AN ELEMENTARY TREATISE ON

FOURIER'S SERIES

AND SPHERICAL, CYLINDRICAL, AND ELLIPSOIDAL HARMONICS,
WITH
APPLICATIONS TO PROBLEMS IN MATHEMATICAL PHYSICS.

WILLIAM ELWOOD BYERLY, Ph.D.,
PROFESSOR OF MATHEMATICS IN HARVARD UNIVERSITY.

HAWK PRESS

Published by

Hawk Press
4836/24, Ansari Road, Daryaganj
New Delhi – 110 002
Phones : 9643330713, 011-23278618, 011-35676207
E-mail: thehawkpress@gmail.com
www.thehawkpress.com

ISBN: 978-93-93971-60-9

PREFACE.

About ten years ago I gave a course of lectures on Trigonometric Series, following closely the treatment of that subject in Riemann's "Partielle Differentialgleichungen," to accompany a short course on The Potential Function, given by Professor B. O. Peirce.

My course has been gradually modified and extended until it has become an introduction to Spherical Harmonics and Bessel's and Lamé's Functions.

Two years ago my lecture notes were lithographed by my class for their own use and were found so convenient that I have prepared them for publication, hoping that they may prove useful to others as well as to my own students. Meanwhile, Professor Peirce has published his lectures on "The Newtonian Potential Function" (Boston, Ginn & Co.), and the two sets of lectures form a course (Math. 10) given regularly at Harvard, and intended as a partial introduction to modern Mathematical Physics.

Students taking this course are supposed to be familiar with so much of the infinitesimal calculus as is contained in my "Differential Calculus" (Boston, Ginn & Co.) and my "Integral Calculus" (second edition, same publishers), to which I refer in the present book as "Dif. Cal." and "Int. Cal." Here, as in the "Calculus," I speak of a "derivative" rather than a "differential coefficient," and use the notation D_x instead of $\frac{\delta}{\delta x}$ for "partial derivative with respect to x."

The course was at first, as I have said, an exposition of Riemann's "Partielle Differentialgleichungen." In extending it, I drew largely from Ferrer's "Spherical Harmonics" and Heine's "Kugelfunctionen," and was somewhat indebted to Todhunter ("Functions of Laplace, Bessel, and Lamé"), Lord Rayleigh ("Theory of Sound"), and Forsyth ("Differential Equations").

In preparing the notes for publication, I have been greatly aided by the criticisms and suggestions of my colleagues, Professor B. O. Peirce and Dr. Maxime Bôcher, and the latter has kindly contributed the brief historical sketch contained in Chapter IX.

W. E. BYERLY.

CAMBRIDGE, MASS., Sept. 1893.

ANALYTICAL TABLE OF CONTENTS.

CHAPTER I.

CHAPTER II.

CHAPTER VI.

CHAPTER VII.

CHAPTER I.

INTRODUCTION.

1. In many important problems in mathematical physics we are obliged to deal with *partial differential equations* of a comparatively simple form.

For example, in the Analytical Theory of Heat we have for the change of temperature of any solid due to the flow of heat within the solid, the equation

$$D_t u = a^2(D_x^2 u + D_y^2 u + D_z^2 u),^{1} \qquad \text{[I]}$$

where u represents the temperature at any point of the solid and t the time.

In the simplest case, that of a slab of infinite extent with parallel plane faces, where the temperature can be regarded as a function of one coördinate, [I] reduces to

$$D_t u = a^2 D_x^2 u, \qquad \text{[II]}$$

a form of considerable importance in the consideration of the problem of the cooling of the earth's crust.

In the problem of the permanent state of temperatures in a thin rectangular plate, the equation [I] becomes

$$D_x^2 u + D_y^2 u = 0. \qquad \text{[III]}$$

In *polar* or *spherical coördinates* [I] is less simple, it is

$$D_t u = \frac{a^2}{r^2}\left[D_r(r^2 D_r u) + \frac{1}{\sin\theta} D_\theta(\sin\theta D_\theta u) + \frac{1}{\sin^2\theta} D_\phi^2 u\right]. \qquad \text{[IV]}$$

In the case where the solid in question is a sphere and the temperature at any point depends merely on the distance of the point from the centre [IV] reduces to

$$D_t(ru) = a^2 D_r^2(ru). \qquad \text{[V]}$$

In *cylindrical coördinates* [I] becomes

$$D_t u = a^2[D_r^2 u + \frac{1}{r} D_r u + \frac{1}{r^2} D_\phi^2 u + D_z^2 u]. \qquad \text{[VI]}$$

In considering the flow of heat in a cylinder when the temperature at any point depends merely on the distance r of the point from the axis [VI] becomes

$$D_t u = a^2(D_r^2 u + \frac{1}{r} D_r u). \qquad \text{[VII]}$$

In Acoustics in several problems we have the equation

$$D_t^2 y = a^2 D_x^2 y; \qquad \text{[VIII]}$$

[1] For the sake of brevity we shall often use the symbol ∇^2 for the operation $D_x^2 + D_y^2 + D_z^2$; and with this notation equation [I] would be written $D_t u = a^2 \nabla^2 u$.

for instance, in considering the transverse or the longitudinal vibrations of a stretched elastic string, or the transmission of plane sound waves through the air.

If in considering the transverse vibrations of a stretched string we take account of the resistance of the air [VIII] is replaced by

$$D_t^2 y + 2kD_t y = a^2 D_x^2 y. \qquad \text{[IX]}$$

In dealing with the vibrations of a stretched elastic membrane, we have the equation

$$D_t^2 z = c^2(D_x^2 z + D_y^2 z), \qquad \text{[X]}$$

or in *cylindrical coördinates*

$$D_t^2 z = c^2(D_r^2 z + \frac{1}{r} D_r z + \frac{1}{r^2} D_\phi^2 z). \qquad \text{[XI]}$$

In the theory of *Potential* we constantly meet Laplace's Equation

$$D_x^2 V + D_y^2 V + D_z^2 V = 0 \qquad \text{[XII]}$$

or

$$\nabla^2 V = 0$$

which in *spherical coördinates* becomes

$$\frac{1}{r^2}\left[r D_r^2(rV) + \frac{1}{\sin\theta} D_\theta(\sin\theta D_\theta V) + \frac{1}{\sin^2\theta} D_\phi^2 V \right] = 0, \qquad \text{[XIII]}$$

and in *cylindrical coördinates*

$$D_r^2 V + \frac{1}{r} D_r V + \frac{1}{r^2} D_\phi^2 V + D_z^2 V = 0. \qquad \text{[XIV]}$$

In *curvilinear coördinates* it is

$$h_1 h_2 h_3 \left[D_{\rho_1}\left(\frac{h_1}{h_2 h_3} D_{\rho_1} V\right) + D_{\rho_2}\left(\frac{h_2}{h_3 h_1} D_{\rho_2} V\right) + D_{\rho_3}\left(\frac{h_3}{h_1 h_2} D_{\rho_3} V\right)\right] = 0; \qquad \text{[XV]}$$

where $f_1(x, y, z) = \rho_1,\ f_2(x, y, z) = \rho_2,\ f_3(x, y, z) = \rho_3$

represent a set of surfaces which cut one another at right angles, no matter what values are given to ρ_1, ρ_2, and ρ_3; and where

$$h_1^2 = (D_x\rho_1)^2 + (D_y\rho_1)^2 + (D_z\rho_1)^2$$
$$h_2^2 = (D_x\rho_2)^2 + (D_y\rho_2)^2 + (D_z\rho_2)^2$$
$$h_3^2 = (D_x\rho_3)^2 + (D_y\rho_3)^2 + (D_z\rho_3)^2,$$

and, of course, must be expressed in terms of ρ_1, ρ_2, and ρ_3.

If it happens that $\nabla^2\rho_1 = 0$, $\nabla^2\rho_2 = 0$, and $\nabla^2\rho_3 = 0$, then Laplace's Equation [XV] assumes the very simple form

$$h_1^2 D_{\rho_1}^2 V + h_2^2 D_{\rho_2}^2 V + h_3^2 D_{\rho_3}^2 V = 0. \qquad \text{[XVI]}$$

2. A *differential equation* is an equation containing derivatives or differentials with or without the primitive variables from which they are derived.

The *general solution* of a differential equation is the equation expressing the most general relation between the primitive variables which is consistent with the given differential equation and which does not involve differentials or derivatives. A general solution will always contain arbitrary (*i. e.*, undetermined) *constants* or *arbitrary functions.*

A *particular solution* of a differential equation is a relation between the primitive variables which is consistent with the given differential equation, but which is less general than the general solution, although included in it.

Theoretically, every particular solution can be obtained from the general solution by substituting in the general solution particular values for the arbitrary constants or particular functions for the arbitrary functions; but in practice it is often easy to obtain particular solutions directly from the differential equation when it would be difficult or impossible to obtain the general solution.

3. If a problem requiring for its solution the solving of a differential equation is *determinate*, there must always be given in addition to the differential equation enough outside conditions for the determination of all the arbitrary constants or arbitrary functions that enter into the general solution of the equation; and in dealing with such a problem, if the differential equation can be readily solved the natural method of procedure is to obtain its general solution, and then to determine the constants or functions by the aid of the given conditions.

It often happens, however, that the general solution of the differential equation in question cannot be obtained, and then, since the problem *if determinate* will be solved if by any means a solution of the equation can be found which will also satisfy the given outside conditions, it is worth while to try to get *particular solutions* and so to combine them as to form a result which shall satisfy the given conditions without ceasing to satisfy the differential equation.

4. A differential equation is *linear* when it would be of the first degree if the dependent variable and all its derivatives were regarded as algebraic unknown quantities. If it is linear and contains no term which does not involve the dependent variable or one of its derivatives, it is said to be linear and *homogeneous.*

All the differential equations collected in Art. 1 are linear and homogeneous.

5. *If a value of the dependent variable has been found which satisfies a given homogeneous, linear, differential equation, the product formed by multiplying this value by any constant will also be a value of the dependent variable which will satisfy the equation.*

For if all the terms of the given equation are transposed to the first member, the substitution of the first-named value must reduce that member to zero; substituting the second value is equivalent to multiplying each term of the result of the first substitution by the same constant factor, which therefore may be taken out as a factor of the whole first member. The remaining factor being zero, the product is zero and the equation is satisfied.

If several values of the dependent variable have been found each of which satisfies the given differential equation, their sum will satisfy the equation; for if the sum of the values in question is substituted in the equation each term of the sum will give rise to a set of terms which must be equal to zero, and therefore the sum of these sets must be zero.

6. It is generally possible to get by some simple device *particular solutions* of such differential equations as those we have collected in Art. 1. The object of the branch of mathematics with which we are about to deal is to find methods of so combining these particular solutions as to satisfy any given conditions which are consistent with the nature of the problem in question.

This often requires us to be able to develop any given function of the variables which enter into the expression of these conditions in terms of *normal forms* suited to the problem with which we happen to be dealing, and suggested by the form of particular solution that we are able to obtain for the differential equation.

These normal forms are frequently sines and cosines, but they are often much more complicated functions known as *Legendre's Coefficients,* or *Zonal Harmonics; Laplace's Coefficients,* or *Spherical Harmonics: Bessel's Functions,* or *Cylindrical Harmonics; Lamé's Functions,* or *Ellipsoidal Harmonics,* &c.

7. As an illustration, let us take Fourier's problem of the permanent state of temperatures in a thin rectangular plate of breadth π and of infinite length whose faces are impervious to heat. We shall suppose that the two long edges of the plate are kept at the constant temperature zero, that one of the short edges, which we shall call the base of the plate, is kept at the temperature unity, and that the temperatures of points in the plate decrease indefinitely as we recede from the base; we shall attempt to find the temperature at any point of the plate.

Let us take the base as the axis of X and one end of the base as the origin. Then to solve the problem we are to find the temperature u of any point from the equation

$$D_x^2 u + D_y^2 u = 0 \qquad \text{[III] Art. 1}$$

subject to the conditions

$$u = 0 \quad \text{when} \quad x = 0 \qquad (1)$$

$$u = 0 \quad \text{“} \quad x = \pi \qquad (2)$$

$$u = 0 \quad \text{“} \quad y = \infty \qquad (3)$$

$$u = 1 \quad \text{“} \quad y = 0. \qquad (4)$$

We shall begin by getting a particular solution of [III], and we shall use a device which always succeeds when the equation is *linear* and *homogeneous* and has *constant coefficients.*

Assume[2] $u = e^{\alpha y+\beta x}$, where α and β are constants, substitute in [III] and divide by $e^{\alpha y+\beta x}$, and we have $\alpha^2+\beta^2=0$. If, then, this condition is satisfied $u = e^{\alpha y+\beta x}$ is a solution.

Hence $u = e^{\alpha y\pm\alpha x i}$ [3] is a solution of [III], no matter what value may be given to α.

This form is objectionable, since it involves an imaginary. We can, however, readily improve it.

Take $u = e^{\alpha y}e^{\alpha x i}$, a solution of [III], and $u = e^{\alpha y}e^{-\alpha x i}$, another solution of [III]; add these values of u and divide the sum by 2 and we have $e^{\alpha y}\cos\alpha x$. (v. Int. Cal. Art. 35, [1].) Therefore by Art. 5

$$u = e^{\alpha y}\cos\alpha x \tag{5}$$

is a solution of [III]. Take $u = e^{\alpha y}e^{\alpha x i}$ and $u = e^{\alpha y}e^{-\alpha x i}$, subtract the second value of u from the first and divide by $2i$ and we have $e^{\alpha y}\sin\alpha x$. (v. Int. Cal. Art. 35, [2]). Therefore by Art. 5

$$u = e^{\alpha y}\sin\alpha x \tag{6}$$

is a solution of [III].

Let us now see if out of these particular solutions we can build up a solution which will satisfy the conditions (1), (2), (3), and (4).

Consider
$$u = e^{\alpha y}\sin\alpha x. \tag{6}$$

It is zero when $x=0$ for all values of α. It is zero when $x=\pi$ if α is a whole number. It is zero when $y=\infty$ if α is negative. If, then, we write u equal to a sum of terms of the form $Ae^{-my}\sin mx$, where m is a positive integer, we shall have a solution of [III] which satisfies conditions (1), (2) and (3). Let this solution be

$$u = A_1e^{-y}\sin x + A_2e^{-2y}\sin 2x + A_3e^{-3y}\sin 3x + A_4e^{-4y}\sin 4x + \cdots \tag{7}$$

A_1, A_2, A_3, A_4, &c., being undetermined constants.

When $y=0$ (7) reduces to

$$u = A_1\sin x + A_2\sin 2x + A_3\sin 3x + A_4\sin 4x + \cdots. \tag{8}$$

If now it is possible to develop unity into a series of the form (8), our problem is solved; we have only to substitute the coefficients of that series for A_1, A_2, A_3, &c. in (7).

It will be proved later that

$$1 = \frac{4}{\pi}\left(\sin x + \frac{1}{3}\sin 3x + \frac{1}{5}\sin 5x + \frac{1}{7}\sin 7x + \cdots\right)$$

[2]This assumption must be regarded as purely tentative. It must be tested by substituting in the equation, and is justified if it leads to a solution.

[3]We shall regularly use the symbol i for $\sqrt{-1}$.

for all values of x between 0 and π; hence our required solution is

$$u = \frac{4}{\pi}\left[e^{-y}\sin x + \frac{1}{3}e^{-3y}\sin 3x + \frac{1}{5}e^{-5y}\sin 5x + \frac{1}{7}e^{-7y}\sin 7x + \cdots\right] \tag{9}$$

for this satisfies the differential equation and all the given conditions.

If the given temperature of the base of the plate instead of being unity is a function of x, we can solve the problem as before if we can express the given function of x as a sum of terms of the form $A\sin mx$, where m is a whole number.

The problem of finding the value of the *potential function* at any point of a long, thin, rectangular conducting sheet, of breadth π, through which an electric current is flowing, when the two long edges are kept at potential zero, and one short edge at potential unity, is mathematically identical with the problem we have just solved.

EXAMPLE.

Taking the temperature of the base of the plate described above as 100° centigrade, and that of the sides of the plate as 0°, compute the temperatures of the points

$$(a)\ \left(\frac{\pi}{6}, 1\right);\ (b)\ \left(\frac{\pi}{3}, 2\right);\ (c)\ \left(\frac{\pi}{2}, 3\right),$$

correct to the nearest degree. *Ans.* (a) 26°; (b) 15°; (c) 6°.

8. As another illustration, we shall take the problem of the transverse vibrations of a stretched string fastened at the ends, initially distorted into some given curve and then allowed to swing.

Let the length of the string be l. Take the position of equilibrium of the string as the axis of X, and one of the ends as the origin, and suppose the string initially distorted into a curve whose equation $y = f(x)$ is given.

We have then to find an expression for y which will be a solution of the equation

$$D_t^2 y = a^2 D_x^2 y \qquad \text{[VIII] Art. 1,}$$

while satisfying the conditions

$$y = 0 \quad \text{when} \quad x = 0 \tag{1}$$
$$y = 0 \quad \text{"} \quad x = l \tag{2}$$
$$y = f(x) \quad \text{"} \quad t = 0 \tag{3}$$
$$D_t y = 0 \quad \text{"} \quad t = 0, \tag{4}$$

the last condition meaning merely that the string starts from rest.

As in the last problem let[4] $y = e^{\alpha x + \beta t}$ and substitute in [VIII]. Divide by $e^{\alpha x + \beta t}$ and we have $\beta^2 = a^2\alpha^2$ as the condition that our assumed value of y shall satisfy the equation.

$$y = e^{\alpha x \pm a\alpha t} \tag{5}$$

[4]See note on page 5.

is, then, a solution of (VIII) whatever the value of α.

It is more convenient to have a trigonometric than an exponential form to deal with, and we can readily obtain one by using an imaginary value for α in (5). Replace α by αi and (5) becomes $y = e^{(x \pm at)\alpha i}$, a solution of [VIII]. Replace α by $-\alpha i$ and (5) becomes $y = e^{-(x \pm at)\alpha i}$, another solution of [VIII]. Add these values of y and divide by 2 and we have $\cos\alpha(x \pm at)$. Subtract the second value of y from the first and divide by $2i$ and we have $\sin\alpha(x \pm at)$.

$$y = \cos\alpha(x + at)$$
$$y = \cos\alpha(x - at)$$
$$y = \sin\alpha(x + at)$$
$$y = \sin\alpha(x - at)$$

are, then, solutions of [VIII]. Writing y successively equal to half the sum of the first pair of values, half their difference, half the sum of the last pair of values, and half their difference, we get the very convenient particular solutions of [VIII].

$$y = \cos\alpha x \cos\alpha at$$
$$y = \sin\alpha x \sin\alpha at$$
$$y = \sin\alpha x \cos\alpha at$$
$$y = \cos\alpha x \sin\alpha at.$$

If we take the third form

$$y = \sin\alpha x \cos\alpha at$$

it will satisfy conditions (1) and (4), no matter what value may be given to α, and it will satisfy (2) if $\alpha = \frac{m\pi}{l}$ where m is an integer.

If then we take

$$y = A_1 \sin\frac{\pi x}{l}\cos\frac{\pi at}{l} + A_2 \sin\frac{2\pi x}{l}\cos\frac{2\pi at}{l} + A_3 \sin\frac{3\pi x}{l}\cos\frac{3\pi at}{l} + \cdots \qquad (6)$$

where A_1, A_2, A_3 $\cdots$ are undetermined constants, we shall have a solution of [VIII] which satisfies (1), (2), and (4). When $t = 0$ it reduces to

$$y = A_1 \sin\frac{\pi x}{l} + A_2 \sin\frac{2\pi x}{l} + A_3 \sin\frac{3\pi x}{l} + \cdots \qquad (7)$$

If now it is possible to develop $f(x)$ into a series of the form (7), we can solve our problem completely. We have only to take the coefficients of this series as values of A_1, A_2, A_3 $\cdots$ in (6), and we shall have a solution of [VIII] which satisfies all our given conditions.

In each of the preceding problems the *normal function*, in terms of which a given function has to be expressed, is the sine of a simple multiple of the variable. It would be easy to modify the problem so that the *normal form* should be a cosine.

We shall now take a couple of problems which are much more complicated and where the normal function is an unfamiliar one.

9. Let it be required to find the potential function due to a circular wire ring of small cross section and of given radius c, supposing the matter of the ring to attract according to the law of nature.

We can readily find, by direct integration, the value of the potential function at any point of the axis of the ring. We get for it

$$V = \frac{M}{\sqrt{c^2 + x^2}} \tag{1}$$

where M is the mass of the ring, and x the distance of the point from the centre of the ring.

Let us use spherical coördinates, taking the centre of the ring as origin and the axis of the ring as the polar axis.

To obtain the value of the potential function at any point in space, we must satisfy the equation

$$rD_r^2(rV) + \frac{1}{\sin\theta}D_\theta(\sin\theta D_\theta V) + \frac{1}{\sin^2\theta}D_\phi^2 V = 0, \qquad \text{[XIII] Art. 1,}$$

subject to the condition

$$V = \frac{M}{(c^2 + r^2)^{\frac{1}{2}}} \quad \text{when} \quad \theta = 0. \tag{1}$$

From the symmetry of the ring, it is clear that the value of the potential function must be independent of ϕ, so that [XIII] will reduce to

$$rD_r^2(rV) + \frac{1}{\sin\theta}D_\theta(\sin\theta D_\theta V) = 0. \tag{2}$$

We must now try to get particular solutions of (2), and as the coefficients are not constant, we are driven to a new device.

Let[5] $V = r^m P$, where P is a function of θ only, and m is a positive integer, and substitute in (2), which becomes

$$m(m+1)r^m P + \frac{r^m}{\sin\theta}D_\theta(\sin\theta D_\theta P) = 0.$$

Divide by r^m and use the notation of ordinary derivatives since P depends upon θ only, and we have the equation

$$m(m+1)P + \frac{1}{\sin\theta}\frac{d\left(\sin\theta\dfrac{dP}{d\theta}\right)}{d\theta} = 0, \tag{3}$$

from which to obtain P.

Equation (3) can be simplified by changing the independent variable. Let $x = \cos\theta$ and (3) becomes

$$\frac{d}{dx}\left[(1-x^2)\frac{dP}{dx}\right] + m(m+1)P = 0. \tag{4}$$

[5]See note on page 5.

Assume[6] now that P can be expressed as a sum or as a series of terms involving whole powers of x multiplied by constant coefficients.

Let $P = \sum a_n x^n$ and substitute this value of P in (4). We get

$$\sum[n(n-1)a_n x^{n-2} - n(n+1)a_n x^n + m(m+1)a_n x^n] = 0, \tag{5}$$

where the symbol $\sum$ indicates that we are to form all the terms we can by taking successive whole numbers for n.

As (5) must be true no matter what the value of x, the coefficient of any given power of x, as for instance x^k, must vanish. Hence

$$(k+2)(k+1)a_{k+2} - k(k+1)a_k + m(m+1)a_k = 0 \tag{6}$$

and

$$a_{k+2} = -\frac{m(m+1) - k(k+1)}{(k+1)(k+2)} a_k. \tag{7}$$

If now any set of coefficients satisfying the relation (7) be taken, $P = \sum a_k x^k$ will be a solution of (4).

If $$k = m, \quad a_{k+2} = 0, \quad a_{k+4} = 0, \quad \&c.$$

Since it will answer our purpose if we pick out the simplest set of coefficients that will obey the condition (7), we can take a set including a_m.

Let us rewrite (7) in the form

$$a_k = -\frac{(k+2)(k+1)}{(m-k)(m+k+1)} a_{k+2}. \tag{8}$$

We get from (8), beginning with $k = m - 2$,

$$a_{m-2} = -\frac{m(m-1)}{2.(2m-1)} a_m$$

$$a_{m-4} = \frac{m(m-1)(m-2)(m-3)}{2.4.(2m-1)(2m-3)} a_m$$

$$a_{m-6} = -\frac{m(m-1)(m-2)(m-3)(m-4)(m-5)}{2.4.6.(2m-1)(2m-3)(2m-5)} a_m, \quad \&c.$$

If m is even we see that the set will end with a_0, if m is odd, with a_1.

$$P = a_m \left[x^m - \frac{m(m-1)}{2.(2m-1)} x^{m-2} + \frac{m(m-1)(m-2)(m-3)}{2.4.(2m-1)(2m-3)} x^{m-4} - \cdots \right]$$

where a_m is entirely arbitrary, is, then, a solution of (4). It is found convenient to take a_m equal to

$$\frac{(2m-1)(2m-3)\cdots 1}{m!}$$

and it can be shown that with this value of a_m $P = 1$ when $x = 1$.

[6]See note on page 5.

P is a function of x and contains no higher powers of x than x^m. It is usual to write it as $P_m(x)$.

We proceed to compute a few values of $P_m(x)$ from the formula

$$P_m(x) = \frac{(2m-1)(2m-3)\cdots 1}{m!}\left[x^m - \frac{m(m-1)}{2.(2m-1)}x^{m-2} + \frac{m(m-1)(m-2)(m-3)}{2.4.(2m-1)(2m-3)}x^{m-4} - \cdots\right]. \tag{9}$$

We have:

$$\left.\begin{array}{lll} P_0(x) = 1 & \text{or} & P_0(\cos\theta) = 1 \\ P_1(x) = x & \text{“} & P_1(\cos\theta) = \cos\theta \\ P_2(x) = \frac{1}{2}(3x^2-1) & \text{“} & P_2(\cos\theta) = \frac{1}{2}(3\cos^2\theta - 1) \\ P_3(x) = \frac{1}{2}(5x^3-3x) & \text{“} & P_3(\cos\theta) = \frac{1}{2}(5\cos^3\theta - 3\cos\theta) \\ P_4(x) = \frac{1}{8}(35x^4 - 30x^2 + 3) \text{ or} & & \\ & & P_4(\cos\theta) = \frac{1}{8}(35\cos^4\theta - 30\cos^2\theta + 3) \\ P_5(x) = \frac{1}{8}(63x^5 - 70x^3 + 15x) \text{ or} & & \\ & & P_5(\cos\theta) = \frac{1}{8}(63\cos^5\theta - 70\cos^3\theta + 15\cos\theta). \end{array}\right\} \tag{10}$$

We have obtained $P = P_m(x)$ as a particular solution of (4) and $P = P_m(\cos\theta)$ as a particular solution of (3). $P_m(x)$ or $P_m(\cos\theta)$ is a new function, known as a *Legendre's Coefficient*, or as a *Surface Zonal Harmonic*, and occurs as a normal form in many important problems.

$V = r^m P_m(\cos\theta)$ is a particular solution of (2) and $r^m P_m(\cos\theta)$ is sometimes called a *Solid Zonal Harmonic*.

We can now proceed to the solution of our original problem.

$$V = A_0 r^0 P_0(\cos\theta) + A_1 r P_1(\cos\theta) + A_2 r^2 P_2(\cos\theta) + A_3 r^3 P_3(\cos\theta) + \cdots \tag{11}$$

where A_0, A_1, A_2, &c., are entirely arbitrary, is a solution of (2) (v. Art. 5). When $\theta = 0$ (11) reduces to

$$V = A_0 + A_1 r + A_2 r^2 + A_3 r^3 + \cdots,$$

since, as we have said, $P_m(x) = 1$ when $x = 1$, or $P_m(\cos\theta) = 1$ when $\theta = 0$.

By our condition (1)

$$V = \frac{M}{(c^2 + r^2)^{\frac{1}{2}}}$$

when $\theta = 0$.

By the Binomial Theorem

$$\frac{M}{(c^2+r^2)^{\frac{1}{2}}} = \frac{M}{c}\left[1 - \frac{1}{2}\frac{r^2}{c^2} + \frac{1.3}{2.4}\frac{r^4}{c^4} - \frac{1.3.5}{2.4.6}\frac{r^6}{c^6} + \cdots\right]$$

provided $r < c$. Hence

$$V = \frac{M}{c}\left[P_0(\cos\theta) - \frac{1}{2}\frac{r^2}{c^2}P_2(\cos\theta) + \frac{1.3}{2.4}\frac{r^4}{c^4}P_4(\cos\theta) - \frac{1.3.5}{2.4.6}\frac{r^6}{c^6}P_6(\cos\theta) + \cdots\right] \tag{12}$$

is our required solution if $r < c$; for it is a solution of equation (2) and satisfies condition (1).

EXAMPLE.

Taking the mass of the ring as one pound and the radius of the ring as one foot, compute to two decimal places the value of the potential function due to the ring at the points

(a) $(r = .2,\ \theta = 0)$; (d) $(r = .6,\ \theta = 0)$; (f) $\left(r = .6,\ \theta = \frac{\pi}{3}\right)$;
(b) $\left(r = .2,\ \theta = \frac{\pi}{4}\right)$; (e) $\left(r = .6,\ \theta = \frac{\pi}{6}\right)$; (g) $\left(r = .6,\ \theta = \frac{\pi}{2}\right)$;
(c) $\left(r = .2,\ \theta = \frac{\pi}{2}\right)$;

Ans. (a) .98; (b) .99; (c) 1.01; (d) .86; (e) .90; (f) 1.00; (g) 1.10.

The unit used is the potential due to a pound of mass concentrated at a point and attracting a second pound of mass concentrated at a point, the two points being a foot apart.

10. A slightly different problem calling for development in terms of Zonal Harmonics is the following:

Required the permanent temperatures within a solid sphere of radius 1, one half of the surface being kept at the constant temperature zero, and the other half at the constant temperature unity.

Let us take the diameter perpendicular to the plane separating the unequally heated surfaces as our axis and let us use spherical coördinates. As in the last problem, we must solve the equation

$$rD_r^2(ru) + \frac{1}{\sin\theta}D_\theta(\sin\theta D_\theta u) + \frac{1}{\sin^2\theta}D_\phi^2 u = 0 \qquad \text{[XIII] Art. 1}$$

which as before reduces to

$$rD_r^2(ru) + \frac{1}{\sin\theta}D_\theta(\sin\theta D_\theta u) = 0 \tag{1}$$

from the consideration that the temperatures must be independent of ϕ.

Our equation of condition is

$$u = 1 \text{ from } \theta = 0 \text{ to } \theta = \frac{\pi}{2} \text{ and } u = 0 \text{ from } \theta = \frac{\pi}{2} \text{ to } \theta = \pi, \tag{2}$$

when $r = 1$.

As we have seen $u = r^m P_m(\cos\theta)$ is a particular solution of (1), m being any positive whole number, and

$$u = A_0 r^0 P_0(\cos\theta) + A_1 r P_1(\cos\theta) + A_2 r^2 P_2(\cos\theta) + A_3 r^3 P_3(\cos\theta) + \cdots \quad (3)$$

where A_0, A_1, A_2, $A_3 \cdots$ are undetermined constants, is a solution of (1).

When $r = 1$ (3) reduces to

$$u = A_0 P_0(\cos\theta) + A_1 P_1(\cos\theta) + A_2 P_2(\cos\theta) + A_3 P_3(\cos\theta) + \cdots \quad (4)$$

If then we can develop our function of θ which enters into equation (2) in a series of the form (4), we have only to take the coefficients of that series as the values of A_0, A_1, A_2, &c., in (3) and we shall have our required solution.

11. As a last example we shall take the problem of the vibration of a stretched circular membrane fastened at the circumference, that is, of an ordinary drumhead. We shall suppose the membrane initially distorted into any given form which has circular symmetry[7] about an axis through the centre perpendicular to the plane of the boundary, and then allowed to vibrate.

Here we have to solve

$$D_t^2 z = c^2 \left(D_r^2 z + \frac{1}{r} D_r z + \frac{1}{r^2} D_\phi^2 z \right) \qquad \text{[XI] Art. 1}$$

subject to the conditions

$$\begin{aligned} z &= f(r) \quad \text{when} \quad t = 0 & (1) \\ D_t z &= 0 \quad\quad \text{"} \quad\quad t = 0 & (2) \\ z &= 0 \quad\quad \text{"} \quad\quad r = a. & (3) \end{aligned}$$

From the symmetry of the supposed initial distortion z must be independent of ϕ, therefore [XI] reduces to

$$D_t^2 z = c^2 \left(D_r^2 z + \frac{1}{r} D_r z \right) \quad (4)$$

and this is the equation for which we wish to find a particular solution.

We shall employ a device not unlike that used in Art. 9.

Assume[8] $z = R.T$ where R is a function of r alone and T is a function of t alone. Substitute this value of z in (4) and we get

$$R D_t^2 T = c^2 T \left(D_r^2 R + \frac{1}{r} D_r R \right)$$

[7]A function of the coördinates of a point has *circular symmetry* about an axis when its value is not affected by rotating the point through any angle about the axis. A surface has circular symmetry about an axis when it is a surface of revolution about the axis.

[8]See note on page 5.

or
$$\frac{1}{c^2T}\frac{d^2T}{dt^2} = \frac{1}{R}\left(\frac{d^2R}{dr^2} + \frac{1}{r}\frac{dR}{dr}\right). \tag{5}$$

The second member of (5) does not involve t, therefore its equal the first member must be independent of t. The first member of (5) does not involve r, and consequently since it contains neither t nor r, it must be constant. Let it equal $-\mu^2$, where μ of course is an undetermined constant.

Then (5) breaks up into the two differential equations

$$\frac{d^2T}{dt^2} + \mu^2c^2T = 0 \tag{6}$$

$$\frac{d^2R}{dr^2} + \frac{1}{r}\frac{dR}{dr} + \mu^2R = 0. \tag{7}$$

(6) can be solved by familiar methods, and we get $T = \cos\mu ct$ and $T = \sin\mu ct$ as simple particular solutions (v. Int. Cal. p. 319, § 21).

To solve (7) is not so easy. We shall first simplify it by a change of independent variable. Let $r = \frac{x}{\mu}$. (7) becomes

$$\frac{d^2R}{dx^2} + \frac{1}{x}\frac{dR}{dx} + R = 0. \tag{8}$$

Assume, as in Art. 9, that R can be expressed in terms of whole powers of x. Let $R = \sum a_nx^n$ and substitute in (8). We get

$$\sum[n(n-1)a_nx^{n-2} + na_nx^{n-2} + a_nx^n] = 0,$$

an equation which must be true no matter what the value of x. The coefficient of any given power of x, as x^{k-2}, must, then, vanish, and

$$k(k-1)a_k + ka_k + a_{k-2} = 0$$

or
$$k^2a_k + a_{k-2} = 0$$

whence we obtain
$$a_{k-2} = -k^2a_k \tag{9}$$

as the only relation that need be satisfied by the coefficients in order that $R = \sum a_kx^k$ shall be a solution of (8).

If $\quad k = 0, \quad a_{k-2} = 0, \quad a_{k-4} = 0, \quad$ &c.

We can then begin with $k = 0$ as our lowest subscript.

From (9)
$$a_k = -\frac{a_{k-2}}{k^2}.$$

Then
$$a_2 = -\frac{a_0}{2^2}$$
$$a_4 = \frac{a_0}{2^2.4^2}$$
$$a_6 = -\frac{a_0}{2^2.4^2.6^2}, \text{ \&c.}$$

Hence
$$R = a_0\left[1 - \frac{x^2}{2^2} + \frac{x^4}{2^2.4^2} - \frac{x^6}{2^2.4^2.6^2} + \cdots\right]$$

where a_0 may be taken at pleasure, is a solution of (8), provided the series is convergent.

Take $a_0 = 1$, and then $R = J_0(x)$ where

$$J_0(x) = 1 - \frac{x^2}{2^2} + \frac{x^4}{2^2.4^2} - \frac{x^6}{2^2.4^2.6^2} + \frac{x^8}{2^2.4^2.6^2.8^2} - \cdots \tag{10}$$

is a solution of (8).

$J_0(x)$ is easily shown to be convergent for all values real or imaginary of x, since the series made up of the moduli of the terms of $J_0(x)$ (v. Int. Cal. Art. 30)

$$1 + \frac{r^2}{2^2} + \frac{r^4}{2^2.4^2} + \frac{r^6}{2^2.4^2.6^2} + \cdots,$$

where r is the modulus of x, is convergent for all values of r. For the ratio of the $n+1$st term of this series to the nth term is $\frac{r^2}{4n^2}$ and approaches zero as its limit as n is indefinitely increased, no matter what the value of r. Therefore $J_0(x)$ is *absolutely convergent.*

$J_0(x)$ is a new and important form. It is called a *Bessel's Function* of the zeroth order, or a *Cylindrical Harmonic.*

Equation (8) was obtained from (7) by the substitution of $x = \mu r$, therefore

$$R = J_0(\mu r) = 1 - \frac{(\mu r)^2}{2^2} + \frac{(\mu r)^4}{2^2.4^2} - \frac{(\mu r)^6}{2^2.4^2.6^2} + \cdots$$

is a solution of (7), no matter what the value of μ, and $z = J_0(\mu r)\cos \mu ct$ or $z = J_0(\mu r)\sin \mu ct$ is a solution of (4).

$z = J_0(\mu r)\cos \mu ct$ satisfies condition (2) whatever the value of μ. In order that it should also satisfy condition (3) μ must be so taken that

$$J_0(\mu a) = 0; \tag{11}$$

that is, μ must be a root of (11) regarded as an equation in μ.

It can be shown that $J_0(x) = 0$ has an infinite number of real positive roots, any one of which can be obtained to any required degree of approximation without serious difficulty. Let x_1, x_2, x_3, $\cdots$ be these roots. Then if

$$\frac{x_1}{a} = \mu_1, \quad \frac{x_2}{a} = \mu_2, \quad \frac{x_3}{a} = \mu_3, \quad \&c.$$

$$z = A_1 J_0(\mu_1 r)\cos \mu_1 ct + A_2 J_0(\mu_2 r)\cos \mu_2 ct + A_3 J_0(\mu_3 r)\cos \mu_3 ct + \cdots, \tag{12}$$

where A_1, A_2, A_3, &c., are any constants, is a solution of (4) which satisfies conditions (2) and (3).

When $t = 0$ (12) reduces to

$$z = A_1 J_0(\mu_1 r) + A_2 J_0(\mu_2 r) + A_3 J_0(\mu_3 r) + \cdots. \tag{13}$$

If then $f(r)$ can be expressed as a series of the form just given, the solution of our problem can be obtained by substituting the coefficients of that series for A_1, A_2, A_3, &c., in (12).

EXAMPLE.

The temperature of a long cylinder is at first unity throughout. The convex surface is then kept at the constant temperature zero. Show that the temperature of any point in the cylinder at the expiration of the time t is

$$u = A_1 e^{-a^2\mu_1^2 t} J_0(\mu_1 r) + A_2 e^{-a^2\mu_2^2 t} J_0(\mu_2 r) + A_3 e^{-a^2\mu_3^2 t} J_0(\mu_3 r) + \cdots$$

where μ_1, μ_2, &c., are the roots of $J_0(\mu c) = 0$, and where

$$1 = A_1 J_0(\mu_1 r) + A_2 J_0(\mu_2 r) + A_3 J_0(\mu_3 r) + \cdots,$$

c being the radius of the cylinder.

12. Each of the five problems which we have taken up forces upon us the consideration of the development of a given function in terms of some *normal form*, and in two of them the normal form suggested is an unfamiliar function. It is clear, then, that a complete treatment of our subject will require the investigation of the properties and relations of certain new and important functions, as well as the consideration of methods of developing in terms of them.

13. In each of the problems just taken up we have to deal with a homogeneous linear partial differential equation involving two independent variables, and we are content if we can obtain particular solutions. In each case the assumption made in the last problem, that there exists a solution of the equation in which the dependent variable is the product of two factors each of which involves but one of the independent variables, will reduce the question to solving two ordinary differential equations which can be treated separately.

If these equations are familiar ones their solutions can be written down at once; if unfamiliar, the device used in problems 3 and 5 is often serviceable, namely, that of assuming that the dependent variable can be expressed as a sum or series of terms involving whole powers of the independent variable, and then determining the coefficients.

Let us consider again the equations used in the first, second and third problems.

(*a*)
$$D_x^2 u + D_y^2 u = 0 \tag{1}$$

Assume $u = X.Y$ where X involves x but not y, and Y involves y but not x.

Substitute in (1),
$$Y D_x^2 X + X D_y^2 Y = 0,$$

or, since we are now dealing with functions of a single variable,

$$\frac{1}{X}\frac{d^2X}{dx^2} + \frac{1}{Y}\frac{d^2Y}{dy^2} = 0,$$

or
$$\frac{1}{Y}\frac{d^2Y}{dy^2} = -\frac{1}{X}\frac{d^2X}{dx^2}. \tag{2}$$

Since the first member of (2) does not contain x, and the second member does not contain y, and the two members must be identically equal, neither of them can contain either x or y, and each must be equal to a constant, say α^2.

Then
$$\frac{d^2Y}{dy^2} - \alpha^2 Y = 0 \tag{3}$$
and
$$\frac{d^2X}{dx^2} + \alpha^2 X = 0; \tag{4}$$

and if (3) and (4) can be solved, we can solve (1). They have for their complete solutions

$$Y = Ae^{\alpha y} + Be^{-\alpha y}$$
and
$$X = C\sin\alpha x + D\cos\alpha x. \qquad \text{(v. Int. Cal. p. 319, § 21.)}$$

Hence $Y = e^{\alpha y}$ and $Y = e^{-\alpha y}$ are particular solutions of (3), $X = \sin\alpha x$ and $X = \cos\alpha x$ are particular solutions of (1), and consequently

$$u = e^{\alpha y}\sin\alpha x, \quad u = e^{\alpha y}\cos\alpha x, \quad u = e^{-\alpha y}\sin\alpha x, \text{ and } u = e^{-\alpha y}\cos\alpha x$$

are particular solutions of (1). These agree with the results of Art. 7.

(b)
$$D_t^2 y = a^2 D_x^2 y \tag{1}$$

Assume $y = T.X$ where T is a function of t only and X a function of x only; substitute in (1) and divide by a^2TX. We get

$$\frac{1}{a^2T}\frac{d^2T}{dt^2} = \frac{1}{X}\frac{d^2X}{dx^2}; \tag{2}$$

hence as in the last case $\frac{1}{X}\frac{d^2X}{dx^2}$ is a constant; call it $-\alpha^2$, and (2) breaks up into

$$\frac{d^2X}{dx^2} + \alpha^2 X = 0 \tag{3}$$
$$\frac{d^2T}{dt^2} + \alpha^2 a^2 T = 0. \tag{4}$$

The complete solutions of (3) and (4) are

$$X = A\sin\alpha x + B\cos\alpha x$$
and
$$T = C\sin\alpha at + D\cos\alpha at, \qquad \text{(v. Int. Cal. p. 319, § 21).}$$

$$y = \sin\alpha x\cos\alpha at, \; y = \sin\alpha x\sin\alpha at, \; y = \cos\alpha x\cos\alpha at, \; y = \cos\alpha x\sin\alpha at$$

are particular solutions of (1), and agree with the results of Art. 8.

(c)
$$rD_r^2(rV) + \frac{1}{\sin\theta}D_\theta(\sin\theta D_\theta V) = 0. \tag{1}$$

Assume $V = R.\Theta$ where R involves r alone, and Θ involves θ alone; substitute in (1), divide by $R.\Theta$, and transpose; we get

$$\frac{r}{R}\frac{d^2(rR)}{dr^2} = -\frac{1}{\Theta\sin\theta}\frac{d\left(\sin\theta\frac{d\Theta}{d\theta}\right)}{d\theta}. \tag{2}$$

Since by the reasoning used in (*a*) and (*b*) each member of (2) must be a constant, say α^2, we have

$$r\frac{d^2(rR)}{dr^2} = \alpha^2 R \tag{3}$$

and

$$\frac{1}{\sin\theta}\frac{d\left(\sin\theta\frac{d\Theta}{d\theta}\right)}{d\theta} + \alpha^2\Theta = 0. \tag{4}$$

(3) can be expanded into

$$r^2\frac{d^2R}{dr^2} + 2r\frac{dR}{dr} - \alpha^2 R = 0. \tag{5}$$

(5) can be solved (v. Int. Cal. p. 321, § 23), and has for its complete solution

$$R = Ar^m + Br^n,$$

where $m = -\frac{1}{2} + \sqrt{\alpha^2 + \frac{1}{4}}$ and $n = -\frac{1}{2} - \sqrt{\alpha^2 + \frac{1}{4}}$.

Hence $n = -m-1$, and α^2 may be written $m(m+1)$, m being wholly arbitrary; and

$$R = Ar^m + Br^{-m-1}.$$

$$R = r^m, \quad \text{and} \quad R = \frac{1}{r^{m+1}}$$

are, then, particular solutions of

$$r^2\frac{d^2R}{dr^2} + 2r\frac{dR}{dr} - m(m+1)R = 0. \tag{6}$$

With the new value of α^2 (4) becomes

$$\frac{1}{\sin\theta}\frac{d\left(\sin\theta\frac{d\Theta}{d\theta}\right)}{d\theta} + m(m+1)\Theta = 0. \tag{7}$$

which has been treated in Art. 9 for the case where m is a positive integer, and the particular solution $\Theta = P_m(\cos\theta)$ has been obtained.

Hence $$V = r^m P_m(\cos\theta)$$

and $$V = \frac{1}{r^{m+1}}P_m(\cos\theta),$$

m being a positive integer, are particular solutions of (1). The first of these was obtained in Art. 9, but the second is new and exceedingly important.

14. The method of obtaining a particular solution of an ordinary linear differential equation, which we have used in Articles 9 and 11, is of very extensive application, and often leads to the general solution of the equation in question.

As a very simple example, let us take the equation Art. 13 (a) (4), which we shall write

$$\frac{d^2z}{dx^2} + \alpha^2 z = 0. \tag{1}$$

Assume that there is a solution which can be expressed in terms of powers of x; that is, let $z = \sum a_n x^n$, where the coefficients are to be determined. Substitute this value for z in (1) and we get

$$\sum [n(n-1)a_n x^{n-2} + \alpha^2 a_n x^n] = 0.$$

Since this equation must be true from its form, without reference to the value of x, that is, since it must be an identical equation, the coefficient of each power of x must equal zero, and we have

$$(n+1)(n+2)a_{n+2} + \alpha^2 a_n = 0;$$

whence
$$a_n = -\frac{(n+1)(n+2)}{\alpha^2} a_{n+2}$$

is the only relation that need hold between the coefficients in order that $z = \sum a_n x^n$ should be a solution of (1).

If $n + 2 = 0$ or $n + 1 = 0$, a_n will be zero and a_{n-2}, a_{n-4}, &c., will be zero. In the first case the series will begin with a_0, in the second with a_1.

$$a_{n+2} = -\frac{\alpha^2}{(n+1)(n+2)} a_n.$$

If we begin with a_0 we have

$$a_2 = -\frac{\alpha^2}{2!} a_0, \qquad a_4 = \frac{\alpha^4}{4!} a_0, \qquad a_6 = -\frac{\alpha^6}{6!} a_0, \ \&\text{c.}, \cdots$$

and
$$z = a_0 \left(1 - \frac{\alpha^2 x^2}{2!} + \frac{\alpha^4 x^4}{4!} - \frac{\alpha^6 x^6}{6!} + \cdots\right) \tag{2}$$

or
$$z = a_0 \cos \alpha x \tag{3}$$

is a particular solution of (1).

If we begin with a_1 we have

$$a_3 = -\frac{\alpha^2}{3!} a_1, \qquad a_5 = \frac{\alpha^4}{5!} a_1, \qquad a_7 = -\frac{\alpha^6}{7!} a_1, \ \&\text{c.}, \cdots$$

and
$$z = a_1 \left(x - \frac{\alpha^2 x^3}{3!} + \frac{\alpha^4 x^5}{5!} - \frac{\alpha^6 x^7}{7!} + \cdots\right) \tag{4}$$

is a solution of (1); a_1 can be taken at pleasure. Let $a_1 = \alpha$, (4) becomes

$$z = \alpha x - \frac{\alpha^3 x^3}{3!} + \frac{\alpha^5 x^5}{5!} - \frac{\alpha^7 x^7}{7!} + \cdots$$

or

$$z = \sin \alpha x$$

which, then, is a particular solution of (1).

$$z = A \sin \alpha x + B \cos \alpha x \tag{5}$$

is, then, a solution of (1), and since it contains two arbitrary constants it is the general solution.

15. As another example we will take the equation

$$x^2 \frac{d^2 z}{dx^2} + 2x \frac{dz}{dx} - m(m+1)z = 0, \tag{1}$$

which is in effect equation (6), Art. 13 (*c*), and let m be a positive integer.

Assume $z = \sum a_n x^n$ and substitute in (1). We get

$$\sum [n(n+1) - m(m+1)] a_n x^n = 0.$$

This is an identical equation, therefore

$$[n(n+1) - m(m+1)] a_n = 0.$$

Hence $a_n = 0$ for all values of n except those which make

$$n(n+1) - m(m+1) = 0,$$

that is, for all values of n except $n = m$ and $n = -m - 1$. Then

$$z = Ax^m + Bx^{-m-1} \tag{2}$$

is the general solution of (1) and

$$z = x^m \quad \text{and} \quad z = \frac{1}{x^{m+1}}$$

are particular solutions. If m is not a positive integer this method will not lead to a result, and we are driven back to that employed in Art. 13 (*c*).

16. Let us now take the equation

$$\frac{d}{dx}\left[(1 - x^2)\frac{dz}{dx}\right] + m(m+1)z = 0 \tag{1}$$

which is in effect equation (4), Art. 9, and is known as *Legendre's Equation.* (1) may be written

$$(1 - x^2)\frac{d^2 z}{dx^2} - 2x \frac{dz}{dx} + m(m+1)z = 0. \tag{2}$$

Assume $z = \sum a_n x^n$ and substitute in (2). We get

$$\sum\{n(n-1)a_n x^{n-2} + [m(m+1) - n(n+1)]a_n x^n\} = 0.$$

Hence $$(n+1)(n+2)a_{n+2} + [m(m+1) - n(n+1)]a_n = 0,$$

or $$a_n = -\frac{(n+1)(n+2)}{m(m+1) - n(n+1)}a_{n+2}. \tag{3}$$

If $a_n = 0$, then $a_{n-2} = 0$, $a_{n-4} = 0$, &c.; but $a_n = 0$ if $n = -2$ or $n = -1$. For the first case we have the sequence of coefficients

$$\begin{aligned} a_2 &= -\frac{m(m+1)}{2!}a_0 \\ a_4 &= \frac{m(m-2)(m+1)(m+3)}{4!}a_0 \\ a_6 &= -\frac{m(m-2)(m-4)(m+1)(m+3)(m+5)}{6!}a_0, \quad \text{\&c.} \end{aligned}$$

Let us take a_0, which is arbitrary, as 1. Then $z = p_m(x)$ where

$$p_m(x) = \left[1 - \frac{m(m+1)}{2!}x^2 + \frac{m(m-2)(m+1)(m+3)}{4!}x^4 - \cdots\right] \tag{4}$$

is a solution of Legendre's Equation if $p_m(x)$ is a finite sum or a convergent series.

For the second case we have the sequence of coefficients

$$\begin{aligned} a_3 &= -\frac{(m-1)(m+2)}{3!}a_1 \\ a_5 &= \frac{(m-1)(m-3)(m+2)(m+4)}{5!}a_1 \\ a_7 &= -\frac{(m-1)(m-3)(m-5)(m+2)(m+4)(m+6)}{7!}a_1, \quad \text{\&c.} \end{aligned}$$

Let us take a_1, which is arbitrary, as 1. Then $z = q_m(x)$ where

$$q_m(x) = \left[x - \frac{(m-1)(m+2)}{3!}x^3 + \frac{(m-1)(m-3)(m+2)(m+4)}{5!}x^5 - \cdots\right] \tag{5}$$

is a solution of Legendre's Equation if $q_m(x)$ is a finite sum or a convergent series.

If m is a positive even whole number, $p_m(x)$ will terminate with the term containing x^m, and is easily seen to be identical with

$$(-1)^{\frac{m}{2}}\frac{2^m\left[\Gamma\left(\frac{m}{2}+1\right)\right]^2}{\Gamma(m+1)}P_m(x). \qquad \text{[v. Art. 9 (9)]}$$

For all other values of m, $p_m(x)$ is a series.

The ratio of the $(n+1)$st term of $p_m(x)$ to the nth, when m is not a positive even integer, is

$$\frac{(2n-2-m)(2n-1+m)}{(2n-1)(2n)}x^2.$$

Its limiting value, as n is increased, is x^2, and the series is therefore convergent if $-1 < x < 1$. It is divergent for all other values of x.

If m is a positive odd whole number $q_m(x)$ will terminate with the term containing x^m, and is easily seen to be identical with

$$(-1)^{\frac{m-1}{2}}\frac{2^{m-1}\left[\Gamma\left(\frac{m+1}{2}\right)\right]^2}{\Gamma(m+1)}P_m(x).$$

For all other values of m, $q_m(x)$ is a series, and can be shown to be convergent if $-1 < x < 1$, and divergent for all other values of x.

$$z = Ap_m(x) + Bq_m(x) \tag{6}$$

is the general solution of Legendre's Equation if $-1 < x < 1$, no matter what the value of m. From Art. 13 (c) it follows that

$$\left.\begin{aligned} V &= r^m p_m(\cos\theta) \\ V &= \frac{1}{r^{m+1}}p_m(\cos\theta) \\ V &= r^m q_m(\cos\theta) \\ V &= \frac{1}{r^{m+1}}q_m(\cos\theta) \end{aligned}\right\} \tag{7}$$

are particular solutions of

$$rD_r^2(rV) + \frac{1}{\sin\theta}D_\theta(\sin\theta D_\theta V) = 0,$$

no matter what the value of m, provided $\cos\theta$ is neither one nor minus one.

In the work we shall have to do with Laplace's and Legendre's Equations, it is generally possible to restrict m to being a positive integer, and hereafter we shall usually confine our attention to that case.

With this understanding let us return to (3), which may be rewritten

$$a_{n+2} = -\frac{(m-n)(m+n+1)}{(n+1)(n+2)}a_n.$$

If $\quad a_{n+2} = 0$, then $a_{n+4} = 0,\ a_{n+6} = 0$, &c.;

but $\quad a_{n+2} = 0$ if $n = m$, or $n = -m-1$.

If in (3) we begin with $n = m - 2$, we get the sequence of coefficients already obtained in Art. 9, and we have $z = P_m(x)$, where

$$P_m(x) = \frac{(2m-1)(2m-3)\cdots 1}{m!}\left[x^m - \frac{m(m-1)}{2(2m-1)}x^{m-2}\right.$$
$$+ \frac{m(m-1)(m-2)(m-3)}{2.4.(2m-1)(2m-3)}x^{m-4}$$
$$\left. - \frac{m(m-1)(m-2)(m-3)(m-4)(m-5)}{2.4.6.(2m-1)(2m-3)(2m-5)}x^{m-6} + \cdots\right], \tag{8}$$

as a particular solution of Legendre's Equation.

If, however, we begin with $n = -m - 3$, we have

$$a_{-m-3} = \frac{(m+1)(m+2)}{2(2m+3)}a_{-m-1}$$
$$a_{-m-5} = \frac{(m+1)(m+2)(m+3)(m+4)}{2.4.(2m+3)(2m+5)}a_{-m-1}$$
$$a_{-m-7} = \frac{(m+1)(m+2)(m+3)(m+4)(m+5)(m+6)}{2.4.6.(2m+3)(2m+5)(2m+7)}a_{-m-1}, \quad \&c.$$

a_{-m-1} may be taken at pleasure, and is usually taken as $\dfrac{m!}{1.3.5.\cdots(2m+1)}$, and $z = Q_m(x)$ where

$$Q_m(x) = \frac{m!}{(2m+1)(2m-1)\cdots 1}\left[\frac{1}{x^{m+1}} + \frac{(m+1)(m+2)}{2.(2m+3)}\frac{1}{x^{m+3}}\right.$$
$$\left. + \frac{(m+1)(m+2)(m+3)(m+4)}{2.4.(2m+3)(2m+5)}\frac{1}{x^{m+5}} + \cdots\right] \tag{9}$$

is a second particular solution of Legendre's Equation, provided the series is convergent. $Q_m(x)$ is called a *Surface Zonal Harmonic* of the *second kind*. It is easily seen to be convergent if $x < -1$ or $x > 1$, and divergent if $-1 < x < 1$.

Hence if m is a positive integer,

$$z = AP_m(x) + BQ_m(x) \tag{10}$$

is the general solution of Legendre's Equation if $x < -1$ or $x > 1$.

We have seen that for $-1 < x < 1$

$$P_m(x) = (-1)^{\frac{m}{2}}\frac{\Gamma(m+1)}{2^m\left[\Gamma\left(\frac{m}{2}+1\right)\right]^2}p_m(x) \tag{11}$$

if m is an even integer, and

$$P_m(x) = (-1)^{\frac{m-1}{2}}\frac{\Gamma(m+1)}{2^{m-1}\left[\Gamma\left(\frac{m+1}{2}\right)\right]^2}q_m(x) \tag{12}$$

if m is an odd integer.

If now we define $Q_m(x)$ as follows when $-1 < x < 1$

$$Q_m(x) = (-1)^{\frac{m+1}{2}} \frac{2^{m-1}\left[\Gamma\left(\frac{m+1}{2}\right)\right]^2}{\Gamma(m+1)} p_m(x) \tag{13}$$

if m is an odd integer, and

$$Q_m(x) = (-1)^{\frac{m}{2}} \frac{2^{m}\left[\Gamma\left(\frac{m}{2}+1\right)\right]^2}{\Gamma(m+1)} q_m(x) \tag{14}$$

if m is an even integer, then (10) will be the general solution of Legendre's Equation if m is a positive integer when $-1 < x < 1$, as well as when $x < -1$ or $x > 1$.

17. Let us last consider the equation

$$\frac{d^2z}{dx^2} + \frac{1}{x}\frac{dz}{dx} + \left(1 - \frac{m^2}{x^2}\right)z = 0 \tag{1}$$

which is known as Bessel's Equation, and which reduces to (8) Art. 11, that is, to

$$\frac{d^2z}{dx^2} + \frac{1}{x}\frac{dz}{dx} + z = 0$$

when $m = 0$;[9] (1) can be simplified by a change of the dependent variable. Let $z = x^m v$ and we get

$$\frac{d^2v}{dx^2} + \frac{2m+1}{x}\frac{dv}{dx} + v = 0 \tag{2}$$

to determine v.

Assume $v = \sum a_n x^n$, and substitute in (2). We get

$$\sum[n(2m+n)a_n x^{n-2} + a_n x^n] = 0;$$

whence

$$a_{n-2} = -n(2m+n)a_n.$$

If we begin with $n = 0$, then $a_{n-2} = 0$, $a_{n-4} = 0$, &c., and we have the set of values

$$a_2 = -\frac{a_0}{2(2m+2)} = -\frac{a_0}{2^2(m+1)}$$

$$a_4 = \frac{a_0}{2.4(2m+2)(2m+4)} = \frac{a_0}{2^4.2!(m+1)(m+2)}$$

[9]This equation was first studied by Fourier in considering the cooling of a cylinder. We shall designate it as "Fourier's Equation."

$$a_6 = -\frac{a_0}{2.4.6(2m+2)(2m+4)(2m+6)} = -\frac{a_0}{2^6.3!(m+1)(m+2)(m+3)};$$

whence
$$z = a_0 x^m \left[1 - \frac{x^2}{2^2(m+1)} + \frac{x^4}{2^4.2!(m+1)(m+2)} - \frac{x^6}{2^6.3!(m+1)(m+2)(m+3)} + \cdots\right] \tag{3}$$

is a solution of Bessel's Equation. a_0 is usually taken as $\frac{1}{2^m m!}$ if m is a positive integer, or as $\frac{1}{2^m\Gamma(m+1)}$ if m is unrestricted in value, and the second member of (3) is represented by $J_m(x)$ and is called a *Bessel's Function* of the mth order, or a *Cylindrical Harmonic* of the mth order.

If $m = 0$, $J_m(x)$ becomes $J_0(x)$ and is the value of z obtained in Art. 11 as the solution of equation (8) of that article.

If in equation (1) we substitute $x^{-m}v$ in place of $x^m v$ for z, we get in place of (2) the equation

$$\frac{d^2v}{dx^2} + \frac{1-2m}{x}\frac{dv}{dx} + v = 0$$

and in place of (3)

$$z = a_0 x^{-m} \left[1 - \frac{x^2}{2^2(1-m)} + \frac{x^4}{2^4.2!(1-m)(2-m)} - \frac{x^6}{2^6.3!(1-m)(2-m)(3-m)} + \cdots\right] \tag{4}$$

If a_0 is taken equal to $\frac{1}{2^{-m}\Gamma(1-m)}$ the second member of (4) is the same function of $-m$ and x that $J_m(x)$ is of $+m$ and x and may be written $J_{-m}(x)$.

Therefore
$$z = AJ_m(x) + BJ_{-m}(x) \tag{5}$$

is the general solution of (1) unless $J_m(x)$ and $J_{-m}(x)$ should prove not to be independent.

It is easily seen that when $m = 0$, $J_{-m}(x)$ and $J_m(x)$ become identical and (5) reduces to

$$z = (A+B)J_0(x)$$

and contains but a single arbitrary constant and is not the general solution of Fourier's Equation (8) Art. (11).

It can be shown that $J_{-m}(x) = (-1)^m J_m(x)$ whenever m is an integer, and consequently that the solution (5) is general only when m if real is fractional or incommensurable.

The general solution for the important case where $m = 0$ is, however, easily obtained. Let $F(m, x)$ be the value which the second member of (3) assumes when $a_0 = 1$; then the value which the second member of (4) assumes when $a_0 = 1$ will be $F(-m, x)$, and it has been shown that $z = F(m, x)$ and $z = F(-m, x)$ are solutions of Bessel's Equation; $z = F(m, x) - F(-m, x)$ is, then, a solution, as is also

$$z = \frac{F(m, x) - F(-m, x)}{2m}, \tag{6}$$

but the limiting value which $\frac{F(m, x) - F(-m, x)}{2m}$ approaches as m approaches zero is $[D_m F(m, x)]_{m=0}$ and consequently

$$z = [D_m F(m, x)]_{m=0} \tag{7}$$

is a solution of the equation

$$\frac{d^2 z}{dx^2} + \frac{1}{x}\frac{dz}{dx} + z = 0, \tag{8}$$

and the general solution of (8) is

$$z = AJ_0(x) + B[D_m F(m, x)]_{m=0}.$$

$$F(m, x) = x^m \left[1 - \frac{x^2}{2^2(m+1)} + \frac{x^4}{2^4.2!(m+1)(m+2)} - \frac{x^6}{2^6.3!(m+1)(m+2)(m+3)} + \cdots\right]$$

$$D_m F(m, x) = x^m \log x \left[1 - \frac{x^2}{2^2(m+1)} + \frac{x^4}{2^4.2!(m+1)(m+2)} - \cdots\right] + x^m D_m \left[1 - \frac{x^2}{2^2(m+1)} + \frac{x^4}{2^4.2!(m+1)(m+2)} + \cdots\right].$$

The general term of the last parenthesis can be written

$$(-1)^k \frac{x^{2k}}{2^{2k}.k!(m+1)(m+2)\cdots(m+k)},$$

and its partial derivative with respect to m is

$$(-1)^k \frac{x^{2k}}{2^{2k}.k!} D_m \frac{1}{(m+1)(m+2)\cdots(m+k)}.$$

$$\log \frac{1}{(m+1)(m+2)\cdots(m+k)} = -[\log(m+1) + \log(m+2) + \cdots + \log(m+k)].$$

Take the D_m of both members and we have

$$D_m \frac{1}{(m+1)(m+2)\cdots(m+k)}$$
$$= -\frac{1}{(m+1)(m+2)\cdots(m+k)}\left[\frac{1}{m+1}+\frac{1}{m+2}+\cdots\frac{1}{m+k}\right].$$
$$D_m\left[1-\frac{x^2}{2^2(m+1)}+\frac{x^4}{2^4.2!(m+1)(m+2)}-\frac{x^6}{2^6.3!(m+1)(m+2)(m+3)}\right.$$
$$\left.+\cdots\right]=\frac{x^2}{2^2}\frac{1}{(m+1)^2}-\frac{x^4}{2^4.2!}\frac{1}{(m+1)(m+2)}\left[\frac{1}{m+1}+\frac{1}{m+2}\right]$$
$$+\frac{x^6}{2^6.3!}\frac{1}{(m+1)(m+2)(m+3)}\left[\frac{1}{m+1}+\frac{1}{m+2}+\frac{1}{m+3}\right]+\cdots$$

and we have

$$[D_mF(m,x)]_{m=0}=J_0(x)\log x+\frac{x^2}{2^2(1!)^2}\frac{1}{1}-\frac{x^4}{2^4(2!)^2}\left(\frac{1}{1}+\frac{1}{2}\right)$$
$$+\frac{x^6}{2^6(3!)^2}\left(\frac{1}{1}+\frac{1}{2}+\frac{1}{3}\right)-\frac{x^8}{2^8(4!)^2}\left(\frac{1}{1}+\frac{1}{2}+\frac{1}{3}+\frac{1}{4}\right)+\cdots;$$

and
$$z=AJ_0(x)+BK_0(x), \tag{9}$$
where
$$K_0(x)=J_0(x)\log x+\frac{x^2}{2^2}-\frac{x^4}{2^4(2!)^2}\left(\frac{1}{1}+\frac{1}{2}\right)$$
$$+\frac{x^6}{2^6(3!)^2}\left(\frac{1}{1}+\frac{1}{2}+\frac{1}{3}\right)-\frac{x^8}{2^8(4!)^2}\left(\frac{1}{1}+\frac{1}{2}+\frac{1}{3}+\frac{1}{4}\right)+\cdots \tag{10}$$

is the general solution of Fourier's Equation (8).

$K_0(x)$ is known as a *Bessel's Function* of the *Second Kind.*

18. It is worth while to confirm the results of the last few articles by getting the general solutions of the equations in question by a different and familiar method.

The general solution of any ordinary linear differential equation of the second order can be obtained when a particular solution of the equation has been found [v. Int. Cal. p. 321, § 24 (a)].

The most general form of a homogeneous ordinary linear differential equation of the second order is

$$\frac{d^2y}{dx^2}+P\frac{dy}{dx}+Qy=0 \tag{1}$$

where P and Q are functions of x. Suppose that

$$y=v \tag{2}$$

is a particular solution of (1). Substitute $y=vz$ in (1) and we get

$$v\frac{d^2z}{dx^2}+\left(2\frac{dv}{dx}+Pv\right)\frac{dz}{dx}=0. \tag{3}$$

Call $\dfrac{dz}{dx} = z'$. Then (3) becomes

$$v\frac{dz'}{dx} + \left(2\frac{dv}{dx} + Pv\right) z' = 0, \tag{4}$$

a differential equation of the first order in which the variables can be separated. Multiply by dx and divide by vz' and (4) reduces to

$$\frac{dz'}{z'} + 2\frac{dv}{v} + Pdx = 0.$$

Integrate and we have

$$\log z' + \log v^2 + \int Pdx = C$$

or

$$z'v^2 = e^{C - \int Pdx} = Be^{-\int Pdx},$$

$$z' = \frac{dz}{dx} = B\frac{e^{-\int Pdx}}{v^2},$$

$$z = A + B\int \frac{e^{-\int Pdx}}{v^2}dx;$$

and

$$y = v\left(A + B\int \frac{e^{-\int Pdx}}{v^2}dx\right) \tag{5}$$

is the general solution of (1), the only arbitrary constants in the second member of (5) being those explicitly written, namely, A and B.

(a) Apply this formula to (1) Art. 14,

$$\frac{d^2z}{dx^2} + \alpha^2 z = 0; \tag{1}$$

given: $z = \cos \alpha x$, as a particular solution. Substituting in (5) we have since $P = 0$

$$\begin{aligned} z &= \cos \alpha x \left(A + B\int \frac{dx}{\cos^2 \alpha x}\right) \\ &= \cos \alpha x \left(A + \frac{B}{\alpha}\tan \alpha x\right) \\ &= A\cos \alpha x + B_1 \sin \alpha x, \end{aligned} \tag{2}$$

as the general solution of (1), and this agrees perfectly with (5) Art. 14.

(b) Take equation (1) Art. 15.

$$x^2\frac{d^2z}{dx^2} + 2x\frac{dz}{dx} - m(m+1)z = 0; \tag{1}$$

given: $z = x^m$, as a particular solution.

Here $P = \frac{2}{x}$, $\int P dx = 2\log x = \log x^2$, and $e^{-\int P dx} = \frac{1}{x^2}$. Hence by (5)

$$z = x^m \left(A + B \int \frac{dx}{x^{2m+2}} \right) = x^m \left(A + \frac{B}{-2m-1} \frac{1}{x^{2m+1}} \right),$$

that is

$$z = Ax^m + \frac{B_1}{x^{m+1}} \tag{2}$$

is the general solution of (1), and agrees with (2) Art. 15.

(*c*) Take Legendre's Equation, (2) Art. 16.

$$(1-x^2)\frac{d^2 z}{dx^2} - 2x\frac{dz}{dx} + m(m+1)z = 0; \tag{1}$$

given: $z = P_m(x)$, as a particular solution.

Here $P = \frac{-2x}{1-x^2}$, $\int P dx = \log(1-x^2)$, and $e^{-\int P dx} = \frac{1}{1-x^2}$.

Hence by (5)

$$z = P_m(x) \left(A + B \int \frac{dx}{(1-x^2)[P_m(x)]^2} \right) \tag{2}$$

is the general solution of (1) and must agree with (10) Art. 16, if m is an integer, and therefore

$$Q_m(x) = CP_m(x) \int \frac{dx}{(1-x^2)[P_m(x)]^2} \tag{3}$$

where C is as yet undetermined, and no constant term is to be understood with the integral in the second member.

(*d*) Take Bessel's Equation, (1) Art. 17.

$$\frac{d^2 z}{dx^2} + \frac{1}{x}\frac{dz}{dx} + \left(1 - \frac{m^2}{x^2}\right) z = 0; \tag{1}$$

given: $z = J_m(x)$, as a particular solution.

Here $P = \frac{1}{x}$, $\int P dx = \log x$, and $e^{-\int P dx} = \frac{1}{x}$. Hence by (5)

$$z = J_m(x) \left(A + B \int \frac{dx}{x[J_m(x)]^2} \right) \tag{2}$$

is the general solution of Bessel's Equation.

If $m = 0$ (2) becomes

$$z = J_0(x) \left(A + B \int \frac{dx}{x[J_0(x)]^2} \right) \tag{3}$$

and must agree with (9) Art. 17. Therefore

$$K_0(x) = CJ_0(x) \int \frac{dx}{x[J_0(x)]^2}, \tag{4}$$

where C is at present undetermined, and no constant term is to be taken with the integral.

The first considerable subject suggested by the problems which we have taken up in this introductory chapter is that of development in Trigonometric Series (v. Arts. 7 and 8).

CHAPTER II.

DEVELOPMENT IN TRIGONOMETRIC SERIES.

19. We have seen in Chapter I. that it is sometimes important to be able to express a given function of a variable x, in terms of the sines or of the cosines of multiples of x. The problem in its general form was first solved by Fourier in his "Analytic Theory of Heat" (1822), and its solution plays a very important part in most branches of modern Physics. Series involving only sines and cosines of whole multiples of x, that is series of the form

$$b_0 + b_1 \cos x + b_2 \cos 2x + \cdots + a_1 \sin x + a_2 \sin 2x + \cdots$$

are generally known as Fourier's series.

Let us endeavor to develop a given function of x in terms of $\sin x$, $\sin 2x$, $\sin 3x$, &c., in such a way that the function and the series shall be equal for all values of x between $x = 0$ and $x = \pi$.

To fix our ideas let us suppose that we have a curve,

$$y = f(x),$$

given, and that we wish to form the equation,

$$y = a_1 \sin x + a_2 \sin 2x + a_3 \sin 3x + \cdots,$$

of a curve which shall coincide with so much of the given curve as lies between the points corresponding to $x = 0$ and $x = \pi$. It is clear that in the equation

$$y = a_1 \sin x \tag{1}$$

a_1 may be determined so that the curve represented shall pass through any given point. For if we substitute in (1) the coördinates of the point in question we shall have an equation of the first degree in which a_1 is the only unknown quantity and which will therefore give us one and only one value for a_1.

In like manner the curve

$$y = a_1 \sin x + a_2 \sin 2x$$

may be made to pass through any two arbitrarily chosen points whose abscissas lie between 0 and π provided that the abscissas are not equal; and

$$y = a_1 \sin x + a_2 \sin 2x + a_3 \sin 3x + \cdots + a_n \sin nx$$

may be made to pass through any n arbitrarily chosen points whose abscissas lie between 0 and π provided as before that their abscissas are all different.

If, then, the given function $f(x)$ is of such a character that for each value of x between $x = 0$ and $x = \pi$ it has one and only one value, and if between

$x = 0$ and $x = \pi$ it is finite and continuous, or if discontinuous has only *finite discontinuities* (v. Int. Cal. Art. 83, p. 78), the coefficients in

$$y = a_1 \sin x + a_2 \sin 2x + a_3 \sin 3x + \cdots + a_n \sin nx \tag{2}$$

can be determined so that the curve represented by (2) will pass through any n arbitrarily chosen points of the curve

$$y = f(x) \tag{3}$$

whose abscissas lie between 0 and π and are all different, and these coefficients will have but one set of values.

For the sake of simplicity suppose that the n points are so chosen that their projections on the axis of X are equidistant.

Call $\dfrac{\pi}{n+1} = \Delta x$; then the coördinates of the n points will be $[\Delta x, f(\Delta x)]$, $[2\Delta x, f(2\Delta x)]$, $[3\Delta x, f(3\Delta x)]$, $\cdots$ $[n\Delta x, f(n\Delta x)]$. Substitute them in (2) and we have

$$\left.\begin{array}{l} f(\Delta x) = \ \ a_1 \sin \Delta x + \ \ a_2 \sin 2\Delta x + \ \ a_3 \sin 3\Delta x + \cdots + \ \ a_n \sin n\Delta x \\ f(2\Delta x) = a_1 \sin 2\Delta x + \ \ a_2 \sin 4\Delta x + \ \ a_3 \sin 6\Delta x + \cdots + a_n \sin 2n\Delta x \\ f(3\Delta x) = a_1 \sin 3\Delta x + \ \ a_2 \sin 6\Delta x + \ \ a_3 \sin 9\Delta x + \cdots + a_n \sin 3n\Delta x \\ \quad \vdots \qquad\qquad \vdots \qquad\qquad \vdots \qquad\qquad \vdots \qquad\qquad \vdots \\ f(n\Delta x) = a_1 \sin n\Delta x + a_2 \sin 2n\Delta x + a_3 \sin 3n\Delta x + \cdots + a_n \sin n^2\Delta x, \end{array}\right\} \tag{4}$$

n equations of the first degree to determine the n coefficients a_1, a_2, a_3, $\cdots$ a_n.

Not only can equations (4) be solved in theory, but they can be actually solved in any given case by a very simple and ingenious method due to Lagrange.

Let us take as an example the simple problem to determine the coefficients a_1, a_2, a_3, a_4, and a_5, so that

$$y = a_1 \sin x + a_2 \sin 2x + a_3 \sin 3x + a_4 \sin 4x + a_5 \sin 5x \tag{5}$$

shall pass through the five points of the line

$$y = x$$

which have the abscissas $\dfrac{\pi}{6}, \dfrac{2\pi}{6}, \dfrac{3\pi}{6}, \dfrac{4\pi}{6}$, and $\dfrac{5\pi}{6}$, $\dfrac{\pi}{6}$ here being Δx.

We must now solve the equations

$$\left.\begin{array}{l} \dfrac{\pi}{6} = a_1 \sin \dfrac{\pi}{6} + a_2 \sin \dfrac{2\pi}{6} + a_3 \sin \dfrac{3\pi}{6} + a_4 \sin \dfrac{4\pi}{6} + a_5 \sin \dfrac{5\pi}{6} \\ \dfrac{2\pi}{6} = a_1 \sin \dfrac{2\pi}{6} + a_2 \sin \dfrac{4\pi}{6} + a_3 \sin \dfrac{6\pi}{6} + a_4 \sin \dfrac{8\pi}{6} + a_5 \sin \dfrac{10\pi}{6} \\ \dfrac{3\pi}{6} = a_1 \sin \dfrac{3\pi}{6} + a_2 \sin \dfrac{6\pi}{6} + a_3 \sin \dfrac{9\pi}{6} + a_4 \sin \dfrac{12\pi}{6} + a_5 \sin \dfrac{15\pi}{6} \\ \dfrac{4\pi}{6} = a_1 \sin \dfrac{4\pi}{6} + a_2 \sin \dfrac{8\pi}{6} + a_3 \sin \dfrac{12\pi}{6} + a_4 \sin \dfrac{16\pi}{6} + a_5 \sin \dfrac{20\pi}{6} \\ \dfrac{5\pi}{6} = a_1 \sin \dfrac{5\pi}{6} + a_2 \sin \dfrac{10\pi}{6} + a_3 \sin \dfrac{15\pi}{6} + a_4 \sin \dfrac{20\pi}{6} + a_5 \sin \dfrac{25\pi}{6}. \end{array}\right\} \tag{6}$$

Multiply the first equation by $2\sin\frac{\pi}{6}$, the second by $2\sin\frac{2\pi}{6}$, the third by $2\sin\frac{3\pi}{6}$, the fourth by $2\sin\frac{4\pi}{6}$, the fifth by $2\sin\frac{5\pi}{6}$ and add the equations.

The coefficient of a_2 is

$$2\sin\frac{\pi}{6}\sin\frac{2\pi}{6}+2\sin\frac{2\pi}{6}\sin\frac{4\pi}{6}+2\sin\frac{3\pi}{6}\sin\frac{6\pi}{6}+2\sin\frac{4\pi}{6}\sin\frac{8\pi}{6}$$
$$+2\sin\frac{5\pi}{6}\sin\frac{10\pi}{6};$$

but
$$2\sin\frac{\pi}{6}\sin\frac{2\pi}{6}=\cos\frac{\pi}{6}-\cos\frac{3\pi}{6}, \text{\&c.}$$

Hence the coefficient of a_2 becomes

$$\left.\begin{aligned}\cos\frac{\pi}{6}+\cos\frac{2\pi}{6}+\cos\frac{3\pi}{6}+\cos\frac{4\pi}{6}+\cos\frac{5\pi}{6}\\ -\cos\frac{3\pi}{6}-\cos\frac{6\pi}{6}-\cos\frac{9\pi}{6}-\cos\frac{12\pi}{6}-\cos\frac{15\pi}{6}\end{aligned}\right\} \quad (7)$$

and this may be reduced by the aid of an important Trigonometric formula which we proceed to establish.

20. LEMMA.

$$\cos\theta+\cos 2\theta+\cos 3\theta+\cdots+\cos n\theta=-\frac{1}{2}+\frac{1}{2}\frac{\sin(2n+1)\frac{\theta}{2}}{\sin\frac{\theta}{2}}. \quad (1)$$

For let $S=\cos\theta+\cos 2\theta+\cos 3\theta+\cdots+\cos n\theta$ and multiply by $2\cos\theta$.

$$\begin{aligned}2S\cos\theta&=2\cos^2\theta+2\cos\theta\cos 2\theta+2\cos\theta\cos 3\theta+\cdots+2\cos\theta\cos n\theta\\ &=1+\cos\theta+\cos 2\theta+\cdots+\cos(n-1)\theta\\ &\quad+\cos 2\theta+\cos 3\theta+\cos 4\theta+\cdots+\cos(n+1)\theta\\ &=2S+1+\cos(n+1)\theta-\cos\theta-\cos n\theta.\end{aligned}$$ Hence

$$S=-\frac{1}{2}+\frac{\cos n\theta-\cos(n+1)\theta}{2(1-\cos\theta)}$$

or
$$S=-\frac{1}{2}+\frac{1}{2}\frac{\sin(2n+1)\frac{\theta}{2}}{\sin\frac{\theta}{2}}.$$ Q.E.D.

21. Applying (1) Art. 20 to (7) Art. 19 the coefficient of a_2 reduces to

$$\frac{\sin\frac{11\pi}{12}}{2\sin\frac{\pi}{12}}-\frac{\sin\frac{33\pi}{12}}{2\sin\frac{3\pi}{12}};$$

but $$\frac{11\pi}{12} = \pi - \frac{\pi}{12}, \text{ and } \frac{33\pi}{12} = 3\pi - \frac{3\pi}{12};$$

therefore $$\frac{\sin\left(\pi - \frac{\pi}{12}\right)}{2\sin\frac{\pi}{12}} - \frac{\sin\left(3\pi - \frac{3\pi}{12}\right)}{2\sin\frac{3\pi}{12}} = \frac{1}{2} - \frac{1}{2} = 0,$$

and a_2 vanishes.

In like manner it may be shown that the coefficients of a_3, a_4, and a_5 vanish.

The coefficient of a_1 is

$$\begin{aligned} & 2\sin^2\frac{\pi}{6} + 2\sin^2\frac{2\pi}{6} + 2\sin^2\frac{3\pi}{6} + 2\sin^2\frac{4\pi}{6} + 2\sin^2\frac{5\pi}{6} \\ = & 1 + 1 + 1 + 1 + 1 \\ & - \cos\frac{2\pi}{6} - \cos\frac{4\pi}{6} - \cos\frac{6\pi}{6} - \cos\frac{8\pi}{6} - \cos\frac{10\pi}{6} \\ = & 5 + \frac{1}{2} - \frac{\sin\frac{11\pi}{6}}{2\sin\frac{\pi}{6}} = 5\tfrac{1}{2} - \frac{\sin\left(2\pi - \frac{\pi}{6}\right)}{2\sin\frac{\pi}{6}} = 6. \end{aligned}$$

The first member of the final equation is

$\frac{2\pi}{6}\sin\frac{\pi}{6} + 2\frac{2\pi}{6}\sin\frac{2\pi}{6} + 2\frac{3\pi}{6}\sin\frac{3\pi}{6} + 2\frac{4\pi}{6}\sin\frac{4\pi}{6} + 2\frac{5\pi}{6}\sin\frac{5\pi}{6}$. Hence

$$a_1 = \frac{2}{6}\sum_{k=1}^{k=5}\frac{k\pi}{6}\sin\frac{k\pi}{6} = \frac{\pi}{6}(2+\sqrt{3}) = 2 \quad \text{approximately.}$$

If we multiply the first equation of (6) Art. 19 by $2\sin\frac{2\pi}{6}$, the second by $2\sin\frac{4\pi}{6}$, the third by $2\sin\frac{6\pi}{6}$, the fourth by $2\sin\frac{8\pi}{6}$, the fifth by $2\sin\frac{10\pi}{6}$, add and reduce as before we shall find

$$a_2 = \frac{2}{6}\sum_{k=1}^{k=5}\frac{k\pi}{6}\sin\frac{2k\pi}{6} = -\frac{\pi}{6}\sqrt{3} = -0.9;$$

and in like manner we get

$$a_3 = \frac{2}{6}\sum_{k=1}^{k=5}\frac{k\pi}{6}\sin\frac{3k\pi}{6} = \frac{\pi}{6} = 0.5$$

$$a_4 = \frac{2}{6}\sum_{k=1}^{k=5}\frac{k\pi}{6}\sin\frac{4k\pi}{6} = -\frac{\pi\sqrt{3}}{18} = -0.3$$

$$a_5 = \frac{2}{6}\sum_{k=1}^{k=5}\frac{k\pi}{6}\sin\frac{5k\pi}{6} = \frac{\pi}{6}(2-\sqrt{3}) = 0.1.$$

Therefore

$$y = 2\sin x - 0.9\sin 2x + 0.5\sin 3x - 0.3\sin 4x + 0.1\sin 5x \qquad (1)$$

cuts the curve $y = x$ at the five points whose abscissas are $\frac{\pi}{6}, \frac{2\pi}{6}, \frac{3\pi}{6}, \frac{4\pi}{6}$, and $\frac{5\pi}{6}$.

22. The equations (4) Art. 19 can be solved by exactly the same device. To find any coefficient a_m multiply the first equation by $2\sin m\Delta x$, the second by $2\sin 2m\Delta x$, the third by $2\sin 3m\Delta x$, &c. and add.

The coefficient of any other a as a_k in the resulting equation will be

$$\begin{aligned}
&2\sin k\Delta x\sin m\Delta x + 2\sin 2k\Delta x\sin 2m\Delta x + 2\sin 3k\Delta x\sin 3m\Delta x + \cdots \\
&\qquad + 2\sin nk\Delta x\sin nm\Delta x \\
&= \cos(m-k)\Delta x + \cos 2(m-k)\Delta x + \cos 3(m-k)\Delta x + \cdots + \cos n(m-k)\Delta x \\
&- \cos(m+k)\Delta x - \cos 2(m+k)\Delta x - \cos 3(m+k)\Delta x - \cdots - \cos n(m+k)\Delta x \\
&= \frac{\sin\frac{2n+1}{2}(m-k)\Delta x}{2\sin\frac{(m-k)\Delta x}{2}} - \frac{\sin\frac{2n+1}{2}(m+k)\Delta x}{2\sin\frac{(m+k)\Delta x}{2}}; \qquad \text{by (1) Art. 20.}
\end{aligned}$$

$$\frac{2n+1}{2} = n + 1 - \frac{1}{2} \quad \text{and} \quad (n+1)\Delta x = \pi.$$

Hence the coefficient of a_k may be written

$$\frac{\sin\left[(m-k)\pi - \frac{(m-k)\Delta x}{2}\right]}{2\sin\frac{(m-k)\Delta x}{2}} - \frac{\sin\left[(m+k)\pi - \frac{(m+k)\Delta x}{2}\right]}{2\sin\frac{(m+k)\Delta x}{2}}$$

but this is equal to $\frac{1}{2} - \frac{1}{2}$ or $-\frac{1}{2} + \frac{1}{2}$ according as $m - k$ is odd or even and so is zero in either case.

The coefficient of a_m will be

$$\begin{aligned}
&2\sin^2 m\Delta x + 2\sin^2 2m\Delta x + 2\sin^2 3m\Delta x + \cdots + 2\sin^2 nm\Delta x \\
&= \quad 1 \quad + \quad 1 \quad + \quad 1 \quad + \quad \cdots \quad + \quad 1 \\
&\quad - \cos 2m\Delta x - \cos 4m\Delta x - \cos 6m\Delta x - \cdots - \cos 2nm\Delta x \\
&= n + \frac{1}{2} - \frac{\sin(2n+1)m\Delta x}{2\sin m\Delta x}, \quad \text{by (1) Art. 20.}
\end{aligned}$$

But $\quad (2n+1)m\Delta x = 2m(n+1)\Delta x - m\Delta x = 2m\pi - m\Delta x,$

therefore $\quad \dfrac{\sin(2n+1)m\Delta x}{2\sin m\Delta x} = \dfrac{\sin(2m\pi - m\Delta x)}{2\sin m\Delta x} = -\dfrac{1}{2},$

and the coefficient of a_m is $n+1$.

The first member of our final equation will be

$$2\sum_{k=1}^{k=n} f(k\Delta x)\sin km\Delta x.$$

Hence

$$a_m = \frac{2}{n+1}\sum_{k=1}^{k=n} f(k\Delta x)\sin km\Delta x, \tag{1}$$

and the curve

$$y = a_1 \sin x + a_2 \sin 2x + \cdots + a_n \sin nx, \tag{2}$$

where the coefficients are given by (1) will pass through the n points of the curve $y = f(x)$ whose abscissas are Δx, $2\Delta x$, $3\Delta x$, $\cdots$ $n\Delta x$. Δx being $\dfrac{\pi}{n+1}$.

It should be noted that since the n equations (4) Art. 19 are all of the first degree there will exist only one set of values for the n quantities a_1, a_2, a_3, $\cdots$ a_n that can satisfy these equations. Consequently the solution which we have obtained is the only solution possible.

23. The result just obtained obviously holds good no matter how great a value of n may be taken.

If now we suppose n indefinitely increased the two curves (2) Art. 22 and $y = f(x)$ will come nearer and nearer to coinciding throughout the whole of their portions between $x = 0$ and $x = \pi$, and consequently the limiting form that equation (2) Art. 22 approaches as n is indefinitely increased will represent a curve absolutely coinciding between the values of x in question with $y = f(x)$.

Let us see what limiting value a_m approaches as n is indefinitely increased.

$$a_m = \frac{2}{n+1}\sum_{k=1}^{k=n} f(k\Delta x)\sin km\Delta x \qquad \text{(1) Art. 22.}$$

$$= \frac{2\Delta x}{\pi}\sum_{k=1}^{k=n} f(k\Delta x)\sin km\Delta x$$

$$= \frac{2}{\pi}\Big[f(\Delta x)\sin m\Delta x.\Delta x + f(2\Delta x)\sin 2m\Delta x.\Delta x + \cdots + f(n\Delta x)\sin nm\Delta x.\Delta x\Big]$$

$$= \frac{2}{\pi}\Big[f(\Delta x)\sin m\Delta x.\Delta x + f(2\Delta x)\sin 2m\Delta x.\Delta x + \cdots + f(\pi-\Delta x)\sin m(\pi-\Delta x).\Delta x\Big]$$

since $\Delta x = \dfrac{\pi}{n+1}$.

As n is increased indefinitely Δx approaches zero as a limit. Hence the limiting value of a_m as n increases indefinitely is

$$\frac{2}{\pi}\lim_{\Delta x \doteq 0}\left[\begin{array}{l} f(\Delta x)\sin m\Delta x.\Delta x + f(2\Delta x)\sin 2m\Delta x.\Delta x + \cdots \\ \qquad\qquad + f(\pi - \Delta x)\sin m(\pi - \Delta x).\Delta x \end{array}\right]^{1}$$

$$= \frac{2}{\pi}\int_0^\pi f(x)\sin mx.dx. \quad \text{[v. Int. Cal. Arts. 80, 81.]}$$

Hence
$$f(x) = a_1\sin x + a_2\sin 2x + a_3\sin 3x + \cdots, \tag{2}$$

where any coefficient a_m is given by the formula

$$a_m = \frac{2}{\pi}\int_0^\pi f(x)\sin mx.dx, \tag{3}$$

is a true development of $f(x)$ for all values of x between $x = 0$ and $x = \pi$ *provided that the series* (2) *is convergent*, for it is in that case only that we can assume that the limiting value of the second member of (2) Art. 22 can be obtained by adding the limiting values of the several terms.

When $x = 0$ and when $x = \pi$ every term in the second member of (2) is zero, and the second member is zero and will not be equal to $f(x)$ unless $f(x)$ is itself zero when $x = 0$ and $x = \pi$; but even when $f(x)$ is not zero for $x = 0$ and $x = \pi$ the development given above holds good for any value of x between zero and π no matter how near it may be taken to either of these values.

24. Instead of actually performing the elimination in equations (4) Art. 19 and getting a formula for a_m in terms of n, and then letting n increase indefinitely, we might have saved labor by the following method.

Return to equations (4) Art. 19 and multiply the first by $\Delta x\sin m\Delta x$, the second by $\Delta x\sin 2m\Delta x$, and so on, that is multiply each equation by Δx times the coefficient of a_m in that equation, and then add the equations.

We get as the coefficient of a_k

$$\sin k\Delta x\sin m\Delta x.\Delta x + \sin 2k\Delta x\sin 2m\Delta x.\Delta x + \cdots + \sin nk\Delta x\sin nm\Delta x.\Delta x.$$

Let us find its limiting value as n is indefinitely increased. It may be written, since $(n + 1)\Delta x = \pi$,

$$\lim_{\Delta x \doteq 0}\left[\begin{array}{l} \sin k\Delta x\sin m\Delta x.\Delta x + \sin 2k\Delta x\sin 2m\Delta x.\Delta x + \cdots \\ \qquad\qquad + \sin k(\pi - \Delta x)\sin m(\pi - \Delta x).\Delta x \end{array}\right]$$

$$= \int_0^\pi \sin kx\sin mx.dx;$$

[1] We shall use the sign $\doteq$ for *approaches*. $\Delta x \doteq 0$ is read Δx approaches zero.

but $$\int_0^\pi \sin kx \sin mx.dx = \tfrac{1}{2}\int_0^\pi [\cos(m-k)x - \cos(m+k)x]dx$$
$$= 0 \text{ if } m \text{ and } k \text{ are not equal.}$$

The coefficient of a_m is

$$\Delta x(\sin^2 m\Delta x + \sin^2 2m\Delta x + \sin^2 3m\Delta x + \cdots + \sin^2 nm\Delta x).$$

Its limiting value

$$\underset{\Delta x \doteq 0}{\text{limit}}\left[\sin^2 m\Delta x.\Delta x + \sin^2 2m\Delta x.\Delta x + \cdots + \sin^2 m(\pi - \Delta x)\Delta x\right]$$
$$= \int_0^\pi \sin^2 mx.dx = \frac{\pi}{2}.$$

The first member is

$$f(\Delta x)\sin m\Delta x.\Delta x + f(2\Delta x)\sin 2m\Delta x.\Delta x + \cdots + f(n\Delta x)\sin mn\Delta x.\Delta x$$

and its limiting value is

$$\int_0^\pi f(x)\sin mx.dx.$$

Hence the limiting form approached by the final equation as n is increased is

$$\int_0^\pi f(x)\sin mx.dx = \frac{\pi}{2}a_m.$$

Whence $$a_m = \frac{2}{\pi}\int_0^\pi f(x)\sin mx.dx$$ as before.

This method is practically the same as *multiplying the equation*

$$f(x) = a_1 \sin x + a_2 \sin 2x + a_3 \sin 3x + \cdots \tag{1}$$

by $\sin mx.dx$ *and integrating both members from zero to* π.

It is exceedingly important to realize that the short method of determining any coefficient a_m of the series (1) which has just been described in the italicized paragraph, is essentially the same as that of obtaining a_m by actual elimination from the equations (4) Art. 19, and then supposing n to increase indefinitely, thus making the curves (3) Art. 19 and (2) Art. 19 absolutely coincide between the values of x which are taken as the limits of the definite integration.

25. We see, then, that any function of x which is single-valued, finite, and continuous between $x = 0$ and $x = \pi$, or if discontinuous has only finite discontinuities each of which is preceded and succeeded by continuous portions, can probably be developed into a series of the form

$$f(x) = a_1 \sin x + a_2 \sin 2x + a_3 \sin 3x + \cdots \tag{1}$$

where
$$a_m = \frac{2}{\pi}\int_0^\pi f(x) \sin mx.dx = \frac{2}{\pi}\int_0^\pi f(\alpha) \sin m\alpha.d\alpha; \tag{2}$$

and the series and the function will be identical for all values of x between $x = 0$ and $x = \pi$, not including the values $x = 0$ and $x = \pi$ unless the given function is equal to zero for those values.

An elaborate investigation of the question of the convergence of the series (1), for which we have not space, entirely confirms the result formulated above[2] and shows in addition that at a point of finite discontinuity the series has a value equal to half the sum of the two values which the function approaches as we approach the point in question from opposite sides.

The investigation which we have made in the preceding sections establishes the fact that the curve represented by $y = f(x)$ need not follow the same mathematical law throughout its length, but may be made up of portions of entirely different curves. For example, a broken line or a locus consisting of finite parts of several different and disconnected straight lines can be represented perfectly well by $y =$ a sine series.

26. Let us obtain a few sine developments.

(*a*) Let
$$f(x) = x. \tag{1}$$

We have
$$x = a_1 \sin x + a_2 \sin 2x + a_3 \sin 3x + \cdots \tag{2}$$

where
$$a_m = \frac{2}{\pi}\int_0^\pi x \sin mx.dx \tag{3}$$

$$\int x \sin mx.dx = \frac{1}{m^2}(\sin mx - mx \cos mx),$$

$$\int_0^\pi x \sin mx.dx = -\frac{(-1)^m \pi}{m},$$

and
$$x = 2\left(\frac{\sin x}{1} - \frac{\sin 2x}{2} + \frac{\sin 3x}{3} - \frac{\sin 4x}{4} + \cdots\right) \tag{4}$$

[2]Provided the function has not an infinite number of maxima and minima in the neighborhood of a point. v. Arts. 37–38.

(*b*) Let
$$f(x) = 1. \tag{1}$$

$$a_m = \frac{2}{\pi}\int_0^\pi \sin mx.dx; \tag{2}$$

$$\int \sin mx.dx = -\frac{\cos mx}{m},$$

$$\int_0^\pi \sin mx.dx = \frac{1}{m}(1-\cos m\pi) = \frac{1}{m}[1-(-1)^m]$$
$$= 0 \text{ if } m \text{ is even}$$
$$= \frac{2}{m} \text{ if } m \text{ is odd.}$$

Hence
$$1 = \frac{4}{\pi}\left(\frac{\sin x}{1} + \frac{\sin 3x}{3} + \frac{\sin 5x}{5} + \frac{\sin 7x}{7} + \cdots\right). \tag{3}$$

It is to be noticed that (3) gives at once a sine development for any constant c. It is,
$$c = \frac{4c}{\pi}\left(\frac{\sin x}{1} + \frac{\sin 3x}{3} + \frac{\sin 5x}{5} + \cdots\right). \tag{4}$$

If we substitute $x = \frac{\pi}{2}$ in (4) (*a*) or (3) (*b*) we get a familiar result, namely
$$\frac{\pi}{4} = \frac{1}{1} - \frac{1}{3} + \frac{1}{5} - \frac{1}{7} + \cdots, \tag{5}$$
a formula usually derived by substituting $x = 1$ in the power series for $\tan^{-1} x$. (v. Dif. Cal. Art. 135.)

(4) (*a*) does not hold good when $x = \pi$, and (3) (*b*) fails when $x = 0$ and when $x = \pi$, for in all these cases the series reduces to zero.

(*c*) Let $f(x) = x$ from $x = 0$ to $x = \frac{\pi}{2}$ and $f(x) = \pi - x$ from $x = \frac{\pi}{2}$ to $x = \pi$. That is, let $y = f(x)$ represent the broken line in the figure.

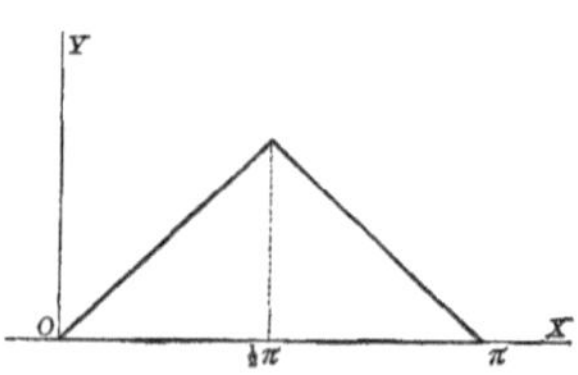

As the mathematical expression for $f(x)$ is different in the two halves of the curve we must break up
$$\int_0^\pi f(x)\sin mx.dx \quad \text{into} \quad \int_0^{\frac{\pi}{2}} f(x)\sin mx.dx + \int_{\frac{\pi}{2}}^\pi f(x)\sin mx.dx.$$

We have, then,

$$a_m = \frac{2}{\pi}\int_0^{\frac{\pi}{2}} x \sin mx.dx + \frac{2}{\pi}\int_{\frac{\pi}{2}}^{\pi} (\pi - x)\sin mx.dx \tag{1}$$

$$= \frac{4}{m^2\pi}\sin m\frac{\pi}{2}.$$

But

$$\begin{aligned} \sin m\frac{\pi}{2} &= 1 \quad \text{if} \quad m = 1 \quad \text{or} \quad 4k+1 \\ &= 0 \quad \text{“} \quad m = 2 \quad \text{“} \quad 4k+2 \\ &= -1 \quad \text{“} \quad m = 3 \quad \text{“} \quad 4k+3 \\ &= 0 \quad \text{“} \quad m = 4 \quad \text{“} \quad 4k. \end{aligned}$$

Hence if $y = f(x)$ represents our broken line,

$$f(x) = \frac{4}{\pi}\left(\frac{\sin x}{1^2} - \frac{\sin 3x}{3^2} + \frac{\sin 5x}{5^2} - \frac{\sin 7x}{7^2} + \cdots\right). \tag{2}$$

When $x = \frac{\pi}{2}$ $f(x) = \frac{\pi}{2}$ and we have

$$\frac{\pi^2}{8} = \frac{1}{1^2} + \frac{1}{3^2} + \frac{1}{5^2} + \frac{1}{7^2} + \cdots \tag{3}$$

(*d*) As a case where the function has a finite discontinuity, let

$$f(x) = 1 \quad \text{from} \quad x = 0 \quad \text{to} \quad x = \frac{\pi}{2} \quad \text{and}$$
$$f(x) = 0 \quad \text{“} \quad x = \frac{\pi}{2} \quad \text{“} \quad x = \pi.$$

$y = f(x)$ will in this case represent the locus in the figure.

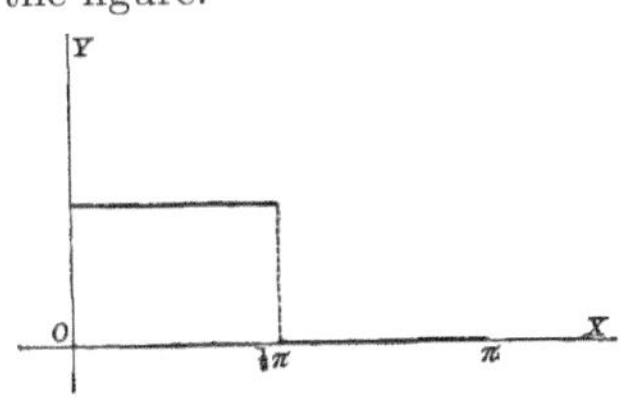

As before

$$\int_0^{\pi} f(x)\sin mx.dx = \int_0^{\frac{\pi}{2}} f(x)\sin mx.dx + \int_{\frac{\pi}{2}}^{\pi} f(x)\sin mx.dx.$$

$$a_m = \frac{2}{\pi}\int_0^{\frac{\pi}{2}} \sin mx.dx + \frac{2}{\pi}\int_{\frac{\pi}{2}}^{\pi} 0.\sin mx.dx. \tag{1}$$

$$a_m = \frac{2}{\pi}\int_0^{\frac{\pi}{2}} \sin mx.dx = \frac{2}{\pi}\frac{1}{m}\left(1 - \cos m\frac{\pi}{2}\right).$$

But
$$\begin{aligned}\cos m\frac{\pi}{2} &= \ \ 0 \text{ if } m=1 \text{ or } 4k+1\\ &= -1 \text{ `` } m=2 \text{ `` } 4k+2\\ &= \ \ 0 \text{ `` } m=3 \text{ `` } 4k+3\\ &= \ \ 1 \text{ `` } m=4 \text{ `` } 4k.\end{aligned}$$

Hence

$$f(x)=\frac{2}{\pi}\left(\frac{\sin x}{1}+\frac{2\sin 2x}{2}+\frac{\sin 3x}{3}+\frac{\sin 5x}{5}+\frac{2\sin 6x}{6}+\frac{\sin 7x}{7}+\cdots\right). \quad (2)$$

If $x=\frac{\pi}{2}$ the second member of (2) reduces to $\frac{1}{2}$, for

$$\frac{2}{\pi}\left(\frac{1}{1}-\frac{1}{3}+\frac{1}{5}-\frac{1}{7}+\cdots\right)=\frac{1}{2} \qquad \text{by (5) } (b);$$

and we see that the series represents the function completely for all values of x between $x=0$ and $x=\pi$ except for $x=\frac{\pi}{2}$ and there it has a value which is the mean of the values approached by the function as x approaches $\frac{\pi}{2}$ from opposite sides.

EXAMPLES.

Obtain the following developments:—

(1) $$x^2=\frac{2}{\pi}\left[\left(\frac{\pi^2}{1}-\frac{4}{1^3}\right)\sin x-\frac{\pi^2}{2}\sin 2x+\left(\frac{\pi^2}{3}-\frac{4}{3^3}\right)\sin 3x-\frac{\pi^2}{4}\sin 4x+\left(\frac{\pi^2}{5}-\frac{4}{5^3}\right)\sin 5x-\cdots\right].$$

(2) $$x^3=\frac{2}{\pi}\left[\left(\frac{\pi^3}{1}-\frac{6\pi}{1^3}\right)\sin x-\left(\frac{\pi^3}{2}-\frac{6\pi}{2^3}\right)\sin 2x+\left(\frac{\pi^3}{3}-\frac{6\pi}{3^3}\right)\sin 3x-\left(\frac{\pi^3}{4}-\frac{6\pi}{4^3}\right)\sin 4x+\cdots\right].$$

(3) $$f(x)=\frac{2}{\pi}\left[\frac{\sin x}{1^2}+\frac{\pi}{2^2}\sin 2x-\frac{\sin 3x}{3^2}-\frac{2\pi}{4^2}\sin 4x+\frac{\sin 5x}{5^2}+\frac{3\pi}{6^2}\sin 6x-\cdots\right],$$

if $f(x)=x$ from $x=0$ to $x=\frac{\pi}{2}$ and $f(x)=0$ from $x=\frac{\pi}{2}$ to $x=\pi$.

(4) $$\sin \mu x = \frac{2}{\pi} \sin \mu\pi \left[\frac{\sin x}{1^2 - \mu^2} - \frac{2 \sin 2x}{2^2 - \mu^2} + \frac{3 \sin 3x}{3^2 - \mu^2} - \frac{4 \sin 4x}{4^2 - \mu^2} + \cdots \right]$$

if μ is a fraction.

(5) $$e^x = \frac{2}{\pi} \left[\frac{1}{2}(1 + e^\pi) \sin x + \frac{2}{5}(1 - e^\pi) \sin 2x + \frac{3}{10}(1 + e^\pi) \sin 3x + \frac{4}{17}(1 - e^\pi) \sin 4x + \cdots \right].$$

(6) $$\sinh x = \frac{2 \sinh \pi}{\pi} \left[\frac{1}{2} \sin x - \frac{2}{5} \sin 2x + \frac{3}{10} \sin 3x - \frac{4}{17} \sin 4x + \cdots \right].$$

(7) $$\cosh x = \frac{2}{\pi} \left[\frac{1}{2}(1 + \cosh \pi) \sin x + \frac{2}{5}(1 - \cosh \pi) \sin 2x + \frac{3}{10}(1 + \cosh \pi) \sin 3x + \cdots \right].$$

27. Let us now try to develop a given function of x in a series of cosines. As before suppose that $f(x)$ has a single value for each value of x between $x = 0$ and $x = \pi$, that it does not become infinite between $x = 0$ and $x = \pi$, and that if discontinuous it has only finite discontinuities.

Assume

$$f(x) = b_0 + b_1 \cos x + b_2 \cos 2x + b_3 \cos 3x + \cdots \tag{1}$$

To determine any coefficient b_m multiply (1) by $\cos mx.dx$ and integrate each term from 0 to π.

$$\int_0^\pi b_0 \cos mx.dx = 0.$$

$$\int_0^\pi b_k \cos kx \cos mx.dx = \frac{b_k}{2} \int_0^\pi [\cos(m - k)x + \cos(m + k)x] dx$$

$$= 0 \text{ if } m \text{ and } k \text{ are not equal.}$$

$$\int b_m \cos^2 mx.dx = \frac{b_m}{2m}(mx + \cos mx \sin mx),$$

$$\int_0^\pi b_m \cos^2 mx.dx = \frac{\pi}{2} b_m, \qquad \text{if } m \text{ is not zero.}$$

Hence $$b_m = \frac{2}{\pi} \int_0^\pi f(x) \cos mx.dx = \frac{2}{\pi} \int_0^\pi f(\alpha) \cos m\alpha.d\alpha, \tag{2}$$

if m is not zero.

To get b_0 multiply (1) by dx and integrate from zero to π.

$$\int_0^\pi b_0 dx = b_0\pi,$$

$$\int_0^\pi b_k \cos kx.dx = 0.$$

Hence
$$b_0 = \frac{1}{\pi}\int_0^\pi f(x)dx = \frac{1}{\pi}\int_0^\pi f(\alpha)d\alpha, \tag{3}$$

which is just half the value that would be given by formula (2) if zero were substituted for m.

To save a separate formula (1) is usually written

$$f(x) = \tfrac{1}{2}b_0 + b_1\cos x + b_2\cos 2x + b_3\cos 3x + \cdots \tag{4}$$

and then the formula

$$b_m = \frac{2}{\pi}\int_0^\pi f(x)\cos mx.dx = \frac{2}{\pi}\int_0^\pi f(\alpha)\cos m\alpha.d\alpha \tag{2}$$

will give b_0 as well as the other coefficients.

It is important to see clearly that what we have just done in determining the coefficients of (1) is equivalent to taking $n+1$ terms of (4), substituting in

$$y = \tfrac{1}{2}b_0 + b_1\cos x + b_2\cos 2x + \cdots + b_n\cos nx \tag{5}$$

in turn the coördinates of the $n+1$ points of the curve

$$y = f(x)$$

whose projections on the axis of X are equidistant, determining $b_0, b_1, b_2, \cdots b_n$ by elimination from the $n+1$ resulting equations, and then taking the limiting values they approach as n is indefinitely increased. (v. Art. 24.)

If $\Delta x = \dfrac{\pi}{n+1}$ the abscissas of the $n+1$ points used are 0, Δx, $2\Delta x$, $3\Delta x$, $\cdots$ $n\Delta x$, so that we should expect our cosine development to hold for $x=0$ as well as for values of x between zero and π.

28. Let us take one or two examples:

(*a*) Let
$$f(x) = x. \tag{1}$$

$$b_0 = \frac{2}{\pi}\int_0^\pi x\,dx = \frac{2}{\pi}\frac{\pi^2}{2} = \pi.$$

$$b_m = \frac{2}{\pi}\int_0^\pi x\cos mx.dx = \frac{2}{m^2\pi}(\cos m\pi - 1) = \frac{2}{m^2\pi}[(-1)^m - 1].$$

$$\text{Hence} \quad x = \frac{\pi}{2} - \frac{4}{\pi}\left(\cos x + \frac{\cos 3x}{3^2} + \frac{\cos 5x}{5^2} + \frac{\cos 7x}{7^2} + \cdots\right). \tag{2}$$

(2) holds good not only for values of x between zero and π but for $x = 0$ and $x = \pi$ as well, since for these values we have

$$0 = \frac{\pi}{2} - \frac{4}{\pi}\left(1 + \frac{1}{3^2} + \frac{1}{5^2} + \frac{1}{7^2} + \cdots\right) \tag{3}$$

and

$$\pi = \frac{\pi}{2} + \frac{4}{\pi}\left(1 + \frac{1}{3^2} + \frac{1}{5^2} + \frac{1}{7^2} + \cdots\right) \tag{4}$$

which are true by Art. 26 $(c)(3)$.

(*b*) Let

$$f(x) = x \sin x. \tag{1}$$

$$b_0 = \frac{2}{\pi}\int_0^\pi x \sin x.dx = \frac{2}{\pi}\pi = 2,$$

$$b_1 = \frac{2}{\pi}\int_0^\pi x \sin x \cos x.dx = \frac{1}{\pi}\int_0^\pi x \sin 2x.dx = -\frac{1}{2},$$

$$\begin{aligned} b_m = \frac{2}{\pi}\int_0^\pi x \sin x \cos mx.dx &= \frac{1}{\pi}\int_0^\pi [x \sin(m+1)x - x \sin(m-1)x]dx \\ &= \frac{2}{(m-1)(m+1)} \quad \text{if } m \text{ is odd} \\ &= -\frac{2}{(m-1)(m+1)} \quad \text{if } m \text{ is even.} \end{aligned}$$

Hence

$$x \sin x = 1 - \frac{\cos x}{2} - \frac{2\cos 2x}{1.3} + \frac{2\cos 3x}{2.4} - \frac{2\cos 4x}{3.5} + \cdots \tag{2}$$

If $x = \dfrac{\pi}{2}$ we have

$$\frac{\pi}{4} = \frac{1}{2} + \frac{1}{1.3} - \frac{1}{3.5} + \frac{1}{5.7} - \cdots. \tag{3}$$

EXAMPLES.

Obtain the following developments:

$$(1)\quad f(x) = \frac{\pi}{4} - \frac{2}{\pi}\left[\frac{\cos 2x}{1^2} + \frac{\cos 6x}{3^2} + \frac{\cos 10x}{5^2} + \frac{\cos 14x}{7^2} + \cdots\right]$$

if $f(x) = x$ from $x = 0$ to $x = \dfrac{\pi}{2}$ and $f(x) = \pi - x$ from $x = \dfrac{\pi}{2}$ to $x = \pi$.

$$(2)\quad f(x) = \frac{1}{2} + \frac{2}{\pi}\left[\frac{\cos x}{1} - \frac{\cos 3x}{3} + \frac{\cos 5x}{5} - \frac{\cos 7x}{7} + \cdots\right],$$

if $f(x) = 1$ from $x = 0$ to $x = \dfrac{\pi}{2}$ and $f(x) = 0$ from $x = \dfrac{\pi}{2}$ to $x = \pi$.

$$(3)\quad x^2 = \frac{\pi^2}{3} - 4\left[\frac{\cos x}{1^2} - \frac{\cos 2x}{2^2} + \frac{\cos 3x}{3^2} - \frac{\cos 4x}{4^2} + \cdots\right],$$

$$(4)\quad x^3 = \frac{\pi^3}{4} - \frac{6}{\pi}\left[\left(\frac{\pi^2}{1^2} - \frac{4}{1^4}\right)\cos x - \frac{\pi^2}{2^2}\cos 2x + \left(\frac{\pi^2}{3^2} - \frac{4}{3^4}\right)\cos 3x \right.$$
$$\left. - \frac{\pi^2}{4^2}\cos 4x + \left(\frac{\pi^2}{5^2} - \frac{4}{5^4}\right)\cos 5x - \cdots\right],$$

$$(5)\quad f(x) = \frac{\pi}{8} + \frac{2}{\pi}\left[\left(\frac{\pi}{2} - 1\right)\cos x - \frac{2}{2^2}\cos 2x - \frac{1}{3^2}\left(\frac{3\pi}{2} + 1\right)\cos 3x \right.$$
$$\left. + \frac{1}{5^2}\left(\frac{5\pi}{2} - 1\right)\cos 5x - \frac{2}{6^2}\cos 6x - \cdots\right],$$

if $f(x) = x$ from $x = 0$ to $x = \dfrac{\pi}{2}$ and $f(x) = 0$ from $x = \dfrac{\pi}{2}$ to $x = \pi$.

$$(6)\quad e^x = \frac{2}{\pi}\left[\frac{1}{2}(e^\pi - 1) - \frac{1}{1+1^2}(e^\pi + 1)\cos x + \frac{1}{1+2^2}(e^\pi - 1)\cos 2x \right.$$
$$\left. - \frac{1}{1+3^2}(e^\pi + 1)\cos 3x + \cdots\right],$$

$$(7)\quad \cosh x = \frac{2\sinh\pi}{\pi}\left[\frac{1}{2} - \frac{1}{2}\cos x + \frac{1}{5}\cos 2x - \frac{1}{10}\cos 3x \right.$$
$$\left. + \frac{1}{17}\cos 4x - \cdots\right],$$

$$(8)\quad \sinh x = \frac{2}{\pi}\left[\frac{1}{2}(\cosh\pi - 1) - \frac{1}{2}(\cosh\pi + 1)\cos x \right.$$
$$\left. + \frac{1}{5}(\cosh\pi - 1)\cos 2x - \frac{1}{10}(\cosh\pi + 1)\cos 3x + \cdots\right],$$

$$(9)\quad \cos\mu x = \frac{2\mu\sin\mu\pi}{\pi}\left[\frac{1}{2\mu^2} - \frac{\cos x}{\mu^2 - 1^2} + \frac{\cos 2x}{\mu^2 - 2^2} - \frac{\cos 3x}{\mu^2 - 3^2} \right.$$
$$\left. + \frac{\cos 4x}{\mu^2 - 4^2} - \cdots\right],$$

if μ is a fraction.

29. Although any function can be expressed both as a sine series and as a cosine series, and the function and either series will be equal for all values of x between zero and π, there is a decided difference in the two series for other values of x.

Both series are periodic functions of x having the period 2π. If then we let y equal the series in question and construct the portion of the corresponding curve which lies between the values $x = -\pi$ and $x = \pi$ the whole curve will consist of repetitions of this portion.

Since $\sin mx = -\sin(-mx)$ the ordinate corresponding to any value of x between $-\pi$ and zero in the sine curve; will be the negative of the ordinate corresponding to the same value of x with the positive sign. In other words the curve

$$y = a_1 \sin x + a_2 \sin 2x + a_3 \sin 3x + \cdots \tag{1}$$

is symmetrical with respect to the origin.

Since $\cos mx = \cos(-mx)$ the ordinate corresponding to any value of x between $-\pi$ and zero in the cosine curve will be the same as the ordinate belonging to the corresponding positive value of x. In other words the curve

$$y = \tfrac{1}{2}b_0 + b_1 \cos x + b_2 \cos 2x + b_3 \cos 3x + \cdots \tag{2}$$

is symmetrical with respect to the axis of Y.

If then $f(x) = -f(-x)$, that is if $f(x)$ is an *odd* function the sine series corresponding to it will be equal to it for all values of x between $-\pi$ and π, except perhaps for the value $x = 0$ for which the series will necessarily be zero.

If $f(x) = f(-x)$, that is if $f(x)$ is an *even* function the cosine series corresponding to it will be equal to it for all values of x between $x = -\pi$ and $x = \pi$, not excepting the value $x = 0$.

As an example of the difference between the sine and cosine developments of the same function let us take the series for x.

$$y = 2\left[\sin x - \frac{\sin 2x}{2} + \frac{\sin 3x}{3} - \frac{\sin 4x}{4} + \cdots\right] \tag{3}$$

$$y = \frac{\pi}{2} - \frac{4}{\pi}\left[\cos x + \frac{\cos 3x}{3^2} + \frac{\cos 5x}{5^2} + \frac{\cos 7x}{7^2} + \cdots\right] \tag{4}$$

[v. Art. 26(a) and Art. 28(a)]. (3) represents the curve

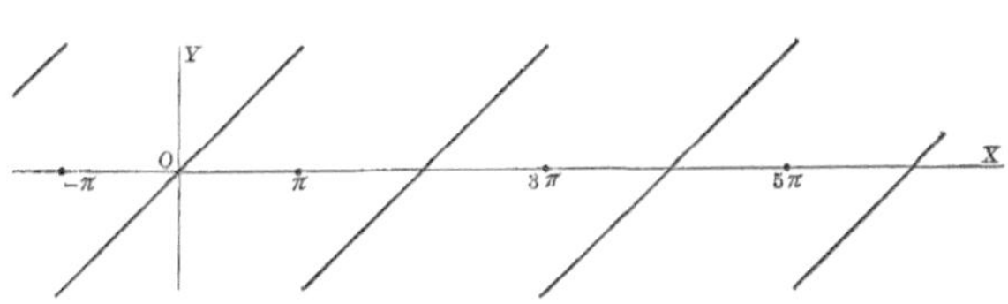

and (4) the curve

Both coincide with $y = x$ from $x = 0$ to $x = \pi$, (3) coincides with $y = x$ from $x = -\pi$ to $x = \pi$, and neither coincides with $y = x$ for values of x less than $-\pi$ or greater than π. Moreover (3), in addition to the continuous portions of the locus represented in the figure, gives the isolated points $(-\pi, 0)$ $(\pi, 0)$ $(3\pi, 0)$ &c.

30. We have seen that if $f(x)$ is an *odd* function its development in sine series holds for all values of x from $-\pi$ to π, as does the development of $f(x)$ in cosine series if $f(x)$ is an *even* function.

Thus the developments of Art. 26(a), Art. 26 Exs. (2), (4), (6); Art. 28(b), Art. 28 Exs. (3), (7), (9) are valid for all values of x between $-\pi$ and π.

Any function of x can be developed into a Trigonometric series to which it is equal for all values of x between $-\pi$ and π.

Let $f(x)$ be the given function of x. It can be expressed as the sum of an even function of x and an odd function of x by the following device.

$$f(x) = \frac{f(x) + f(-x)}{2} + \frac{f(x) - f(-x)}{2} \tag{1}$$

identically; but $\dfrac{f(x) + f(-x)}{2}$ is not changed by reversing the sign of x and is therefore an *even* function of x; and when we reverse the sign of x, $\dfrac{f(x) - f(-x)}{2}$ is affected only to the extent of having its sign reversed and is consequently an *odd* function of x.

Therefore for all values of x between $-\pi$ and π

$$\frac{f(x) + f(-x)}{2} = \frac{1}{2}b_0 + b_1 \cos x + b_2 \cos 2x + b_3 \cos 3x + \cdots$$

where $$b_m = \frac{2}{\pi}\int_0^\pi \frac{f(x) + f(-x)}{2} \cos mx.dx;$$ and

$$\frac{f(x) - f(-x)}{2} = a_1 \sin x + a_2 \sin 2x + a_3 \sin 3x + \cdots$$

where $$a_m = \frac{2}{\pi}\int_0^\pi \frac{f(x) - f(-x)}{2} \sin mx.dx.$$

b_m and a_m can be simplified a little.

$$b_m = \frac{2}{\pi}\int_0^\pi \frac{f(x)+f(-x)}{2}\cos mx.dx$$

$$= \frac{1}{\pi}\left[\int_0^\pi f(x)\cos mx.dx + \int_0^\pi f(-x)\cos mx.dx\right],$$

but if we replace x by $-x$, we get

$$\int_0^\pi f(-x)\cos mx.dx = -\int_0^{-\pi} f(x)\cos mx.dx = \int_{-\pi}^0 f(x)\cos mx.dx,$$

and we have $$b_m = \frac{1}{\pi}\int_{-\pi}^{\pi} f(x)\cos mx.dx.$$

In the same way we can reduce the value of a_m to

$$\frac{1}{\pi}\int_{-\pi}^{\pi} f(x)\sin mx.dx.$$

Hence

$$\left\{\begin{aligned} f(x) = \frac{1}{2}b_0 + b_1\cos x + b_2\cos 2x + b_3\cos 3x + \cdots \\ + a_1\sin x + a_2\sin 2x + a_3\sin 3x + \cdots \end{aligned}\right\} \tag{2}$$

where $$b_m = \frac{1}{\pi}\int_{-\pi}^{\pi} f(x)\cos mx.dx = \frac{1}{\pi}\int_{-\pi}^{\pi} f(\alpha)\cos m\alpha.d\alpha. \tag{3}$$

and $$a_m = \frac{1}{\pi}\int_{-\pi}^{\pi} f(x)\sin mx.dx = \frac{1}{\pi}\int_{-\pi}^{\pi} f(\alpha)\sin m\alpha.d\alpha. \tag{4}$$

and this development holds for all values of x between $-\pi$ and π.

The second member of (2) is known as a Fourier's Series.

EXAMPLES.

1. Obtain the following developments, all of which are valid from $x = -\pi$ to $x = \pi$:—

(1) $$e^x = \frac{2\sinh\pi}{\pi}\left[\frac{1}{2} - \frac{1}{2}\cos x + \frac{1}{5}\cos 2x - \frac{1}{10}\cos 3x + \frac{1}{17}\cos 4x + \cdots\right]$$
$$+ \frac{2\sinh\pi}{\pi}\left[\frac{1}{2}\sin x - \frac{2}{5}\sin 2x + \frac{3}{10}\sin 3x - \frac{4}{17}\sin 4x + \cdots\right].$$

$$(2)\quad f(x)=\frac{\pi}{4}-\frac{2}{\pi}\left[\cos x+\frac{\cos 3x}{3^2}+\frac{\cos 5x}{5^2}+\frac{\cos 7x}{7^2}+\cdots\right]$$
$$+\frac{\sin x}{1}-\frac{\sin 2x}{2}+\frac{\sin 3x}{3}-\frac{\sin 4x}{4}+\cdots,$$

where $f(x)=0$ from $x=-\pi$ to $x=0$ and $f(x)=x$ from $x=0$ to $x=\pi$.

$$(3)\quad f(x)=-\frac{3\pi}{16}+\frac{1}{\pi}\left[\frac{1}{1^2}\cos x+\frac{2}{2^2}\cos 2x+\frac{1}{3^2}\cos 3x+\frac{1}{5^2}\cos 5x\right.$$
$$\left.+\frac{2}{6^2}\cos 6x+\cdots\right]$$
$$+\frac{1}{\pi}\left[\left(\frac{3\pi}{2}-1\right)\sin x-\frac{3\pi}{4}\sin 2x+\left(\frac{3\pi}{6}+\frac{1}{3^2}\right)\sin 3x\right.$$
$$\left.-\frac{3\pi}{8}\sin 4x+\left(\frac{3\pi}{10}-\frac{1}{5^2}\right)\sin 5x-\cdots\right],$$

where $\quad f(x)=x$ from $x=-\pi$ to $x=0$, $\quad f(x)=0$ from $x=0$ to $x=\dfrac{\pi}{2}$,

and $\quad f(x)=x-\dfrac{\pi}{2}$ from $x=\dfrac{\pi}{2}$ to $x=\pi$.

2. Show that formula (2) Art. 30 can be written

$$f(x)=\frac{1}{2}c_0\cos\beta_0+c_1\cos(x-\beta_1)+c_2\cos(2x-\beta_2)+c_3\cos(3x-\beta_3)+\cdots$$

where $$c_m=(a_m^2+b_m^2)^{\frac{1}{2}}\text{ and }\beta_m=\tan^{-1}\frac{a_m}{b_m}.$$

3. Show that formula (2) Art. 30 can be written

$$f(x)=\frac{1}{2}c_0\sin\beta_0+c_1\sin(x+\beta_1)+c_2\sin(2x+\beta_2)+c_3\sin(3x+\beta_3)+\cdots$$

where $$c_m=(a_m^2+b_m^2)^{\frac{1}{2}}\text{ and }\beta_m=\tan^{-1}\frac{b_m}{a_m}.$$

31. In developing a function of x into a Trigonometric series it is often inconvenient to be held within the narrow boundaries $x=-\pi$ and $x=\pi$. Let us see if we cannot widen them.

Let it be required to develop a function of x into a Trigonometric series which shall be equal to $f(x)$ for all values of x between $x=-c$ and $x=c$.

Introduce a new variable

$$z=\frac{\pi}{c}x,$$

which is equal to $-\pi$ when $x=-c$ and to π when $x=c$.

$f(x)=f\left(\dfrac{c}{\pi}z\right)$ can be developed in terms of z by Art. 30 (2), (3), and (4).

We have

$$f\left(\frac{c}{\pi}z\right) = \frac{1}{2}b_0 + b_1 \cos z + b_2 \cos 2z + b_3 \cos 3z + \cdots \\ + a_1 \sin z + a_2 \sin 2z + a_3 \sin 3z + \cdots \tag{1}$$

where
$$b_m = \frac{1}{\pi}\int_{-\pi}^{\pi} f\left(\frac{c}{\pi}z\right) \cos mz.dz. \tag{2}$$

and
$$a_m = \frac{1}{\pi}\int_{-\pi}^{\pi} f\left(\frac{c}{\pi}z\right) \sin mz.dz. \tag{3}$$

and (1) holds good from $z = -\pi$ to $z = \pi$.

Replace z by its value in terms of x and (1) becomes

$$f(x) = \frac{1}{2}b_0 + b_1 \cos\frac{\pi x}{c} + b_2 \cos\frac{2\pi x}{c} + b_3 \cos\frac{3\pi x}{c} + \cdots \\ + a_1 \sin\frac{\pi x}{c} + a_2 \sin\frac{2\pi x}{c} + a_3 \sin\frac{3\pi x}{c} + \cdots \tag{4}$$

The coefficients in (4) are the same as in (1), and (4) holds good from $x = -c$ to $x = c$.

Formulas (2) and (3) can be put into more convenient shape.

$$b_m = \frac{1}{\pi}\int_{-\pi}^{\pi} f\left(\frac{c}{\pi}z\right) \cos mz.dz = \frac{1}{\pi}\int_{-c}^{c} f(x) \cos\frac{m\pi x}{c}\frac{\pi}{c}dx$$

or
$$b_m = \frac{1}{c}\int_{-c}^{c} f(x) \cos\frac{m\pi x}{c}dx = \frac{1}{c}\int_{-c}^{c} f(\lambda) \cos\frac{m\pi\lambda}{c}d\lambda. \tag{5}$$

In like manner we can transform (3) into

$$a_m = \frac{1}{c}\int_{-c}^{c} f(x) \sin\frac{m\pi x}{c}dx = \frac{1}{c}\int_{-c}^{c} f(\lambda) \sin\frac{m\pi\lambda}{c}d\lambda. \tag{6}$$

By treating in like fashion formulas (1) and (2) Art. 25 and formulas (4) and (2) Art. 27 we get

$$f(x) = a_1 \sin\frac{\pi x}{c} + a_2 \sin\frac{2\pi x}{c} + a_3 \sin\frac{3\pi x}{c} + \cdots \tag{7}$$

where
$$a_m = \frac{2}{c}\int_{0}^{c} f(x) \sin\frac{m\pi x}{c}dx = \frac{2}{c}\int_{0}^{c} f(\lambda) \sin\frac{m\pi\lambda}{c}d\lambda. \tag{8}$$

and
$$f(x) = \frac{1}{2}b_0 + b_1 \cos\frac{\pi x}{c} + b_2 \cos\frac{2\pi x}{c} + b_3 \cos\frac{3\pi x}{c} + \cdots \tag{9}$$

where
$$b_m = \frac{2}{c}\int_{0}^{c} f(x) \cos\frac{m\pi x}{c}dx = \frac{2}{c}\int_{0}^{c} f(\lambda) \cos\frac{m\pi\lambda}{c}d\lambda. \tag{10}$$

and (7) and (9) hold good from $x = 0$ to $x = c$.

EXAMPLES.

1. Obtain the following developments:

(1) $$1 = \frac{4}{\pi}\left[\sin\frac{\pi x}{c} + \frac{1}{3}\sin\frac{3\pi x}{c} + \frac{1}{5}\sin\frac{5\pi x}{c} + \cdots\right]$$

from $x = 0$ to $x = c$.

(2) $$x = \frac{2c}{\pi}\left[\sin\frac{\pi x}{c} - \frac{1}{2}\sin\frac{2\pi x}{c} + \frac{1}{3}\sin\frac{3\pi x}{c} - \frac{1}{4}\sin\frac{4\pi x}{c} + \cdots\right]$$

from $x = -c$ to $x = c$.

$$x = \frac{c}{2} - \frac{4c}{\pi^2}\left[\cos\frac{\pi x}{c} + \frac{1}{3^2}\cos\frac{3\pi x}{c} + \frac{1}{5^2}\cos\frac{5\pi x}{c} + \frac{1}{7^2}\cos\frac{7\pi x}{c} + \cdots\right]$$

from $x = -c$ to $x = c$.

(3) $$x^2 = \frac{2c^2}{\pi^3}\left[\left(\frac{\pi^2}{1} - \frac{4}{1^3}\right)\sin\frac{\pi x}{c} - \frac{\pi^2}{2}\sin\frac{2\pi x}{c} + \left(\frac{\pi^2}{3} - \frac{4}{3^2}\right)\sin\frac{3\pi x}{c}\right.$$
$$\left. - \frac{\pi^2}{4}\sin\frac{4\pi x}{c} + \left(\frac{\pi^2}{5} - \frac{4}{5^3}\right)\sin\frac{5\pi x}{c} + \cdots\right]$$

from $x = 0$ to $x = c$.

$$x^2 = \frac{c^2}{3} - \frac{4c^2}{\pi^2}\left[\cos\frac{\pi x}{c} - \frac{1}{2^2}\cos\frac{2\pi x}{c} + \frac{1}{3^2}\cos\frac{3\pi x}{c} - \frac{1}{4^2}\cos\frac{4\pi x}{c} + \cdots\right]$$

from $x = -c$ to $x = c$.

(4) $$e^x = 2\pi\left[\frac{1 + e^c}{c^2 + \pi^2}\sin\frac{\pi x}{c} + \frac{2(1 - e^c)}{c^2 + 4\pi^2}\sin\frac{2\pi x}{c}\right.$$
$$\left. + \frac{3(1 + e^c)}{c^2 + 9\pi^2}\sin\frac{3\pi x}{c} + \frac{4(1 - e^c)}{c^2 + 16\pi^2}\sin\frac{4\pi x}{c} + \cdots\right],$$

$$e^x = 2c\left[\frac{1}{2}\frac{e^c - 1}{c^2} - \frac{e^c + 1}{c^2 + \pi^2}\cos\frac{\pi x}{c} + \frac{e^c - 1}{c^2 + 4\pi^2}\cos\frac{2\pi x}{c}\right.$$
$$\left. - \frac{e^c + 1}{c^2 + 9\pi^2}\cos\frac{3\pi x}{c} + \cdots\right]$$

from $x = 0$ to $x = c$.

(5) $$f(x) = \frac{4c}{\pi^2}\left[\sin\frac{\pi x}{c} - \frac{1}{3^2}\sin\frac{3\pi x}{c} + \frac{1}{5^2}\sin\frac{5\pi x}{c} + \cdots\right]$$

from $x = 0$ to $x = c$,

where $f(x) = x$ from $x = 0$ to $x = \frac{c}{2}$ and $f(x) = c - x$ from $x = \frac{c}{2}$ to $x = c$.

2. Show that formula (4) Art. 31 can be written

$$f(x) = \frac{1}{2}c_0 \cos\beta_0 + c_1 \cos\left(\frac{\pi x}{c} - \beta_1\right) + c_2 \cos\left(\frac{2\pi x}{c} - \beta_2\right) + c_3 \cos\left(\frac{3\pi x}{c} - \beta_3\right) + \cdots$$

where $$c_m = (a_m^2 + b_m^2)^{\frac{1}{2}} \quad \text{and} \quad \beta_m = \tan^{-1}\frac{a_m}{b_m}.$$

3. Show that formula (4) Art. 31 can be written

$$f(x) = \frac{1}{2}c_0 \sin\beta_0 + c_1 \sin\left(\frac{\pi x}{c} + \beta_1\right) + c_2 \sin\left(\frac{2\pi x}{c} + \beta_2\right) + c_3 \sin\left(\frac{3\pi x}{c} + \beta_3\right) + \cdots$$

where $$c_m = (a_m^2 + b_m^2)^{\frac{1}{2}} \quad \text{and} \quad \beta_m = \tan^{-1}\frac{a_m}{b_m}.$$

32. In the formulas of Art. 31 c may have as great a value as we please, so that we can obtain a Trigonometric Series for $f(x)$ that will represent the given function through as great an interval as we may choose to take. If, then, we can obtain the limiting form approached by the series (4) Art. 31 as c is indefinitely increased the expression in question ought to be equal to the given function of x for all values of x. Equation (4) Art. 31 can be written as follows if we replace $b_0, b_1, b_2, \cdots a_1, a_2, \cdots$ by their values given in Art. 31 (5) and (6).

$$\begin{aligned} f(x) = \frac{1}{c}\Bigg[\frac{1}{2}\int_{-c}^{c} f(\lambda)\,d\lambda &+ \int_{-c}^{c} f(\lambda)\cos\frac{\pi\lambda}{c}\cos\frac{\pi x}{c}d\lambda + \int_{-c}^{c} f(\lambda)\cos\frac{2\pi\lambda}{c}\cos\frac{2\pi x}{c}d\lambda + \cdots \\ &+ \int_{-c}^{c} f(\lambda)\sin\frac{\pi\lambda}{c}\sin\frac{\pi x}{c}d\lambda + \int_{-c}^{c} f(\lambda)\sin\frac{2\pi\lambda}{c}\sin\frac{2\pi x}{c}d\lambda + \cdots\Bigg] \\ = \frac{1}{c}\int_{-c}^{c} f(\lambda)\,d\lambda\Bigg[\frac{1}{2} &+ \cos\frac{\pi\lambda}{c}\cos\frac{\pi x}{c} + \sin\frac{\pi\lambda}{c}\sin\frac{\pi x}{c} \\ &+ \cos\frac{2\pi\lambda}{c}\cos\frac{2\pi x}{c} + \sin\frac{2\pi\lambda}{c}\sin\frac{2\pi x}{c} + \cdots\Bigg] \end{aligned}$$

$$f(x) = \frac{1}{c}\int_{-c}^{c} f(\lambda)\,d\lambda\left[\frac{1}{2} + \cos\frac{\pi}{c}(\lambda - x) + \cos\frac{2\pi}{c}(\lambda - x) + \cdots\right]$$

$$= \frac{1}{2c}\int_{-c}^{c} f(\lambda)\,d\lambda\left[1 + \cos\frac{\pi}{c}(\lambda - x) + \cos\frac{2\pi}{c}(\lambda - x) + \cdots \right.$$
$$\left. + \cos\left(-\frac{\pi}{c}\right)(\lambda - x) + \cos\left(-\frac{2\pi}{c}\right)(\lambda - x) + \cdots\right]$$

since $\cos(-\phi) = \cos\phi$.

$$f(x) = \frac{1}{2\pi}\int_{-c}^{c} f(\lambda)\,d\lambda\left[\cdots + \frac{\pi}{c}\cos\left(-\frac{2\pi}{c}\right)(\lambda - x) + \frac{\pi}{c}\cos\left(-\frac{\pi}{c}\right)(\lambda - x) \right.$$
$$+ \frac{\pi}{c}\cos\frac{0\pi}{c}(\lambda - x) + \frac{\pi}{c}\cos\frac{\pi}{c}(\lambda - x)$$
$$\left. + \frac{\pi}{c}\cos\frac{2\pi}{c}(\lambda - x) + \cdots\right] \tag{1}$$

As c is indefinitely increased the limiting value approached by the parenthesis in (1) is

$$\int_{-\infty}^{\infty} \cos\alpha(\lambda - x).d\alpha.$$

Hence the limiting form approached by (1) is

$$f(x) = \frac{1}{2\pi}\int_{-\infty}^{\infty} f(\lambda)\,d\lambda\int_{-\infty}^{\infty} \cos\alpha(\lambda - x).d\alpha, \tag{2}$$

and the second member of (2) must be equal to $f(x)$ for all values of x.

The double integral in (2) is known as *Fourier's Integral*, and since it is a limiting form of *Fourier's Series* it is subject to the same limitations as the series.

That is, in order that (2) should be true $f(x)$ must be finite, continuous, and single valued for all values of x, or if discontinuous, must have only finite discontinuities.[3]

(2) is sometimes given in a slightly different form.

Since $\displaystyle\int_{-\infty}^{\infty} \cos\alpha(\lambda - x).d\alpha = \int_{-\infty}^{0} \cos\alpha(\lambda - x).d\alpha + \int_{0}^{\infty} \cos\alpha(\lambda - x).d\alpha$

and

$$\int_{-\infty}^{0} \cos\alpha(\lambda - x).d\alpha = \int_{\infty}^{0} \cos(-\alpha)(\lambda - x).d(-\alpha) = -\int_{\infty}^{0} \cos\alpha(\lambda - x).d\alpha$$

$$\int_{-\infty}^{\infty} \cos\alpha(\lambda - x).d\alpha = 2\int_{0}^{\infty} \cos\alpha(\lambda - x).d\alpha$$

[3]See note on page 38.

and (2) may be written

$$f(x) = \frac{1}{\pi}\int_{-\infty}^{\infty} f(\lambda)\,d\lambda\int_{0}^{\infty}\cos\alpha(\lambda - x).d\alpha. \tag{3}$$

If $f(x)$ is an *even* function or an *odd* function (3) can be still further simplified.

Let $$f(x) = -f(-x).$$

Since the limits of integration in (3) do not contain α or λ the integrations may be performed in whichever order we choose. That is

$$\int_{-\infty}^{\infty} f(\lambda)\,d\lambda\int_{0}^{\infty}\cos\alpha(\lambda - x).d\alpha = \int_{0}^{\infty} d\alpha\int_{-\infty}^{\infty} f(\lambda)\cos\alpha(\lambda - x).d\lambda.$$

Now

$$\int_{-\infty}^{\infty} f(\lambda)\cos\alpha(\lambda - x).d\lambda = \int_{-\infty}^{0} f(\lambda)\cos\alpha(\lambda - x).d\lambda + \int_{0}^{\infty} f(\lambda)\cos\alpha(\lambda - x).d\lambda.$$

$$\int_{-\infty}^{0} f(\lambda)\cos\alpha(\lambda - x).d\lambda = \int_{\infty}^{0} f(-\lambda)\cos\alpha(-\lambda - x).d(-\lambda)$$

$$= -\int_{0}^{\infty} f(\lambda)\cos\alpha(\lambda + x).d\lambda$$

and (3) becomes

$$f(x) = \frac{1}{\pi}\int_{0}^{\infty} d\alpha\int_{0}^{\infty} f(\lambda)[\cos\alpha(\lambda - x) - \cos\alpha(\lambda + x).d\lambda$$

$$= \frac{2}{\pi}\int_{0}^{\infty} d\alpha\int_{0}^{\infty} f(\lambda)\sin\alpha\lambda\sin\alpha x.d\lambda$$

or

$$f(x) = \frac{2}{\pi}\int_{0}^{\infty} f(\lambda)d\lambda\int_{0}^{\infty}\sin\alpha\lambda\sin\alpha x.d\alpha. \tag{4}$$

If $f(x) = f(-x)$ (3) can be reduced in like manner to

$$f(x) = \frac{2}{\pi}\int_{0}^{\infty} f(\lambda)\,d\lambda\int_{0}^{\infty}\cos\alpha\lambda\cos\alpha x.d\alpha. \tag{5}$$

Although (4) holds for all values of x only in case $f(x)$ is an *odd* function, and (5) only in case $f(x)$ is an *even*, function, both (4) and (5) hold for all *positive* values of x in the case of any function.

EXAMPLE.

(1) Obtain formulas (4) and (5) directly from (7) and (9) Art. 31.

CHAPTER III.

CONVERGENCE OF FOURIER'S SERIES.

33. The question of the *convergence* of a Fourier's Series is altogether too large to be completely handled in an elementary treatise. We will, however, consider at some length one of the most important of the series we have obtained, namely

$$\frac{4}{\pi}\left[\sin x + \frac{\sin 3x}{3} + \frac{\sin 5x}{5} + \frac{\sin 7x}{7} + \cdots\right], \qquad \text{[v. (3) Art. 26(}b\text{).]}$$

and prove that for all values of x between zero and π its sum is absolutely equal to unity; that is, that the limit approached by the sum of n terms of the series

$$\frac{2}{\pi}\left[\sin x\int_0^\pi \sin\alpha.d\alpha + \sin 2x\int_0^\pi \sin 2\alpha.d\alpha + \sin 3x\int_0^\pi \sin 3\alpha.d\alpha + \cdots\right],$$

as n is indefinitely increased, is 1, provided that x lies between zero and π.

Let

$$S_n = \frac{2}{\pi}\left[\sin x\int_0^\pi \sin\alpha.d\alpha + \sin 2x\int_0^\pi \sin 2\alpha.d\alpha + \sin 3x\int_0^\pi \sin 3\alpha.d\alpha + \cdots\right.$$
$$\left. + \sin nx\int_0^\pi \sin n\alpha.d\alpha\right]. \qquad (1)$$

Then

$$S_n = \frac{2}{\pi}\int_0^\pi [\sin\alpha\sin x + \sin 2\alpha\sin 2x + \sin 3\alpha\sin 3x + \cdots + \sin n\alpha\sin nx]d\alpha$$
$$= \frac{1}{\pi}\int_0^\pi [\cos(\alpha - x) - \cos(\alpha + x) + \cos 2(\alpha - x) - \cos 2(\alpha + x) + \cdots$$
$$+ \cos n(\alpha - x) - \cos n(\alpha + x)]\,d\alpha$$
$$= \frac{1}{\pi}\int_0^\pi [\cos(\alpha - x) + \cos 2(\alpha - x) + \cos 3(\alpha - x) + \cdots + \cos n(\alpha - x)]d\alpha$$
$$- \frac{1}{\pi}\int_0^\pi [\cos(\alpha + x) + \cos 2(\alpha + x) + \cos 3(\alpha + x) + \cdots + \cos n(\alpha + x)]d\alpha.$$

Therefore by Art. 20 (1)

$$S_n = \frac{1}{\pi}\int\limits_0^\pi \left[-\frac{1}{2} + \frac{1}{2}\frac{\sin(2n+1)\dfrac{\alpha - x}{2}}{\sin\dfrac{\alpha - x}{2}}\right] d\alpha$$

$$- \frac{1}{\pi}\int\limits_0^\pi \left[-\frac{1}{2} + \frac{1}{2}\frac{\sin(2n+1)\dfrac{\alpha + x}{2}}{\sin\dfrac{\alpha + x}{2}}\right] d\alpha.$$

$$S_n = \frac{1}{2\pi}\int\limits_0^\pi \frac{\sin(2n+1)\dfrac{\alpha - x}{2}}{\sin\dfrac{\alpha - x}{2}} d\alpha - \frac{1}{2\pi}\int\limits_0^\pi \frac{\sin(2n+1)\dfrac{\alpha + x}{2}}{\sin\dfrac{\alpha + x}{2}} d\alpha.$$

In the first integral substitute β for $\dfrac{\alpha - x}{2}$, and in the second integral substitute β for $\dfrac{\alpha + x}{2}$.

We get

$$S_n = \frac{1}{\pi}\int\limits_{-\frac{x}{2}}^{\frac{\pi}{2}-\frac{x}{2}} \frac{\sin(2n+1)\beta}{\sin\beta} d\beta - \frac{1}{\pi}\int\limits_{\frac{x}{2}}^{\frac{\pi}{2}+\frac{x}{2}} \frac{\sin(2n+1)\beta}{\sin\beta} d\beta. \tag{2}$$

It remains to find the limit approached by S_n as n is indefinitely increased.

34.

$$\int\limits_0^{\frac{\pi}{2}} \frac{\sin(2n+1)\beta}{\sin\beta} d\beta = \frac{\pi}{2}. \tag{1}$$

For

$$\frac{\sin(2n+1)\beta}{2\sin\beta} = \tfrac{1}{2} + \cos 2\beta + \cos 4\beta + \cdots + \cos 2n\beta, \qquad \text{by Art. 20.}$$

and

$$\int\limits_0^{\frac{\pi}{2}} \cos 2k\beta . d\beta = 0.$$

Let us construct the curve

$$y = \frac{\sin(2n+1)x}{\sin x}.$$

We have only to draw the curve $y = \sin(2n+1)x$ and then to divide the length of each ordinate by the value of the sine of the corresponding abscissa.

In $y = \sin(2n+1)x$ the successive arches into which the curve is divided by the axis of X are equal, and consequently their areas are equal.

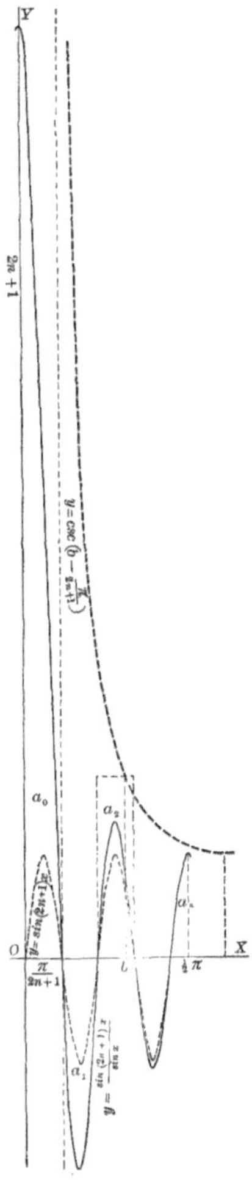

Each arch has for its altitude unity and for its base $\dfrac{\pi}{2n+1}$ and is symmetrical with respect to the ordinate of its highest or lowest point.

If now we form the curve $y = \dfrac{\sin(2n+1)x}{\sin x}$ from the curve $y = \sin(2n+1)x$, it is clear that, since $\sin x$ increases as x increases from 0 to $\dfrac{\pi}{2}$, the ordinate of any point of the new curve will be shorter than the ordinate of the corresponding point in the preceding arch, and that consequently the area of each arch $y = \dfrac{\sin(2n+1)x}{\sin x}$ will be less than that of the arch before it.

If $a_0, a_1, a_2, \cdots a_{n-1}$ are the areas of the successive arches and a_n that of the incomplete arch terminated by the ordinate corresponding to $x = \dfrac{\pi}{2}$

$$\int_0^{\frac{\pi}{2}} \frac{\sin(2n+1)x}{\sin x}\,dx = a_0 - a_1 + a_2 - a_3 + \cdots .$$

But

$$\int_0^{\frac{\pi}{2}} \frac{\sin(2n+1)x}{\sin x}\,dx = \int_0^{\frac{\pi}{2}} \frac{\sin(2n+1)\beta}{\sin \beta}\,d\beta = \frac{\pi}{2} \quad \text{by (1).}$$

Hence

$$\frac{\pi}{2} = a_0 - a_1 + a_2 - a_3 + a_4 - \cdots + a_n \quad \text{if } n \text{ is even,}$$

or

$$\frac{\pi}{2} = a_0 - a_1 + a_2 - a_3 + a_4 - \cdots - a_n \quad \text{if } n \text{ is odd.}$$

These equations can be written

$$\frac{\pi}{2} = a_0 + (-a_1 + a_2) + (-a_3 + a_4) + (-a_5 + a_6) + \cdots + (-a_{n-1} + a_n)$$

if n is even, and

$$\frac{\pi}{2} = a_0 + (-a_1 + a_2) + (-a_3 + a_4) + (-a_5 + a_6) + \cdots + (-a_{n-2} + a_{n-1}) + (-a_n)$$

if n is odd.

In either case each parenthesis is a negative quantity since

$$a_0 > a_1 > a_2 > a_3 \cdots > a_n,$$

and it follows that a_0 is greater than $\frac{\pi}{2}$.

Again

$$\frac{\pi}{2} = a_0 - a_1 + (a_2 - a_3) + (a_4 - a_5) + \cdots + (a_{n-2} - a_{n-1}) + a_n$$

if n is even and

$$\frac{\pi}{2} = a_0 - a_1 + (a_2 - a_3) + (a_4 - a_5) + \cdots + (a_{n-1} - a_n)$$

if n is odd.

In either case each parenthesis is positive and it follows that $a_0 - a_1$ is is less than $\frac{\pi}{2}$.

Since

$$a_0 > \frac{\pi}{2} > a_0 - a_1,$$

a_0 and $a_0 - a_1$ differ from $\frac{\pi}{2}$ by less than they differ from each other, that is, by less than a_1.

In like manner we can show that $a_0 - a_1$ and $a_0 - a_1 + a_2$ differ from $\frac{\pi}{2}$ by less than a_2; and in general that $a_0 - a_1 + a_2 - a_3 + \cdots \pm a_k$ differs from $\frac{\pi}{2}$ by less than a_k; or even that

$$a_0 - a_1 + a_2 - a_3 + \cdots \pm \frac{a_k}{p}$$

differs from $\frac{\pi}{2}$ by less than a_k no matter the value of p, provided p is greater than unity.

35. From what has been proved in the last article it follows that

$$\int_0^b \frac{\sin(2n+1)x}{\sin x}\,dx,$$

where b is some value between $\frac{\pi}{2n+1}$ and $\frac{\pi}{2}$, differs from $\frac{\pi}{2}$ by less than the area of the arch in which the ordinate of $y = \frac{\sin(2n+1)x}{\sin x}$ corresponding to $x = b$ falls if this ordinate divides an arch, or by less than the area of the arch next beyond the point $(b, 0)$ if the curve crosses the axis of X at that point.

The area of the arch in question is less than $\frac{\pi}{2n+1}$, its base, multiplied by $\frac{1}{\sin\left(b - \frac{\pi}{2n+1}\right)}$, a value greater than the length of its longest ordinate.

Therefore $$\int_0^b \frac{\sin(2n+1)x}{\sin x}\,dx$$

differs from $\frac{\pi}{2}$ by less than $\frac{\pi}{2n+1}\frac{1}{\sin\left(b-\frac{\pi}{2n+1}\right)}$.

If now n is indefinitely increased $\frac{\pi}{2n+1}\frac{1}{\sin\left(b-\frac{\pi}{2n+1}\right)}$ approaches zero as its limit, and we get the very important result

$$\operatorname*{limit}_{n=\infty}\left[\int_0^b \frac{\sin(2n+1)x}{\sin x}\,dx\right]=\frac{\pi}{2} \tag{1}$$

if $0<b<\frac{\pi}{2}$.

36. $$S_n=\frac{1}{\pi}\int_{-\frac{x}{2}}^{\frac{\pi}{2}-\frac{x}{2}}\frac{\sin(2n+1)\beta}{\sin\beta}\,d\beta-\frac{1}{\pi}\int_{\frac{x}{2}}^{\frac{\pi}{2}+\frac{x}{2}}\frac{\sin(2n+1)\beta}{\sin\beta}\,d\beta.\quad\text{[Art. 33. (2)]}$$

$$=\frac{1}{\pi}\int_{-\frac{x}{2}}^{0}\frac{\sin(2n+1)\beta}{\sin\beta}\,d\beta+\frac{1}{\pi}\int_{0}^{\frac{\pi}{2}-\frac{x}{2}}\frac{\sin(2n+1)\beta}{\sin\beta}\,d\beta$$

$$-\frac{1}{\pi}\int_{0}^{\frac{\pi}{2}}\frac{\sin(2n+1)\beta}{\sin\beta}\,d\beta+\frac{1}{\pi}\int_{0}^{\frac{x}{2}}\frac{\sin(2n+1)\beta}{\sin\beta}\,d\beta$$

$$-\frac{1}{\pi}\int_{\frac{\pi}{2}}^{\frac{\pi}{2}+\frac{x}{2}}\frac{\sin(2n+1)\beta}{\sin\beta}\,d\beta.$$

This last value for S_n can be somewhat simplified.
Substituting $\gamma=-\beta$ we get

$$\int_{-\frac{x}{2}}^{0}\frac{\sin(2n+1)\beta}{\sin\beta}\,d\beta=-\int_{\frac{x}{2}}^{0}\frac{\sin(2n+1)\gamma}{\sin\gamma}\,d\gamma=\int_{0}^{\frac{x}{2}}\frac{\sin(2n+1)\beta}{\sin\beta}\,d\beta.$$

Substituting $\gamma=\pi-\beta$ in

$$\int_{\frac{\pi}{2}}^{\frac{\pi}{2}+\frac{x}{2}}\frac{\sin(2n+1)\beta}{\sin\beta}\,d\beta$$ we have

$$\int_{\frac{\pi}{2}}^{\frac{\pi}{2}+\frac{x}{2}} \frac{\sin(2n+1)\beta}{\sin\beta} d\beta = -\int_{\frac{\pi}{2}}^{\frac{\pi}{2}-\frac{x}{2}} \frac{\sin(2n+1)\gamma}{\sin\gamma} d\gamma = \int_{\frac{\pi}{2}-\frac{x}{2}}^{\frac{\pi}{2}} \frac{\sin(2n+1)\beta}{\sin\beta} d\beta$$

$$= \int_0^{\frac{\pi}{2}} \frac{\sin(2n+1)\beta}{\sin\beta} d\beta - \int_0^{\frac{\pi}{2}-\frac{x}{2}} \frac{\sin(2n+1)\beta}{\sin\beta} d\beta.$$

Hence

$$S_n = \frac{2}{\pi}\int_0^{\frac{x}{2}} \frac{\sin(2n+1)\beta}{\sin\beta} d\beta + \frac{2}{\pi}\int_0^{\frac{\pi}{2}-\frac{x}{2}} \frac{\sin(2n+1)\beta}{\sin\beta} d\beta - \frac{2}{\pi}\int_0^{\frac{\pi}{2}} \frac{\sin(2n+1)\beta}{\sin\beta} d\beta.$$

$$\int_0^{\frac{\pi}{2}} \frac{\sin(2n+1)\beta}{\sin\beta} d\beta = \frac{\pi}{2} \qquad \text{by (1) Art. 34.}$$

$$\underset{n=\infty}{\text{limit}}\left[\int_0^{\frac{x}{2}} \frac{\sin(2n+1)\beta}{\sin\beta} d\beta\right] = \frac{\pi}{2} \quad \text{if} \quad 0 < x < \pi \qquad \text{by (1) Art. 35}$$

and

$$\underset{n=\infty}{\text{limit}}\left[\int_0^{\frac{\pi}{2}-\frac{x}{2}} \frac{\sin(2n+1)\beta}{\sin\beta} d\beta\right] = \frac{\pi}{2} \quad \text{if} \quad 0 < x < \pi \qquad \text{by (1) Art. 35.}$$

Therefore $\underset{n=\infty}{\text{limit}}[S_n] = 1 + 1 - 1 = 1 \quad \text{if} \quad 0 < x < \pi$ and

$$\frac{4}{\pi}\left[\sin x + \frac{\sin 3x}{3} + \frac{\sin 5x}{5} + \frac{\sin 7x}{7} + \cdots\right] = 1$$

for all values of x between zero and π.

37. By a somewhat long but not especially difficult extension of the reasoning just given it can be shown that if $f(x)$ is *single-valued* and *finite* between $x = -\pi$ and $x = \pi$, and has only a *finite number* of *discontinuities* and of *maxima* and *minima* between $x = -\pi$ and $x = \pi$ the *Fourier's Series*

$$\frac{1}{2}b_0 + b_1 \cos x + b_2 \cos 2x + b_3 \cos 3x + \cdots$$
$$+ a_1 \sin x + a_2 \sin 2x + a_3 \sin 3x + \cdots$$

where $$a_m = \frac{1}{\pi}\int_{-\pi}^{\pi} f(\alpha) \sin m\alpha . d\alpha$$

and $$b_m = \frac{1}{\pi}\int_{-\pi}^{\pi} f(\alpha) \cos m\alpha . d\alpha,$$

and that Fourier's Series only is equal to $f(x)$ for all values of x between $x = -\pi$ and $x = \pi$, *excepting the values of x corresponding to the discontinuities* of $f(x)$, and the values π and $-\pi$ if $f(\pi)$ is not equal to $f(-\pi)$; and that if c is a value of x corresponding to a discontinuity of $f(x)$, the value of the series when $x = c$ is

$$\frac{1}{2} \operatorname*{limit}_{\epsilon \doteq 0} [f(c - \epsilon) + f(c + \epsilon)];$$

and that if $f(\pi)$ is not equal to $f(-\pi)$ the value of the series when $x = -\pi$ and when $x = \pi$ is

$$\frac{1}{2}[f(-\pi) + f(\pi)].$$

If $f(x)$ while satisfying the conditions named in the preceding paragraph except for a finite number of values of x, becomes infinite for those values, the series is equal to the function except for the values of x in question provided that $\int\limits_{-\pi}^{\pi} f(x)\,dx$ is finite and determinate. (v. Int. Cal. Arts. 83 and 84.)

38. The question of the convergency of a Fourier's Series and the conditions under which a function may be developed in such a series was first attacked successfully by Dirichlet in 1829, and his conclusions have been criticised and extended by later mathematicians, notably by Riemann, Heine, Lipschitz, and du Bois Reymond. It may be noted that the criticisms relate not to the sufficiency but to the necessity of Dirichlet's conditions.

An excellent résumé of the literature of the subject is given by Arnold Sachse in a short dissertation published by Gauthier–Villars, Paris, 1880, entitled "Essai Historique sur la Représentation d'une Fonction Arbitraire d'une seule variable par une Série Trigonométrique."

39. A good deal of light is thrown on the peculiarities of trigonometric series by the attempt to construct approximately the curves corresponding to them.

If we construct $y = a_1 \sin x$ and $y = a_2 \sin 2x$ and add the ordinates of the points having the same abscissas we shall obtain points on the curve

$$y = a_1 \sin x + a_2 \sin 2x.$$

If now we construct $y = a_3 \sin 3x$ and add the ordinates to those of $y = a_1 \sin x + a_2 \sin 2x$ we shall get the curve

$$y = a_1 \sin x + a_2 \sin 2x + a_3 \sin 3x.$$

By continuing this process we get successive approximations to

$$y = a_1 \sin x + a_2 \sin 2x + a_3 \sin 3x + a_4 \sin 4x + \cdots$$

Let us apply this method to a few of the series which we have obtained in Chapter II.

Take

$$y = \sin x + \frac{1}{3}\sin 3x + \frac{1}{5}\sin 5x + \cdots \tag{1}$$
$$= 0 \text{ when } x = 0,\ \frac{\pi}{4} \text{ from } x = 0 \text{ to } x = \pi,\ \text{and } 0 \text{ when } x = \pi,$$

v. Art. 26 [b](3).

$$y = 2\left(\sin x - \frac{1}{2}\sin 2x + \frac{1}{3}\sin 3x - \frac{1}{4}\sin 4x + \cdots\right) \tag{2}$$
$$= x \text{ from } x = 0 \text{ to } x = \pi,\ \text{and } 0 \text{ when } x = \pi,$$

Art. 26 [a](4).

$$y = \frac{4}{\pi}\left[\frac{1}{1^2}\sin x - \frac{1}{3^2}\sin 3x + \frac{1}{5^2}\sin 5x - \frac{1}{7^2}\sin 7x + \cdots\right] \tag{3}$$
$$= x \text{ from } x = 0 \text{ to } x = \frac{\pi}{2},\ \text{and } \pi - x \text{ from } x = \frac{\pi}{2} \text{ to } x = \pi,$$

Art. 26 [c](2).

$$y = \frac{1}{1}\sin x + \frac{2}{2}\sin 2x + \frac{1}{3}\sin 3x + \frac{1}{5}\sin 5x - \frac{2}{6}\sin 6x + \frac{1}{7}\sin 7x + \cdots \tag{4}$$
$$= 0 \text{ when } x = 0,\ \frac{\pi}{2} \text{ from } x = 0 \text{ to } x = \frac{\pi}{2},\ \text{and } 0 \text{ from } x = \frac{\pi}{2} \text{ to } x = \pi,$$

v. Art. 26 [d](2).

It must be borne in mind that each of these curves is periodic having the period 2π, and is symmetrical with respect to the origin.

The following figures I, II, III, and IV represent the first four approximations to each of these curves.

In each figure the curve $y =$ the series, and the approximation in question are drawn in continuous lines, and the preceding approximation and the curve corresponding to the term to be added are drawn in dotted lines.

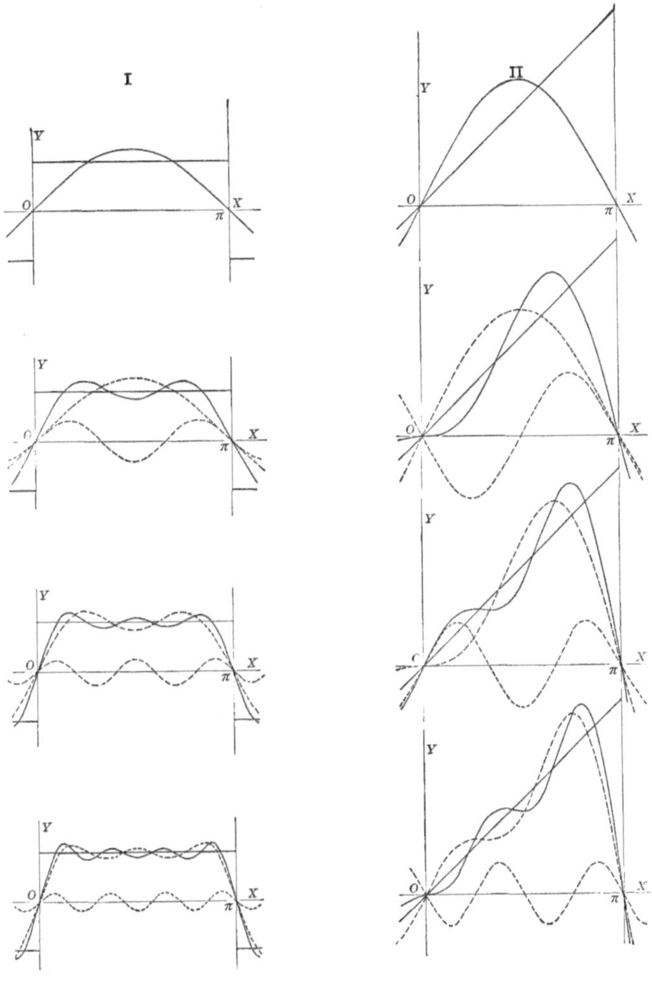

Figs. I, II, III, and IV immediately suggest the following facts:

(a) The curve representing each approximation is continuous even when the curve representing the series is discontinuous.

(b) When the curve representing the series is discontinuous the portion of each successive approximate curve in the neighborhood of the point whose abscissa is a value of x for which the series curve is discontinuous approaches more and more nearly a straight line perpendicular to the axis of X and connecting the separate portions of the series curve.

III

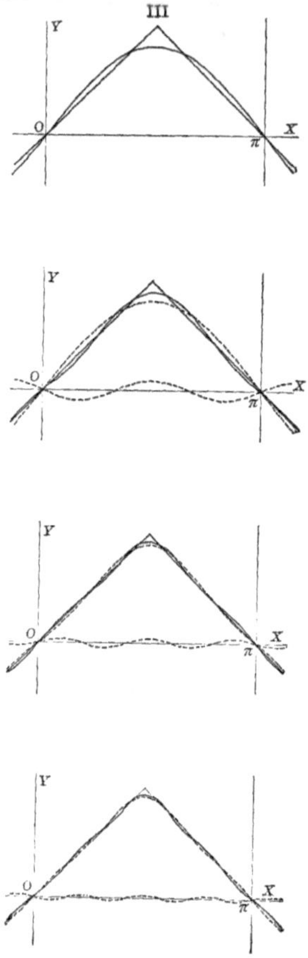

IV

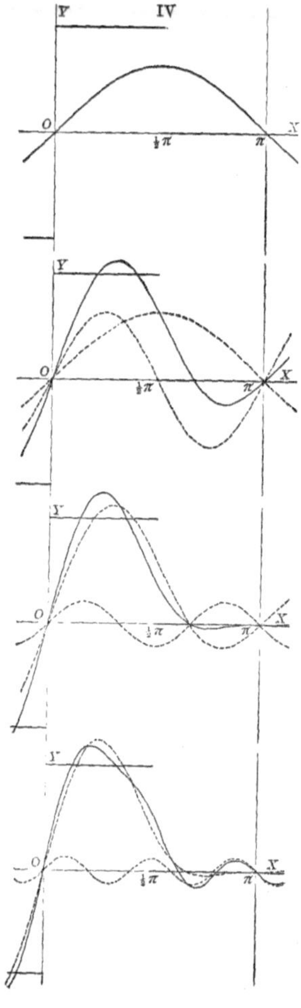

(c) The curves representing successive approximations do not necessarily tend to lose their wavy character, since each is obtained from the preceding one by superposing upon it a wave line whose waves are shorter each time but do not necessarily lose their sharpness of pitch. This is the case in Figures I, II, and IV. In Fig. III the waves of the superposed curves grow rapidly flatter.

It follows from this that in such cases as those represented in Figures I, II, and IV the direction of the approximate curve at a point having a given abscissa does not in general approach the direction of the series curve at the corresponding point, or indeed, approach any limiting value, as the approximation is made closer and closer; and that the length of any portion of the approximate curve will not in general approach the length of the corresponding portion of the series curve.

Analytically this amounts to saying that the derivative of a function of x cannot in general be obtained by differentiating term by term the Fourier's Series which represents the function.

(d) The area bounded by a given ordinate, the approximate curve, the axis of X, and any second ordinate will approach as its limit the corresponding area of the series curve if the series curve is continuous between the ordinates in question; and will approach the area bounded by the given ordinate, the series curve, the axis of X, any second ordinate, and a line perpendicular to the axis of X, and joining the separate portions of the series curve if the latter has a discontinuity between the ordinates in question.

Analytically this amounts to saying that the Fourier's Series corresponding to any given function can be integrated term by term and the resulting series will represent the integral of the function even when the function is discontinuous (v. Int. Cal. Art. 83).

We may note in passing that if the function curve is continuous a curve representing the integral of the function will be continuous and will not change its direction abruptly at any point; while if the function curve is discontinuous the curve representing the integral will still be continuous but will change its direction abruptly at points corresponding to the discontinuities of the given function.

40. The facts that the derivative of a Fourier's Series cannot in general be obtained by differentiating the series term by term and that its integral can be obtained by integrating the series term by term are so important that it is worth while to look at the matter a little more closely. Let us consider the differentiation of the series represented in Art. 39 Figure I.

Let

$$S_n = \sin x + \frac{1}{3}\sin 3x + \frac{1}{5}\sin 5x + \cdots + \frac{1}{2n+1}\sin(2n+1)x.$$

Then $$\frac{dS_n}{dx} = \cos x + \cos 3x + \cos 5x + \cdots + \cos(2n+1)x.$$

If $x = \dfrac{\pi}{2}$

$$\frac{dS_n}{dx} = 0$$

and the curve is parallel to the axis of X for $x = \dfrac{\pi}{2}$ no matter what the value of n.

If $x = 0$ or $x = \pi$

$$\frac{dS_n}{dx} = 1 + 1 + 1 + 1 + \cdots + 1 = n + 1$$

and the curve $y = S_n$ becomes more nearly perpendicular to the axis of X at the origin and for $x = \pi$ as we increase n.

If $x = \dfrac{\pi}{3}$

$$\frac{dS_n}{dx} = \frac{1}{2} - 1 + \frac{1}{2} + \frac{1}{2} - 1 + \frac{1}{2} + \cdots$$

That is

$$\begin{aligned} \frac{dS_n}{dx} &= \frac{1}{2} \quad \text{if} \quad n = 0 \quad \text{or} \quad n = 3k \\ &= -\frac{1}{2} \quad \text{“} \quad n = 1 \quad \text{“} \quad n = 3k + 1 \\ &= 0 \quad \text{“} \quad n = 2 \quad \text{“} \quad n = 3k + 2. \end{aligned}$$

Consequently when $x = \dfrac{\pi}{3}$ $\dfrac{dS_n}{dx}$ does not approach any limiting value as n is indefinitely increased. Indeed, in the successive approximations the point whose abscissa is $\dfrac{\pi}{3}$ is successively on the rear, on the front, and on the crest or in the trough of a wave, and although the waves are getting smaller they do not lose their sharpness of pitch.

If x has any other value between 0 and π $\dfrac{dS_n}{dx}$ will change abruptly as n is changed and will not approach any limiting value as n is increased.

41. In general if we differentiate a Fourier's Series

$$\begin{aligned} S = \frac{1}{2}b_0 &+ b_1 \cos x + b_2 \cos 2x + b_3 \cos 3x + \cdots \\ &+ a_1 \sin x + a_2 \sin 2x + a_3 \sin 3x + \cdots \end{aligned}$$

we get

$$\begin{aligned} &- b_1 \sin x - 2b_2 \sin 2x - 3b_3 \sin 3x - \cdots \\ &+ a_1 \cos x + 2a_2 \cos 2x + 3a_3 \cos 3x + \cdots . \end{aligned}$$

Differentiate again and we get

$$\begin{aligned} &- b_1 \cos x - 2^2 b_2 \cos 2x - 3^2 b_3 \cos 3x - \cdots \\ &- a_1 \sin x - 2^2 a_2 \sin 2x - 3^2 b_3 \sin 3x - \cdots . \end{aligned}$$

We see that each time we differentiate we multiply the coefficient of $\sin kx$ and of $\cos kx$ by k while the term still involves $\cos kx$ or $\sin kx$.

Since the series

$$\begin{aligned} \cos x + \cos 2x + \cos 3x + \cdots \\ + \sin x + \sin 2x + \sin 3x + \cdots \end{aligned}$$

is not convergent, and a Fourier's Series converges only because its coefficients decrease as we advance in the series, the differentiation of a Fourier's Series must make its convergence less rapid if it does not actually destroy it, and repetitions of the process will usually eventually make the derived series diverge.

It is to be observed that the derived series are Fourier's Series, but of somewhat special form, that is they lack the constant term. (v. Art. 30.)

If now we integrate a Fourier's Series

$$\frac{1}{2}b_0 + b_1 \cos x + b_2 \cos 2x + b_3 \cos 3x + \cdots$$
$$+ a_1 \sin x + a_2 \sin 2x + a_3 \sin 3x + \cdots$$

we get
$$C + \frac{1}{2}b_0 x + b_1 \sin x + \frac{1}{2}b_2 \sin 2x + \frac{1}{3}b_3 \sin 3x + \cdots$$
$$- a_1 \cos x - \frac{1}{2}a_2 \cos 2x - \frac{1}{3}a_3 \cos 3x - \cdots,$$

a Trigonometric Series which converges more rapidly than the given series.

It is to be observed that the series obtained by integrating a Fourier's Series is not in general a Fourier's Series owing to the presence of the term $\frac{1}{2}b_0 x$. (v. Art. 30.)

42. We are now ready to consider the conditions under which a function of x can be developed into a Fourier's Series whose term by term derivative shall be equal to the derivative of the function.

Let the function $f(x)$ satisfy the conditions stated in Art. 37. Then there is one Fourier's Series and but one which is equal to it. Call this series S.

Let the derivative $f'(x)$[1] of the given function also satisfy the conditions stated in Art. 37. Then $f'(x)$ can be expressed as a Fourier's Series. By Art. 39 (d) the integral of this latter series will be equal to the integral of $f'(x)$, that is to $f(x)$ plus a constant, and one integral will be equal to $f(x)$.

If this integral which is necessarily a Trigonometric Series is a Fourier's Series it must be identical with S. It will be a Fourier's Series only in case the Fourier's Series for $f'(x)$ lacks the constant term $\frac{1}{2}b_0$.

But
$$b_0 = \frac{1}{\pi}\int_{-\pi}^{\pi} f'(x)dx \qquad \text{by (3) Art. 30.}$$

Therefore
$$b_0 = \frac{1}{\pi}[f(\pi) - f(-\pi)];$$

and will be zero if $f(\pi) = f(-\pi)$.

In order that $f'(x)$ shall satisfy the conditions stated in Art. 37 $f(x)$ while satisfying the same conditions must in addition be finite and continuous between $x = -\pi$ and $x = \pi$.

If, then, $f(x)$ is *single-valued, finite, and continuous, and has only a finite number of maxima and minima*, between $x = -\pi$ and $x = \pi$, (the values $x = -\pi$ and $x = \pi$ being included), *and if* $f(\pi) = f(-\pi)$ $f(x)$ can be developed into a

[1]We shall regularly use the notation $f'(x)$ for $\dfrac{df(x)}{dx}$. v. Dif. Cal. Art. 124.

Fourier's Series whose term by term derivative will be equal to the derivative of the function.

It will be observed that in this case the periodic curve $y = S$ is continuous throughout its whole extent.

43. Since a Fourier's Integral is a limiting case of a Fourier's Series the conclusions stated in this chapter hold, *mutatis mutandis* for a Fourier's Integral.

For example if a function of x is finite and single-valued for all values of x and has not an infinite number of discontinuities or of maxima and minima in the neighborhood of any value of x it will be equal to the Fourier's Integral

$$\frac{1}{\pi}\int_0^\infty d\alpha \int_{-\infty}^{\infty} f(\lambda)\cos\alpha(\lambda - x).d\lambda$$

and to that Fourier's Integral only, and the integral with respect to x of this Fourier's Integral will be equal to $\int f(x)dx$.

If in addition $f(x)$ is finite and continuous for all values of x the derivative of the Fourier's Integral with respect to x will be equal to $\frac{df(x)}{dx}$.

CHAPTER IV.

SOLUTION OF PROBLEMS IN PHYSICS BY THE AID OF FOURIER'S INTEGRALS AND FOURIER'S SERIES.

44. In Art. 7 we have already considered at some length a problem in Heat Conduction which required the use of a Fourier's Series. We shall begin the present chapter with a problem closely analogous in its treatment to that of Art. 7, but calling for the use of a Fourier's Integral.

Suppose that electricity is flowing in a thin plane sheet of infinite extent and that the value of the potential function is given for every point in some straight line in the sheet, required the value of the potential function at any point of the sheet.

Let us take the line as the axis of X and consider at first only those points for which y is positive:

We have, then, to satisfy the equation

$$D_x^2 V + D_y^2 V = 0 \tag{1}$$

subject to the conditions

$$V = 0 \qquad \text{when} \qquad y = \infty \tag{2}$$

$$V = f(x) \qquad \text{"} \qquad y = 0 \tag{3}$$

where $f(x)$ is a given function, and we are not concerned with negative values of y.

As in Art. 7 we have $e^{-\alpha y}\sin\alpha x$ and $e^{-\alpha y}\cos\alpha x$ as particular values of V which satisfy (1) and (2). We must multiply them by constant coefficients and so combine them as to satisfy condition (3).

By (3) Art. 32

$$f(x) = \frac{1}{\pi}\int_0^\infty d\alpha \int_{-\infty}^{\infty} f(\lambda)\cos\alpha(\lambda - x).d\lambda. \tag{4}$$

We wish to build up a value of V which will reduce to (4) when $y = 0$. This requires a little care but not much ingenuity.

Take $e^{-\alpha y}\cos\alpha x$ and $e^{-\alpha y}\sin\alpha x$ and multiply the first by $\cos\alpha\lambda$, and the second by $\sin\alpha\lambda$; they are still values of V which satisfy (1). Add these and we get

$$e^{-\alpha y}\cos\alpha(\lambda - x),$$

still a value of V which satisfies (1), no matter what the values of α and λ. Multiply by $f(\lambda)d\lambda$ and we have

$$e^{-\alpha y} f(\lambda)\cos\alpha(\lambda - x).d\lambda \tag{5}$$

as a value of V which satisfies (1).

$$V = \int_{-\infty}^{\infty} e^{-\alpha y} f(\lambda) \cos \alpha(\lambda - x).d\lambda \tag{6}$$

is still a solution of (1) since it is the limit of the sum of terms covered by the form (5); and finally

$$V = \frac{1}{\pi} \int_0^{\infty} d\alpha \int_{-\infty}^{\infty} e^{-\alpha y} f(\lambda) \cos \alpha(\lambda - x).d\lambda \tag{7}$$

is a solution of (1) as it is $\frac{1}{\pi}$ multiplied by the limit of the sum of terms formed by multiplying the second member of (6) by $d\alpha$ and giving different values to α.

But (7) must be our required solution since while it satisfies (1) and (2), it reduces to (4) when $y = 0$ and therefore satisfies condition (3).

If $f(x)$ is an *even* function we can reduce (7) to the form

$$V = \frac{2}{\pi} \int_0^{\infty} d\alpha \int_0^{\infty} e^{-\alpha y} f(\lambda) \cos \alpha x \cos \alpha\lambda.d\lambda \tag{8}$$

and if $f(x)$ is an *odd* function to the form

$$V = \frac{2}{\pi} \int_0^{\infty} d\alpha \int_0^{\infty} e^{-\alpha y} f(\lambda) \sin \alpha x \sin \alpha\lambda.d\lambda. \tag{9}$$

(7), (8), and (9) are valid only for positive values of y, but as the problem is obviously symmetrical with respect to the axis of X, (7), (8), and (9) enable us to get the value of the potential function at any point of the plane.

EXAMPLES.

1. Obtain forms (8) and (9) directly by the aid of (5) and (4) Art. 32.

2. State a problem in statical electricity of which the solution given in Art. 44 is the solution.

45. As a special case under Art. 44 let us consider the problem:—To find the value of the potential function at any point of a thin plane sheet of infinite extent where all points of a given line which lie to the left of the origin are kept at potential zero, and all points which lie to the right of the origin are kept at potential unity.

Here $f(x) = 0$ if $x < 0$ and $f(x) = 1$ if $x > 0$.

(7) Art. 44 gives us the required solution. It is

$$V = \frac{1}{\pi}\int_0^\infty d\alpha \int_0^\infty e^{-\alpha y}\cos\alpha(\lambda - x).d\lambda; \tag{1}$$

but this can be much simplified.

We have

$$V = \frac{1}{\pi}\int_0^\infty d\lambda \int_0^\infty e^{-\alpha y}\cos\alpha(\lambda - x).d\alpha.$$

Now
$$\int_0^\infty e^{-ax}\cos mx.dx = \frac{a}{a^2+m^2}$$

if $a > 0$. (Int. Cal. Art. 82, Ex. 8.)

Hence
$$\int_0^\infty e^{-\alpha y}\cos\alpha(\lambda - x).d\alpha = \frac{y}{y^2+(\lambda-x)^2},$$

and
$$V = \frac{1}{\pi}\int_0^\infty \frac{y\,d\lambda}{y^2+(\lambda-x)^2} = \frac{1}{\pi}\left(\frac{\pi}{2}+\tan^{-1}\frac{x}{y}\right).$$

$$\tan\left(\frac{\pi}{2}-\tan^{-1}\frac{x}{y}\right) = \text{ctn}\left(\tan^{-1}\frac{x}{y}\right) = \frac{y}{x};$$

and consequently

$$V = \frac{1}{\pi}\left(\frac{\pi}{2}+\tan^{-1}\frac{x}{y}\right) = 1 - \frac{1}{\pi}\tan^{-1}\frac{y}{x}. \tag{2}$$

Since
$$\log z = \log(x+yi) = \frac{1}{2}\log(x^2+y^2) + i\tan^{-1}\frac{y}{x},$$
[Int. Cal. Art. 33 (2)],

$$i - \frac{1}{\pi}\log z = i - \frac{1}{\pi}\log(x+yi) = -\frac{1}{2\pi}\log(x^2+y^2) + i\left(1-\frac{1}{\pi}\tan^{-1}\frac{y}{x}\right)$$

and $1 - \frac{1}{\pi}\tan^{-1}\frac{y}{x}$ and $-\frac{1}{2\pi}\log(x^2+y^2)$ are *conjugate functions.* (v. Int. Cal. Arts. 209 and 210.) Hence

$$V_1 = -\frac{1}{2\pi}\log(x^2+y^2) \tag{3}$$

is a solution of the equation

$$D_x^2 V_1 + D_y^2 V_1 = 0; \tag{4}$$

and the curves

$$\frac{1}{\pi}\left(\frac{\pi}{2} + \tan^{-1}\frac{x}{y}\right) = a \tag{5}$$

and

$$-\frac{1}{2\pi}\log(x^2 + y^2) = b \tag{6}$$

cut each other at right angles.

If we construct the curves obtained by giving different values to a in (5) we get a set of *equipotential lines* for the conducting sheet described at the beginning of this article, and the curves obtained by giving different values to (b) in (6) will be the *lines of flow*.

Moreover since

$$V_1 = -\frac{1}{2\pi}\log(x^2 + y^2) \tag{3}$$

is a solution of Laplace's Equation (4), the lines of flow just mentioned will be equipotential lines for a certain distribution of potential, for which the equipotential lines above mentioned will be lines of flow.

$V = a$, that is

$$\frac{1}{\pi}\left(\frac{\pi}{2} + \tan^{-1}\frac{x}{y}\right) = a, \tag{5}$$

reduces to

$$y = -x\tan a\pi. \tag{7}$$

If now we give to a values differing by a constant amount we get a set of straight lines radiating from the origin and at equal angular intervals.

$V_1 = b$, that is

$$-\frac{1}{2\pi}\log(x^2 + y^2) = b, \tag{6}$$

reduces to

$$x^2 + y^2 = e^{-2\pi b}. \tag{8}$$

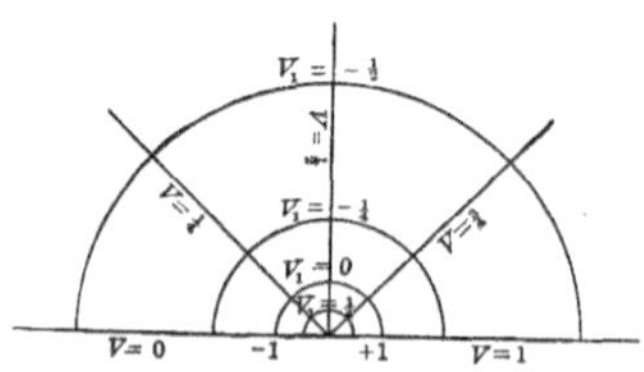

If we give to b a set of values differing by a constant amount we get a set of circles whose centres are at the origin and whose radii form a geometrical progression. They are the equipotential lines for a thin plane sheet of infinite extent where the potential function is kept equal to given different constant values on the circumferences of two given concentric circles or where we have a *source* at the origin; and for this system the lines (7) are lines of flow, and (3) is the complete solution.

The figure gives the equipotential lines and lines of flow for either system, but only for positive values of y. The complete figure has the axis of X as an axis of symmetry.

EXAMPLES.

1. Solve the problem of Art. 44 for the case where

$$f(x) = -1 \quad \text{if} \quad x < 0 \quad \text{and} \quad f(x) = 1 \quad \text{if} \quad x > 0.$$

Ans., $V = \frac{2}{\pi} \tan^{-1} \frac{x}{y}$.

2. Solve the problem of Art. 44 for the case where

$$f(x) = a \quad \text{if} \quad x < 0 \quad \text{and} \quad f(x) = b \quad \text{if} \quad x > 0.$$

Ans., $V = \frac{1}{2}(a+b) + \frac{1}{\pi}(b-a) \tan^{-1} \frac{x}{y}$.

3. Reduce (7), (8), and (9) Art. 44 to the forms

$$V = \frac{1}{\pi} \int_{-\infty}^{\infty} \frac{y f(\lambda) d\lambda}{y^2 + (\lambda - x)^2},$$

$$V = \frac{1}{\pi} \int_0^{\infty} y f(\lambda) d\lambda \left[\frac{1}{y^2 + (\lambda - x)^2} + \frac{1}{y^2 + (\lambda + x)^2} \right],$$

$$V = \frac{1}{\pi} \int_0^{\infty} y f(\lambda) d\lambda \left[\frac{1}{y^2 + (\lambda - x)^2} - \frac{1}{y^2 + (\lambda + x)^2} \right],$$

respectively.

46. An especially interesting case of Art. 44 is the following where

$$f(x) = 0 \text{ if } x < -1, \quad f(x) = 1 \text{ if } -1 < x < 1, \quad \text{and} \quad f(x) = 0 \text{ if } x > 1.$$

Here
$$V = \frac{1}{\pi} \left[\tan^{-1} \frac{1+x}{y} + \tan^{-1} \frac{1-x}{y} \right]. \tag{1}$$

Now
$$\frac{1}{\pi} \log[(1-z)i] = \frac{1}{\pi} \log[(1-x-yi)i] = \frac{1}{\pi} \log[y + (1-x)i]$$
$$= \frac{1}{2\pi} \log[(1-x)^2 + y^2] + \frac{i}{\pi} \tan^{-1} \frac{1-x}{y},$$

and

$$-\frac{1}{\pi} \log[(-1-z)i] = -\frac{1}{\pi} \log[(-1-x-yi)i] = -\frac{1}{\pi} \log[y - (1+x)i]$$
$$= -\frac{1}{2\pi} \log[(1+x)^2 + y^2] + \frac{i}{\pi} \tan^{-1} \frac{1+x}{y}.$$

$$\frac{1}{\pi}\log\frac{1-z}{-1-z} = \frac{1}{2\pi}\log\frac{(1-x)^2+y^2}{(1+x)^2+y^2} + \frac{i}{\pi}\left[\tan^{-1}\frac{1+x}{y} + \tan^{-1}\frac{1-x}{y}\right].$$

Hence

$$\frac{1}{\pi}\left(\tan^{-1}\frac{1+x}{y} + \tan^{-1}\frac{1-x}{y}\right) \quad \text{and} \quad \frac{1}{2\pi}\log\frac{(1-x)^2+y^2}{(1+x)^2+y^2}$$

are *conjugate functions:*[1] and

$$\frac{1}{\pi}\left(\tan^{-1}\frac{1+x}{y} + \tan^{-1}\frac{1-x}{y}\right) = a \tag{2}$$

is any equipotential line, and

$$\frac{1}{2\pi}\log\frac{(1-x)^2+y^2}{(1+x)^2+y^2} = b \tag{3}$$

any line of flow for the system described at the beginning of this article; and

$$V_1 = \frac{1}{2\pi}\log\frac{(1-x)^2+y^2}{(1+x)^2+y^2} \tag{4}$$

is the solution of a new problem for which (3) represents any equipotential line and (2) any line of flow.

---

[1]The function conjugate to

$$\frac{1}{\pi}\left[\tan^{-1}\frac{1+x}{y} + \tan^{-1}\frac{1-x}{y}\right]$$

might have been found as follows. If ϕ is the required function and ψ the given function we have by Int. Cal. Arts. 211, 212, and 213 the relations

$$D_x\phi = D_y\psi \quad \text{and} \quad D_y\phi = -D_x\psi.$$

Here
$$D_y\psi = -\frac{1}{\pi}\left[\frac{1+x}{(1+x)^2+y^2} + \frac{1-x}{(1-x)^2+y^2}\right]$$

and
$$-D_x\psi = -\frac{1}{\pi}\left[\frac{y}{(1+x)^2+y^2} - \frac{y}{(1-x)^2+y^2}\right].$$

If now we integrate $D_y\psi$ with respect to x treating y as a constant and add an arbitrary function of y we shall have ϕ. So that

$$\phi = -\frac{1}{2\pi}\left\{\log[(1+x)^2+y^2] - \log[(1-x)^2+y^2]\right\} + f(y).$$

$$D_y\phi = -\frac{1}{\pi}\left[\frac{y}{(1+x)^2+y^2} - \frac{y}{(1-x)^2+y^2}\right] + \frac{df(y)}{dy}$$

Comparing this with its equal $-D_x\psi$ above we find $\frac{df(y)}{dy} = 0$ and $f(y) = C$ a constant

therefore
$$\frac{1}{2\pi}\log\frac{(1-x)^2+y^2}{(1+x)^2+y^2} + C,$$

where C may be taken at pleasure, is our required conjugate function.

(2) reduces to

$$\frac{2y}{x^2+y^2-1} = \tan a\pi$$

or

$$x^2 + (y - \operatorname{ctn} a\pi)^2 = \csc^2 a\pi; \tag{5}$$

and (3) to

$$x^2 + y^2 + 2\frac{e^{2b\pi}+1}{e^{2b\pi}-1}x + 1 = 0$$

or

$$\left(x + \frac{e^{b\pi}+e^{-b\pi}}{e^{b\pi}-e^{-b\pi}}\right)^2 + y^2 = \left(\frac{e^{b\pi}+e^{-b\pi}}{e^{b\pi}-e^{-b\pi}}\right)^2 - 1$$

or

$$(x + \operatorname{ctnh} b\pi)^2 + y^2 = \operatorname{csch}^2 b\pi. \tag{6}$$

(5) and (6) are circles. The circles (5) have their centres in the axis of Y, and pass through the points $(-1,0)$ and $(1,0)$; and the circles (6) have their centres in the axis of X.

(4) is the complete solution, (6) is any equipotential line and (5) any line of flow for a plane sheet in which the points in the circumferences of two given circles whose centres are further apart than the sum of their radii are kept at different constant potentials, or where a source and a sink of equal intensity are placed at the points $(-1,0)$ and $(1,0)$. An important practical example is where two wires connected with the poles of a battery are placed with their free ends in contact with a thin plane sheet of conducting material. The figure shows the equipotential lines and lines of flow of either system.

The complete figure would have the axis of X for an axis of symmetry.

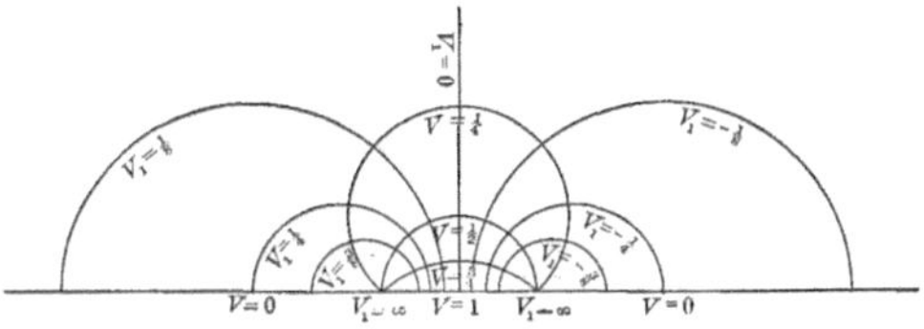

EXAMPLES.

1. Show that if $f(x) = a_1$ when $x < -b$, $f(x) = a_2$ when $-b < x < b$, $f(x) = a_3$ when $x > b$,

$$V = \frac{a_1 + a_3}{2} + \frac{1}{\pi}\left[(a_2 - a_1)\tan^{-1}\frac{b+x}{y} + (a_2 - a_3)\tan^{-1}\frac{b-x}{y}\right].$$

2. Show that if $f(x) = 0$ if $x < 0$, $f(x) = a_1$ if $0 < x < b_1$, $f(x) = a_2$ if $b_1 < x < b_2$, $f(x) = a_3$ if $b_2 < x < b_3$, &c.,

$$V = \frac{1}{\pi}\left[a_1 \tan^{-1}\frac{x}{y} + (a_1 - a_2)\tan^{-1}\frac{b_1 - x}{y} + (a_2 - a_3)\tan^{-1}\frac{b_2 - x}{y} \right.$$
$$\left. + (a_3 - a_4)\tan^{-1}\frac{b_3 - x}{y} + \cdots\right].$$

3. Show that if $f(x) = -1$ if $x < -1$, $f(x) = x$ if $-1 < x < 1$, $f(x) = 1$ if $x > 1$,

$$V = \frac{1}{\pi}\left[(1 + x)\tan^{-1}\frac{1 + x}{y} - (1 - x)\tan^{-1}\frac{1 - x}{y} + \frac{y}{2}\log\frac{(1 - x)^2 + y^2}{(1 + x)^2 + y^2}\right].$$

4. Show that if $f(x) = -1$ if $x < -1$, $f(x) = 0$ if $-1 < x < 1$, $f(x) = 1$ if $x > 1$,

$$V = \frac{1}{\pi}\left[\tan^{-1}\frac{1 + x}{y} - \tan^{-1}\frac{1 - x}{y}\right].$$

Show that the equipotential lines are equilateral hyperbolas passing through the points $(-1, 0)$ and $(1, 0)$, and that the lines of flow are Cassinian ovals having $(-1, 0)$ and $(1, 0)$ as foci. The lines of flow are equipotential lines and the equipotential lines are lines of flow for the case where the points $(-1, 0)$ and $(1, 0)$ are kept at the same infinite potential, or where very small ovals surrounding these points are kept at the same finite potential. The case is approximately that of a pair of wires connected with the same pole of a battery whose other pole is grounded, and then placed with their ends in contact with a thin plane conducting sheet.

5. Show that if $f(x) = 0$ if $x < 0$, $f(x) = -1$ if $0 < x < a$, $f(x) = 0$ if $a < x < b$, and $f(x) = 1$ if $x > b$,

$$V = \frac{1}{\pi}\left[\frac{\pi}{2} - \tan^{-1}\frac{a - x}{y} - \tan^{-1}\frac{b - x}{y} - \tan^{-1}\frac{x}{y}\right].$$

The conjugate function

$$V = \frac{1}{2\pi}\log\frac{x^2 + y^2}{[(a - x)^2 + y^2][(b - x)^2 + y^2]}$$

is the solution for the case where a sink and two sources of equal intensity lie on the axis of X, the sink at the origin and the sources at the distances a and b to the right of the origin. One of the lines of flow is easily seen to be the circle $x^2 + y^2 = ab$.

47. If the plane conducting sheet has two straight edges at right angles with each other and one is kept at potential zero while the value of the potential function is given at each point of the second, that is if $V = 0$ when $x = 0$ and $V = f(x)$ when $y = 0$, the solution is readily obtained. It is

$$V = \frac{2}{\pi}\int_0^\infty d\alpha \int_0^\infty e^{-\alpha y} f(\lambda) \sin \alpha x \sin \alpha\lambda . d\lambda. \tag{1}$$

v. (9) Art. 44.
This reduces to

$$V = \frac{1}{\pi}\int_0^\infty f(\lambda) d\lambda \left[\frac{y}{y^2 + (\lambda - x)^2} - \frac{y}{y^2 + (\lambda + x)^2}\right]. \tag{2}$$

v. Ex. 3 Art. 45.

EXAMPLES.

1. If $V = 0$ when $y = 0$ and $V = F(y)$ when $x = 0$ show that

$$\begin{aligned} V &= \frac{2}{\pi}\int_0^\infty d\alpha \int_0^\infty e^{-\alpha x} F(\lambda) \sin \alpha y \sin \alpha\lambda . d\lambda \\ &= \frac{1}{\pi}\int_0^\infty F(\lambda) d\lambda \left[\frac{x}{x^2 + (\lambda - y)^2} - \frac{x}{x^2 + (\lambda + y)^2}\right]. \end{aligned}$$

2. If $V = f(x)$ when $y = 0$ and $V = F(y)$ when $x = 0$ show that

$$\begin{aligned} V = \frac{1}{\pi}\int_0^\infty \Bigg[& f(\lambda)\left(\frac{y}{y^2 + (\lambda - x)^2} - \frac{y}{y^2 + (\lambda + x)^2}\right) \\ & + F(\lambda)\left(\frac{x}{x^2 + (\lambda - y)^2} - \frac{x}{x^2 + (\lambda + y)^2}\right)\Bigg] d\lambda. \end{aligned}$$

3. If $F(y) = b$ the result of Ex. 2 reduces to

$$V = \frac{2b}{\pi}\tan^{-1}\frac{y}{x} + \frac{1}{\pi}\int_0^\infty f(\lambda) d\lambda \left[\frac{y}{y^2 + (\lambda - x)^2} - \frac{y}{y^2 + (\lambda + x)^2}\right].$$

4. If $F(y) = 1$ for $0 < y < 1$ and $F(y) = 0$ for $y > 1$ while $f(x) = 1$ for $0 < x < 1$ and $f(x) = 0$ for $x > 1$

$$\begin{aligned} V = \frac{1}{\pi}\Bigg[& \tan^{-1}\frac{1-x}{y} - \tan^{-1}\frac{1+x}{y} + 2\tan^{-1}\frac{y}{x} \\ & + \tan^{-1}\frac{1-y}{x} - \tan^{-1}\frac{1+y}{x} + 2\tan^{-1}\frac{x}{y}\Bigg]. \end{aligned}$$

5. If one edge of the conducting sheet treated in Art. 47 is insulated, so that $D_x V = 0$ if $x = 0$ and $V = f(x)$ when $y = 0$

$$V = \frac{2}{\pi}\int_0^\infty d\alpha \int_0^\infty e^{-\alpha y} f(\lambda)\cos\alpha x \cos\alpha\lambda . d\lambda$$

$$= \frac{1}{\pi}\int_0^\infty f(\lambda)d\lambda\left[\frac{y}{y^2+(\lambda+x)^2}+\frac{y}{y^2+(\lambda-x)^2}\right].$$

48. If the conducting sheet is a long strip with parallel edges one of which is at potential zero while the value of the potential function is given at all points of the other, that is if $V = 0$ when $y = 0$ and $V = F(x)$ when $y = b$ the problem is not a very difficult one.

Since we are no longer concerned with the value of V when $y = \infty$ $V = e^{\alpha y}\sin ax$ and $V = e^{\alpha y}\cos ax$ are available as particular solutions of the equation

$$D_x^2 V + D_y^2 V = 0 \tag{1}$$

as well as $V = e^{-\alpha y}\sin\alpha x$ and $V = e^{-\alpha y}\cos\alpha x$.

Consequently $\dfrac{e^{\alpha y}+e^{-\alpha y}}{2}\sin\alpha x = \cosh\alpha y \sin\alpha x$ [Int. Cal. Art. 43 (2)]

and $\dfrac{e^{\alpha y}-e^{-\alpha y}}{2}\sin\alpha x = \sinh\alpha y\sin\alpha x$ [Int. Cal. Art. 43 (1)]

and $\cosh\alpha y\cos\alpha x$ and $\sinh\alpha y\cos\alpha x$

are now available values of V and can be used precisely as $e^{-\alpha y}\cos\alpha x$ and $e^{-\alpha y}\sin\alpha x$ are used in Art. 44.

Following the same course as in Art. 44 we get

$$V = \frac{1}{\pi}\int_0^\infty d\alpha\int_{-\infty}^\infty \frac{\sinh\alpha y}{\sinh\alpha b}F(\lambda)\cos\alpha(\lambda - x).d\lambda \tag{2}$$

as a solution of (1) which will reduce to $V = F(x)$ when $y = b$

and to $V = 0$ when $y = 0$, since $\sinh 0 = \dfrac{1-1}{2} = 0$,

and (2) is therefore our required solution.

If V is to be equal to zero when $y = b$ and to $f(x)$ when $y = 0$ we have only to replace y by $b - y$ and $F(x)$ by $f(x)$ in (2). We get

$$V = \frac{1}{\pi}\int_0^\infty d\alpha\int_{-\infty}^\infty \frac{\sinh\alpha(b-y)}{\sinh\alpha b}f(\lambda)\cos\alpha(\lambda - x).d\lambda. \tag{3}$$

If $V = f(x)$ when $y = 0$ and $V = F(x)$ when $y = b$ then

$$V = \frac{1}{\pi}\int_0^\infty d\alpha \int_{-\infty}^{\infty} \frac{\sinh\alpha(b-y)}{\sinh\alpha b} f(\lambda)\cos\alpha(\lambda - x).d\lambda$$
$$+ \frac{1}{\pi}\int_0^\infty d\alpha \int_{-\infty}^{\infty} \frac{\sinh\alpha y}{\sinh\alpha b} F(\lambda)\cos\alpha(\lambda - x).d\lambda.$$

This can be considerably simplified by the aid of the formula

$$\int_0^\infty \frac{\sinh px}{\sinh qx}\cos rx.dx = \frac{\pi}{2q}\frac{\sin\frac{p\pi}{q}}{\cos\frac{p\pi}{q} + \cosh\frac{r\pi}{q}}$$

if $p^2 < q^2$. [Bierens de Haan, Tables of Def. Int. (7) 265] and becomes

$$V = \frac{1}{2b}\sin\frac{\pi}{b}(b-y)\int_{-\infty}^{\infty} f(\lambda)\frac{d\lambda}{\cos\frac{\pi(b-y)}{b} + \cosh\frac{\pi}{b}(\lambda - x)}$$
$$+ \frac{1}{2b}\sin\frac{\pi y}{b}\int_{-\infty}^{\infty} F(\lambda)\frac{d\lambda}{\cos\frac{\pi y}{b} + \cosh\frac{\pi}{b}(\lambda - x)} \qquad \text{or}$$

$$V = \frac{1}{2b}\sin\frac{\pi y}{b}\int_{-\infty}^{\infty}\left[\frac{f\lambda}{\cosh\frac{\pi}{b}(\lambda - x) - \cos\frac{\pi y}{b}} + \frac{F\lambda}{\cosh\frac{\pi}{b}(\lambda - x) + \cos\frac{\pi y}{b}}\right]d\lambda. \quad (5)$$

EXAMPLES.

1. Given the formula

$$\int\frac{dx}{a + b\cosh x} = \frac{2}{\sqrt{b^2 - a^2}}\tan^{-1}\left(\sqrt{\frac{b-a}{b+a}}\tanh\frac{x}{2}\right) \qquad \text{if } b > a,$$

show that if $V = 1$ when $y = 0$ and $V = 0$ when $y = b$ $V = \frac{1}{b}(b - y)$.

2. Show that if $V = 0$ when $y = b$, $V = -1$ when $y = 0$ and $x < 0$, and $V = 1$ when $y = 0$ and $x > 0$

$$V = \frac{2}{\pi}\tan^{-1}\left[\frac{\tanh\frac{\pi x}{2b}}{\tan\frac{\pi y}{2b}}\right]$$

The solution for the conjugate system, that is, for a strip having a source at $(0, 0)$ and an infinitely distant sink is

$$V = -\frac{1}{\pi}\log\left[\cosh^2\frac{\pi x}{2b} - \cos^2\frac{\pi y}{2b}\right].$$

3. Show that if $V = -1$ when $y = 0$ and $x < 0$, $V = 1$ when $y = 0$ and $x > 0$, $V = -1$ when $y = b$ and $x < 0$, and $V = 1$ when $y = b$ and $x > 0$,

$$V = \frac{2}{\pi}\tan^{-1}\left(\tan\frac{\pi}{2b}(b-y)\tanh\frac{\pi x}{2b}\right) + \frac{2}{\pi}\tan^{-1}\left(\tan\frac{\pi}{2b}y\tanh\frac{\pi x}{2b}\right)$$
$$= \frac{2}{\pi}\tan^{-1}\left[\frac{\sinh\frac{\pi x}{b}}{\sin\frac{\pi y}{b}}\right].$$

The solution for the conjugate system, that is, for a strip having a source and a sink at the points $(0,0)$ and $(0,b)$ is

$$V = \frac{1}{\pi}\log\left[\frac{\cosh\frac{\pi x}{b} + \cos\frac{\pi y}{b}}{\cosh\frac{\pi x}{b} - \cos\frac{\pi y}{b}}\right].$$

4. If $V = 0$ when $x = 0$, $V = f(x)$ when $y = 0$ and $x > 0$, and $V = 0$ when $y = b$ and $x > 0$.

$$V = \frac{1}{\pi}\int_0^\infty d\alpha\int_0^\infty \frac{\sinh\alpha(b-y)}{\sinh\alpha b}[\cos\alpha(\lambda - x) - \cos\alpha(\lambda + x)]f(\lambda)d\lambda$$
$$= \frac{1}{2b}\sin\frac{\pi y}{b}\int_0^\infty\left[\frac{1}{\cosh\frac{\pi}{b}(\lambda - x) - \cos\frac{\pi y}{b}} - \frac{1}{\cosh\frac{\pi}{b}(\lambda + x) - \cos\frac{\pi y}{b}}\right]f(\lambda)d\lambda$$

for positive values of x and for values of y between 0 and b.

5. If $V_1 = 0$ when $x = 0$, $V_1 = F(x)$ when $y = b$ and $x > 0$, and $V_1 = 0$ when $y = 0$ and $x > 0$

$$V_1 = \frac{1}{2b}\sin\frac{\pi y}{b}\int_0^\infty\left[\frac{1}{\cosh\frac{\pi}{b}(\lambda - x) + \cos\frac{\pi y}{b}} - \frac{1}{\cosh\frac{\pi}{b}(\lambda + x) - \cos\frac{\pi y}{b}}\right]F(\lambda)d\lambda$$

for positive values of x and values of y between 0 and b.

6. If $V_2 = 0$ when $x = 0$, $V_2 = f(x)$ when $y = 0$ and $x > 0$, and $V_2 = F(x)$ when $y = b$ and $x > 0$

$$V_2 = V + V_1 \quad \text{for} \quad x > 0 \quad \text{and} \quad 0 < y < b. \text{ (v. Exs. 4 and 5)}$$

7. If one edge of the strip described in Art. 48 is insulated so that we have $V = f(x)$ when $y = 0$ and $D_yV = 0$ when $y = b$ show that

$$V = \frac{1}{\pi}\int_0^\infty d\alpha\int_{-\infty}^\infty \frac{\cosh\alpha(b-y)}{\cosh\alpha b}f(\lambda)\cos\alpha(\lambda - x).d\lambda.$$

By the aid of the formula

$$\int_0^\infty \frac{\cosh px}{\cosh qx}\cos rx.dx = \frac{\pi}{q}\,\frac{\cosh\dfrac{r\pi}{2q}\cos\dfrac{p\pi}{2q}}{\cos\dfrac{p\pi}{q}+\cosh\dfrac{r\pi}{q}} \qquad \text{if } p<q,$$

[Bierens de Haan, Def. Int. Tables (6) 265], reduce this to

$$V = \frac{1}{b}\sin\frac{\pi y}{2b}\int_{-\infty}^{\infty}\frac{f(\lambda)\cosh\dfrac{\pi}{2b}(\lambda-x)}{\cosh\dfrac{\pi}{b}(\lambda-x)-\cos\dfrac{\pi y}{b}}d\lambda.$$

8. If $V=0$ when $y=0$ or b and $x<-a$, $V=1$ when $y=0$ or b and $-a<x<a$, and $V=0$ when $y=0$ or b and $x>a$

$$V=\frac{1}{\pi}\left[\tan^{-1}\frac{\sinh\dfrac{\pi(a-x)}{b}}{\sin\dfrac{\pi y}{b}}+\tan^{-1}\frac{\sinh\dfrac{\pi(a+x)}{b}}{\sin\dfrac{\pi y}{b}}\right].$$

9. If $V=0$ when $y=0$ or b and $x<-a$, $V=1$ when $y=0$ and $-a<x<a$, $V=0$ when $y=0$ or b and $x>a$, and $V=-1$ when $y=b$ and $-a<x<a$

$$V=\frac{1}{\pi}\left[\tan^{-1}\frac{\tanh\dfrac{\pi(a-x)}{b}}{\tan\dfrac{\pi y}{b}}+\tan^{-1}\frac{\tanh\dfrac{\pi(a+x)}{b}}{\tan\dfrac{\pi y}{b}}\right].$$

10. A system conjugate to that of Ex. 9 is $V=+\infty$ when $y=0$ or b and $x=-a$, $V=-\infty$ when $y=0$ or b and $x=a$. In this case

$$V=\frac{1}{2\pi}\log\frac{\sin^2\dfrac{\pi y}{b}+\sinh^2\dfrac{\pi(a-x)}{b}}{\sin^2\dfrac{\pi y}{b}+\sinh^2\dfrac{\pi(a+x)}{b}}.$$

49. Let us take now a problem in the flow of heat. Suppose we have an infinite solid in which heat flows only in one direction, and that at the start the temperature of each point of the solid is given. Let it be required to find the temperature of any point of the solid at the end of the time t.

Here we have to solve the equation

$$D_t u = a^2 D_x^2 u \tag{1}$$

[v. Art. 1 (II)] subject to the condition

$$u=f(x) \quad \text{when} \quad t=0. \tag{2}$$

As the equation (1) is linear with constant coefficients we can get a particular solution by the device used in Arts. 7 and 8.

Let $u = e^{\beta t + \alpha x}$ and substitute in (1). We get

$$\beta = a^2\alpha^2$$

as the only relation which need hold between β and α.

Hence $$u = e^{\alpha x + a^2\alpha^2 t} = e^{a^2\alpha^2 t}e^{\alpha x} \tag{3}$$

is a solution of (1) no matter what value is given to α.

To get a trigonometric form replace α by αi.

Then $$u = e^{-a^2\alpha^2 t}e^{\alpha x i}.$$

If in (3) we replace α by $-\alpha i$ we get

$$u = e^{-a^2\alpha^2 t}e^{-\alpha x i}.$$

As in Arts. 7 and 8 we get from these values

$$u = e^{-a^2\alpha^2 t}\sin \alpha x \quad \text{and} \quad u = e^{-a^2\alpha^2 t}\cos \alpha x$$

as particular solutions of (1), α being wholly unrestricted.

From these values we wish to build up a value of u which shall reduce to $f(x)$ when $t = 0$ and shall still be a solution of (1).

We have $$f(x) = \frac{1}{\pi}\int_0^\infty d\alpha \int_{-\infty}^\infty f(\lambda)\cos\alpha(\lambda - x).d\lambda \tag{4}$$

v. Art. 32 (3), and by proceeding as in Art. 44 we get

$$u = \frac{1}{\pi}\int_0^\infty d\alpha \int_{-\infty}^\infty e^{-a^2\alpha^2 t} f(\lambda)\cos\alpha(\lambda - x).d\lambda \tag{5}$$

as our required value of u.

This can be considerably simplified.

Changing the order of integration

$$u = \frac{1}{\pi}\int_{-\infty}^\infty f(\lambda)d\lambda \int_0^\infty e^{-a^2\alpha^2 t}\cos\alpha(\lambda - x).d\alpha. \tag{6}$$

$$\int_0^\infty e^{-a^2\alpha^2 t}\cos\alpha(\lambda - x).d\alpha = \frac{1}{2a}\sqrt{\frac{\pi}{t}}.e^{-\frac{(\lambda - x)^2}{4a^2 t}} \tag{7}$$

by the formula

$$\int_0^\infty e^{-a^2x^2}\cos bx.dx = \frac{\sqrt{\pi}}{2a}e^{-\frac{b^2}{4a^2}} \qquad \text{[Int. Cal. Art. 94 (2)]}$$

Hence $$u = \frac{1}{2a\sqrt{\pi t}}\int_{-\infty}^\infty f(\lambda)e^{-\frac{(\lambda - x)^2}{4a^2 t}}d\lambda. \tag{8}$$

Let now $$\beta = \frac{\lambda - x}{2a\sqrt{t}},$$

then $$\lambda = x + 2a\sqrt{t}.\beta$$

and $$u = \frac{1}{\sqrt{\pi}} \int_{-\infty}^{\infty} f(x + 2a\sqrt{t}.\beta) e^{-\beta^2} d\beta. \tag{9}$$

EXAMPLES.

1. Let the solid be of infinite extent and let the temperature be equal to a constant c at the time $t = 0$.

Then $$u = \frac{c}{\sqrt{\pi}} \int_{-\infty}^{\infty} e^{-\beta^2} d\beta = \frac{2c}{\sqrt{\pi}} \int_{0}^{\infty} e^{-\beta^2} d\beta = c.$$

v. Int. Cal. Art. 92 (2).

2. Let $u = x$ when $t = 0$.

Then $$u = \frac{1}{\sqrt{\pi}} \int_{-\infty}^{\infty} (x + 2a\sqrt{t}.\beta) e^{-\beta^2} d\beta = x.$$

3. Let $u = x^2$ when $t = 0$.

Then $$u = x^2 + 2a^2 t.$$

4. Let $u = 0$ if $x < -b$, $u = 1$ if $-b < x < b$, and $u = 0$ if $x > b$, when $t = 0$.

Then

$$u = \frac{1}{\sqrt{\pi}} \int_{-\frac{b+x}{2a\sqrt{t}}}^{\frac{b-x}{2a\sqrt{t}}} e^{-\beta^2} d\beta = \frac{2}{\sqrt{\pi}} \left[\frac{b}{2a\sqrt{t}} - \frac{b^3 + 3bx^2}{3(2a\sqrt{t})^3} + \frac{b^5 + 10b^3x^2 + 5bx^4}{5.2!(2a\sqrt{t})^5} - \cdots \right].$$

5. Let $u = 0$ if $x < 0$ and $u = 1$ if $x > 0$ when $t = 0$.
Then

$$u = \frac{1}{\sqrt{\pi}} \int_{-\frac{x}{2a\sqrt{t}}}^{\infty} e^{-\beta^2} d\beta = \frac{1}{\sqrt{\pi}} \left[\int_{0}^{\frac{x}{2a\sqrt{t}}} e^{-\beta^2} d\beta + \int_{0}^{\infty} e^{-\beta^2} d\beta \right] = \frac{1}{\sqrt{\pi}} \int_{0}^{\frac{x}{2a\sqrt{t}}} e^{-\beta^2} d\beta + \frac{1}{2}$$

$$= \frac{1}{2} + \frac{1}{\sqrt{\pi}} \left[\frac{x}{2a\sqrt{t}} - \frac{x^3}{3.(2a\sqrt{t})^3} + \frac{x^5}{5.2!(2a\sqrt{t})^5} - \frac{x^7}{7.3!(2a\sqrt{t})^7} + \cdots \right].$$

6. An iron slab 10 c. m. thick is placed between and in contact with two very thick iron slabs. The initial temperature of the middle slab is 100°, and of

each of the outer slabs 0°. Required the temperature of a point in the middle of the inner slab fifteen minutes after the slabs have been put together. Given $a^2 = 0.185$ in C.G.S. units. *Ans.*, 21°.6.

7. Two very thick iron slabs one of which is at the temperature 0° and the other at the temperature 100° throughout are placed together face to face. Find the temperature of each slab 10 c. m. from their common face fifteen minutes after they have been placed together. *Ans.*, 70°.8, 29°.2.

8. Find a particular solution of $D_t u = a^2 D_x^2 u$ on the assumption that it is of the form $u = T.X$ where T is a function of t alone and X is a function of x alone.

50. If our solid has one plane face which is kept at the constant temperature zero, and we start with any given distribution of heat, the problem is somewhat modified.

Take the origin of coördinates in the plane face. Then we have as before the equation

$$D_t u = a^2 D_x^2 u, \tag{1}$$

but our conditions are

$$u = 0 \quad \text{when} \quad x = 0 \tag{2}$$

$$u = f(x) \quad \text{"} \quad t = 0 \tag{3}$$

and we are concerned only with positive values of x.

We may then use the form (4) Art. 32

$$f(x) = \frac{2}{\pi}\int_0^\infty d\alpha \int_0^\infty f(\lambda) \sin \alpha x \sin \alpha\lambda . d\lambda, \tag{4}$$

and proceeding as in the last section we get

$$u = \frac{2}{\pi}\int_0^\infty d\alpha \int_0^\infty e^{-a^2\alpha^2 t} f(\lambda) \sin \alpha x \sin \alpha\lambda . d\lambda \tag{5}$$

as our required solution. This may be reduced considerably.

$$u = \frac{1}{\pi}\int_0^\infty f(\lambda) d\lambda \int_0^\infty e^{-a^2\alpha^2 t}[\cos \alpha(\lambda - x) - \cos \alpha(\lambda + x)] d\alpha,$$

$$\text{or} \quad u = \frac{1}{2a\sqrt{\pi t}}\int_0^\infty f(\lambda)\left(e^{-\frac{(\lambda - x)^2}{4a^2 t}} - e^{-\frac{(\lambda + x)^2}{4a^2 t}}\right) d\lambda \tag{6}$$

by (7) Art. 49, and this may be reduced to the form

$$u = \frac{1}{\sqrt{\pi}}\left[\int_{-\frac{x}{2a\sqrt{t}}}^\infty e^{-\beta^2} f(x + 2a\sqrt{t}.\beta) d\beta - \int_{\frac{x}{2a\sqrt{t}}}^\infty e^{-\beta^2} f(-x + 2a\sqrt{t}.\beta) d\beta\right]. \tag{7}$$

EXAMPLES.

1. Let the initial temperature be constant and equal to c.
Then

$$u = \frac{c}{\sqrt{\pi}}\left[\int_{-\frac{x}{2a\sqrt{t}}}^{\infty} e^{-\beta^2}d\beta - \int_{\frac{x}{2a\sqrt{t}}}^{\infty} e^{-\beta^2}d\beta\right]$$

$$= \frac{2c}{\sqrt{\pi}}\int_0^{\frac{x}{2a\sqrt{t}}} e^{-\beta^2}d\beta$$

$$= \frac{2c}{\sqrt{\pi}}\left[\frac{x}{2a\sqrt{t}} - \frac{x^3}{3.(2a\sqrt{t})^3} + \frac{x^5}{5.2!(2a\sqrt{t})^5} - \frac{x^7}{7.3!.(2a\sqrt{t})^7} + \cdots\right].$$

2. Assuming that the earth was originally at the temperature 7000° Fahrenheit throughout, and that the surface was kept at the constant temperature 0°, find (1) the temperature 10 miles below the surface 10,000,000 years after the cooling began; (2) the temperature 1 mile below the surface at the same epoch; (3) the temperature 10 miles below the surface 100,000,000 years after the cooling began; (4) the temperature 1 mile below the surface at the same epoch; (5) the rate at which the temperature was increasing with the distance from the surface at each point at each epoch.

Neglect the convexity of the earth's surface and take Sir Wm. Thomson's value of $a^2(400)$ the foot, the Fahrenheit degree, and the year being taken as units. (Thomson and Tait's Nat. Phil. Vol. II. Appendix.)

Ans., (1) 3114°; (2) 329°.5; (3) 1036°; (4) 103°; (5) 1° for every 20 feet, 3° for every 50 feet, 1° for every 50 feet, 1° for every 50 feet.

3. Let the initial temperature be constant and equal to $-b$, then by Ex. 1

$$u = -\frac{2b}{\sqrt{\pi}}\int_0^{\frac{x}{2a\sqrt{t}}} e^{-\beta^2}d\beta.$$

4. Let the temperature of the plane face be b instead of zero, and let the initial temperature be zero.

Then we have only to add b to the second member of the solution in Ex. 3, as we may since $u = b$ is a solution of (1) Art. 49, and we get

$$u = b\left(1 - \frac{2}{\sqrt{\pi}}\int_0^{\frac{x}{2a\sqrt{t}}} e^{-\beta^2}d\beta\right).$$

5. Let $u = b$ when $x = 0$ and $u = f(x)$ when $t = 0$.
Then

$$u = b\left(1 - \frac{2}{\sqrt{\pi}}\int_0^{\frac{x}{2a\sqrt{t}}} e^{-\beta^2}d\beta\right) + \frac{1}{2a\sqrt{\pi t}}\int_0^{\infty} f(\lambda)[e^{-\frac{(\lambda-x)^2}{4a^2t}} - e^{-\frac{(\lambda+x)^2}{4a^2t}}]d\lambda$$

by (6) Art. 50.

6. Let $u = b$ when $x = 0$ and $u = c$ when $t = 0$.

Then
$$u = b + (c - b)\frac{2}{\sqrt{\pi}}\int_0^{\frac{x}{2a\sqrt{t}}} e^{-\beta^2} d\beta.$$

7. If the earth has been cooling for 200,000,000 years from a uniform temperature, prove that the rate of cooling is greatest at a depth of about 76 miles, and that at a depth of about 130 miles the rate of cooling has reached its maximum value for all time. Let $a^2 = 400$.

8. Show that if the plane face of the solid considered in Art. 50 instead of being kept at temperature zero is impervious to heat

$$u = \frac{1}{2a\sqrt{\pi t}}\int_0^{\infty} f(\lambda)(e^{-\frac{(\lambda-x)^2}{4a^2t}} + e^{-\frac{(\lambda+x)^2}{4a^2t}})d\lambda. \qquad \text{v. (6) Art. 50.}$$

51. If the temperature of the plane face of the solid described in Art. 50 is a given function of the time and the initial temperature is zero, the solution of the problem can be obtained by a very ingenious method due to Riemann.

Here we have to solve the equation

$$D_t u = a^2 D_x^2 u \tag{1}$$

subject to the conditions

$$\left.\begin{aligned} u &= F(t) \quad \text{when} \quad x = 0 \\ u &= 0 \qquad\ \ \text{''} \qquad\ t = 0. \end{aligned}\right\} \tag{2}$$

We know that

$$u = \frac{2}{\sqrt{\pi}}\int_0^{\frac{x}{2a\sqrt{t}}} e^{-\beta^2} d\beta$$

is a solution of (1), v. Ex. 1 Art. 50. It is easily shown that

$$u = \frac{2}{\sqrt{\pi}}\int_0^{\frac{x}{2a\sqrt{t-c}}} e^{-\beta^2} d\beta, \tag{3}$$

where c is any constant, is a solution of (1).

For

$$D_t u = -\frac{2}{\sqrt{\pi}}\frac{x}{2a}\frac{1}{2(t-c)^{\frac{3}{2}}}e^{-\frac{x^2}{4a^2(t-c)}} = -\frac{x}{2a\sqrt{\pi}}(t-c)^{-\frac{3}{2}}e^{-\frac{x^2}{4a^2(t-c)}}$$

$$D_x u = \frac{2}{\sqrt{\pi}}\frac{1}{2a\sqrt{t-c}}e^{-\frac{x^2}{4a^2(t-c)}}$$

$$D_x^2 u = -\frac{2}{\sqrt{\pi}}\frac{1}{2a\sqrt{t-c}}\frac{2x}{4a^2(t-c)}e^{-\frac{x^2}{4a^2(t-c)}} = -\frac{x}{2a^3\sqrt{\pi}}(t-c)^{-\frac{3}{2}}e^{-\frac{x^2}{4a^2(t-c)}}$$

and $$D_t u = a^2 D_x^2 u.$$

Let $\phi(x,t)$ be a function of x and t which shall be equal to zero if t is negative and shall be equal to

$$1 - \frac{2}{\sqrt{\pi}} \int_0^{\frac{x}{2a\sqrt{t}}} e^{-\beta^2} d\beta$$

if t is equal to or greater than zero; so that if $x = 0$ $\phi(x,t) = 1$ and if $t = 0$ $\phi(x,t) = 0$.

We shall now attack the following problem, to solve equation (1) subject to the conditions

$$\begin{array}{llll} u = 0 & \text{if} & t = 0 & \\ u = F(0) & \text{"} & x = 0 & \text{and} \quad 0 < t < \tau \\ u = F(k\tau) & \text{"} & x = 0 & \text{"} \quad k\tau < t < (k+1)\tau, \end{array}$$

where k is any whole number and τ is any arbitrarily chosen interval of time.

If we form the value

$$u = F(k\tau)[\phi(x, t - k\tau) - \phi(x, t - (k+1)\tau)] \tag{4}$$

u will satisfy equation (1) since zero, unity and

$$\frac{2}{\sqrt{\pi}} \int_0^{\frac{x}{2a\sqrt{t-k\tau}}} e^{-\beta^2} d\beta$$

are values of u which satisfy (1). u will be zero if $t < k\tau$ by the definition of the function $\phi(x,t)$; if $x = 0$ $u = 0$ if $t > (k+1)\tau$ and $u = F(k\tau)$ if $k\tau < t < (k+1)\tau$.

Therefore

$$u = \sum_{k=0}^{k=\infty} F(k\tau)[\phi(x, t - k\tau) - \phi(x, t - (k+1)\tau)] \tag{5}$$

is the solution of the problem stated above.

(5) can be simplified somewhat from the consideration that for a given value of t $\phi(x, t - k\tau) = 0$ if $k\tau > t$. If, then, $n\tau$ is the greatest whole multiple of τ not exceeding t,

$$u = \sum_{k=0}^{k=n} F(k\tau)[\phi(x, t - k\tau) - \phi(x, t - (k+1)\tau)]. \tag{6}$$

If now we decrease τ indefinitely the limiting form of (6) will be the solution of the problem stated at the beginning of this article.

(6) may be written

$$u = \sum_{k=0}^{k=n} F(k\tau) \left[\frac{\phi(x, t - k\tau) - \phi(x, t - (k+1)\tau)}{\tau}\right] \tau \tag{7}$$

and if τ is indefinitely decreased the limiting form of (7) is

$$u = -\int_0^t F(\lambda) D_\lambda \phi(x, t-\lambda) d\lambda. \tag{8}$$

Since $t - \lambda$ is positive between the limits of integration

$$\phi(x, t-\lambda) = 1 - \frac{2}{\sqrt{\pi}} \int_0^{\frac{x}{2a\sqrt{t-\lambda}}} e^{-\beta^2} d\beta,$$

and $$D_\lambda \phi(x, t-\lambda) = -\frac{x}{2a\sqrt{\pi}} e^{-\frac{x^2}{4a^2(t-\lambda)}} (t-\lambda)^{-\frac{3}{2}};$$

and (8) may be written

$$u = \frac{x}{2a\sqrt{\pi}} \int_0^t F(\lambda) e^{-\frac{x^2}{4a^2(t-\lambda)}} (t-\lambda)^{-\frac{3}{2}} d\lambda, \tag{9}$$

or if we let $$\beta = \frac{x}{2a\sqrt{t-\lambda}}$$

$$u = \frac{2}{\sqrt{\pi}} \int_{\frac{x}{2a\sqrt{t}}}^{\infty} e^{-\beta^2} F\left(t - \frac{x^2}{4a^2\beta^2}\right) d\beta. \tag{10}$$

EXAMPLES.

1. If $u = nt$ when $x = 0$ and $u = 0$ when $t = 0$

$$u = n\left(t + \frac{x^2}{2a^2}\right)\left[1 - \frac{2}{\sqrt{\pi}} \int_0^{\frac{x}{2a\sqrt{t}}} e^{-\beta^2} d\beta\right] - \frac{nx\sqrt{t}}{a\sqrt{\pi}} e^{-\frac{x^2}{4a^2t}}.$$

2. A thick iron slab is at the temperature zero throughout, one of its plane faces is then kept at the temperature 100° Centigrade for 5 minutes, then at the temperature zero for the next 5 minutes, then at the temperature 100° for the next 5 minutes, and then at the temperature zero. Required the temperature of a point in the slab 5 c.m. from the face at the expiration of 18 minutes. Given; $a^2 = .185$.
Ans., 20°.1.

3. If $u = F(t)$ when $x = 0$ and $u = f(x)$ when $t = 0$, then

$$u = \frac{2}{\sqrt{\pi}} \int_{\frac{x}{2a\sqrt{t}}}^{\infty} e^{-\beta^2} F\left(t - \frac{x^2}{4a^2\beta^2}\right) d\beta + \frac{1}{2a\sqrt{\pi t}} \int_0^{\infty} (e^{-\frac{(\lambda-x)^2}{4a^2t}} - e^{-\frac{(\lambda+x)^2}{4a^2t}}) f(\lambda) d(\lambda).$$

v. (6) Art. 50.

4. If in Art. (51) $F(t)$ is a periodic function of the time of period T it can be expressed by a Fourier's series of the form

$$F(t) = \frac{1}{2}b_0 + \sum_{m=1}^{m=\infty} [a_m \sin m\alpha t + b_m \cos m\alpha t], \quad \text{where} \quad \alpha = \frac{2\pi}{T},$$

or $$F(t) = \frac{1}{2b_0} + \sum_{m=1}^{m=\infty} \rho_m \sin(m\alpha t + \lambda_m),$$

where $$\rho_m \cos \lambda_m = a_m \quad \text{and} \quad \rho_m \sin \lambda_m = b_m.$$ v. Art. 31 Ex. 3.

Show that with this value of $F(t)$ (10) Art. 51. becomes

$$u = \frac{1}{\sqrt{\pi}} b_0 \int_{\frac{x}{2a\sqrt{t}}}^{\infty} e^{-\beta^2} d\beta + \frac{2}{\sqrt{\pi}} \sum_{m=1}^{m=\infty} \rho_m \left[\sin(m\alpha t + \lambda_m) \int_{\frac{x}{2a\sqrt{t}}}^{\infty} e^{-\beta^2} \cos \frac{m\alpha x^2}{4a^2\beta^2} d\beta \right.$$
$$\left. - \cos(m\alpha t + \lambda_m) \int_{\frac{x}{2a\sqrt{t}}}^{\infty} e^{-\beta^2} \sin \frac{m\alpha x^2}{4a^2\beta^2} d\beta \right]$$

and that as t increases u approaches the value

$$u = \frac{1}{2} b_0 + \sum_{m=1}^{m=\infty} \rho_m e^{-\frac{x}{a}\sqrt{\frac{m\alpha}{2}}} \sin(m\alpha t - \frac{x}{a}\sqrt{\frac{m\alpha}{2}} + \lambda_m).$$

Given that

$$\int_0^{\infty} e^{-x^2} \sin \frac{b^2}{x^2} dx = \frac{\sqrt{\pi}}{2} e^{-b\sqrt{2}} \sin b\sqrt{2}; \quad \int_0^{\infty} e^{-x^2} \cos \frac{b^2}{x^2} dx = \frac{\sqrt{\pi}}{2} e^{-b\sqrt{2}} \cos b\sqrt{2}.$$

v. *Riemann, Lin. par. dif. gl.* § 54.

5. If we are dealing with a bar of small cross-section where the heat not only flows along the bar but at the same time escapes at the surface of the bar into air at the temperature zero we have to solve the differential equation

$$D_t u = a^2 D_x^2 u - b^2 u.$$ v. Fourier, Heat § 105.

Show that for this case

$$u = e^{-(b^2+a^2\alpha^2)t} \sin \alpha x \quad \text{and} \quad u = e^{-(b^2+a^2\alpha^2)t} \cos \alpha x$$

are particular solutions, and that if $u = f(x)$ when $t = 0$

$$u = \frac{e^{-b^2 t}}{2a\sqrt{\pi t}} \int_{-\infty}^{\infty} e^{-\frac{(\lambda - x)^2}{4a^2 t}} f(\lambda) d\lambda = \frac{e^{-b^2 t}}{\sqrt{\pi}} \int_{-\infty}^{\infty} e^{-\beta^2} f(x + 2a\sqrt{t}.\beta) d\beta.$$

cf. (8) and (9) Art. 49.

If $u = 0$ when $x = 0$ and $u = f(x)$ when $t = 0$

$$u = \frac{e^{-b^2 t}}{\sqrt{\pi}}\left[\int_{-\frac{x}{2a\sqrt{t}}}^{\infty} e^{-\beta^2} f(x + 2a\sqrt{t}.\beta)d\beta - \int_{\frac{x}{2a\sqrt{t}}}^{\infty} e^{-\beta^2} f(-x + 2a\sqrt{t}.\beta)d\beta\right].$$

cf. (7) Art. 50.

If $u = -e^{-\frac{bx}{a}}$ when $t = 0$ and $u = 0$ when $x = 0$

$$u = \frac{1}{\sqrt{\pi}}\left[e^{\frac{bx}{a}} \int_{\frac{x}{2a\sqrt{t}}}^{\infty} e^{-(b\sqrt{t}+\beta)^2} d\beta - e^{-\frac{bx}{a}} \int_{-\frac{x}{2a\sqrt{t}}}^{\infty} e^{-(b\sqrt{t}+\beta)^2} d\beta\right],$$

and if $u = 1$ when $x = 0$ and $u = 0$ when $t = 0$ we have only to add $e^{-\frac{bx}{a}}$ to the second member of the last equation, since $u = e^{-\frac{bx}{a}}$ satisfies the equation

$$D_t u = a^2 D_x^2 u - b^2 u.$$

If $u = F(t)$ when $x = 0$ and $u = 0$ when $t = 0$ we can employ the method of Art. 51.

$$\phi(x, t-\lambda) = e^{-\frac{bx}{a}} + \frac{1}{\sqrt{\pi}}\left[e^{\frac{bx}{a}} \int_{\frac{x}{2a\sqrt{t-\lambda}}}^{\infty} e^{-(b\sqrt{t-\lambda}+\beta)^2} d\beta - e^{-\frac{bx}{a}} \int_{-\frac{x}{2a\sqrt{t-\lambda}}}^{\infty} e^{-(b\sqrt{t-\lambda}+\beta)^2} d\beta\right],$$

$$-D_\lambda \phi(x, t-\lambda) = \frac{x(t-\lambda)^{-\frac{3}{2}}}{2a\sqrt{\pi}} e^{-b^2(t-\lambda) - \frac{x^2}{4a^2(t-\lambda)}};$$

and
$$u = \frac{x}{2a\sqrt{\pi}} \int_0^t (t-\lambda)^{-\frac{3}{2}} e^{-b^2(t-\lambda) - \frac{x^2}{4a^2(t-\lambda)}} F(\lambda) d\lambda,$$

cf. (9) Art. 51,

or
$$u = \frac{2}{\sqrt{\pi}} \int_{\frac{x}{2a\sqrt{t}}}^{\infty} e^{-\beta^2 - \frac{b^2 x^2}{4a^2\beta^2}} F\left(t - \frac{x^2}{4a^2\beta^2}\right) d\beta,$$

cf. (10) Art. 51.

If $F(t)$ is periodic and has the value taken in Ex. 4, show that the value approached by u as t increases is

$$u = \frac{1}{2} b_0 e^{-\frac{bx}{a}} + \sum_{m=1}^{m=\infty} \rho_m e^{-\frac{x\sqrt{2}}{2a} p} \sin\left(m\alpha t - \frac{x\sqrt{2}}{2a} q + \lambda_m\right),$$

where $p = (b^2 + \sqrt{b^4 + m^2\alpha^2})^{\frac{1}{2}}$ and $q = (-b^2 + \sqrt{b^4 + m^2\alpha^2})^{\frac{1}{2}}$.

Given
$$\int_0^\infty e^{-x^2-\frac{a^2}{x^2}}dx = \frac{\sqrt{\pi}}{2}e^{-2a}$$
$$\int_0^\infty e^{-x^2-\frac{a^2}{x^2}}\sin\frac{b^2}{x^2}dx = \frac{\sqrt{\pi}}{2}e^{-2c}\sin 2d$$
and
$$\int_0^\infty e^{-x^2-\frac{a^2}{x^2}}\cos\frac{b^2}{x^2}dx = \frac{\sqrt{\pi}}{2}e^{-2c}\cos 2d,$$
where
$$c = \frac{\sqrt{2}}{2}(a^2+\sqrt{a^4+b^4})^{\frac{1}{2}} \quad \text{and} \quad d = \frac{\sqrt{2}}{2}(-a^2+\sqrt{a^4+b^4})^{\frac{1}{2}}.$$

Ängstrom's method of determining the conductivity of a metal is based on the result just given (v. Phil. Mag. Feb. 1863), and is described by Sir Wm. Thomson (Encyc. Brit. Article "Heat") as by far the best that has yet been devised.

52. If u is a periodic function of the time when $x = 0$ as in Art. 51 Ex. 4 and we are concerned with the limiting value approached by u as t increases we can avoid evaluating a complicated definite integral if we take the following course.

Since as we have seen in Art. 49 $u = e^{\beta t+\alpha x}$ is a solution of
$$D_t u = a^2 D_x^2 u \tag{1}$$
provided only that $\beta = a^2\alpha^2$ we have
$$u = e^{\beta t \pm \frac{x}{a}\sqrt{\beta}}$$
as a solution.

Replacing β by $\pm\beta i$ this becomes
$$u = e^{\pm\beta t i \pm \frac{x}{a}\sqrt{\beta}\sqrt{\pm i}}$$
or
$$u = e^{\pm\beta t i \pm \frac{x}{a}\sqrt{\frac{\beta}{2}}(1\pm i)}$$
since
$$\sqrt{i} = \pm\frac{1}{2}\sqrt{2}(1+i)$$
and
$$\sqrt{-i} = \pm\frac{1}{2}\sqrt{2}(1-i).$$

Hence
$$u = e^{-\frac{x}{a}\sqrt{\frac{\beta}{2}}}\sin\left(\beta t - \frac{x}{a}\sqrt{\frac{\beta}{2}}\right), \quad u = e^{-\frac{x}{a}\sqrt{\frac{\beta}{2}}}\cos\left(\beta t - \frac{x}{a}\sqrt{\frac{\beta}{2}}\right), \tag{2}$$
$$u = e^{\frac{x}{a}\sqrt{\frac{\beta}{2}}}\sin\left(\beta t + \frac{x}{a}\sqrt{\frac{\beta}{2}}\right), \quad u = e^{\frac{x}{a}\sqrt{\frac{\beta}{2}}}\cos\left(\beta t + \frac{x}{a}\sqrt{\frac{\beta}{2}}\right), \tag{3}$$

are particular solutions of (1).

From these we get readily

$$u = \rho_m e^{-\frac{x}{a}\sqrt{\frac{m\alpha}{2}}} \sin\left(m\alpha t - \frac{x}{a}\sqrt{\frac{m\alpha}{2}} + \lambda_m\right) \tag{4}$$

as a solution. (4) reduces to

$$u = \rho_m \sin(m\alpha t + \lambda_m) \quad \text{when} \quad x = 0$$

and to

$$u = \rho_m e^{-\frac{x}{a}\sqrt{\frac{m\alpha}{2}}} \sin\left(\lambda_m - \frac{x}{a}\sqrt{\frac{m\alpha}{2}}\right) \quad \text{when} \quad t = 0.$$

If we add a term which satisfies (1) and which is equal to zero when $x = 0$ and to $-\rho_m e^{-\frac{x}{a}\sqrt{\frac{m\alpha}{2}}} \sin\left(\lambda_m - \frac{x}{a}\sqrt{\frac{m\alpha}{2}}\right)$ when $t = 0$ (v. Art. 50) we shall have a solution of (1) which is zero when $t = 0$ and which is

$$\rho_m \sin(m\alpha t + \lambda_m) \quad \text{when} \quad x = 0.$$

The term in question approaches zero as t increases [v. (7) Art. 50] and we have at once the solution given in Art. 51 Ex. 4, as our required result.

EXAMPLE.

Show that $u = e^{\beta t + \alpha x}$ is a solution of $D_t u = a^2 D_x^2 u - b^2 u$ if $\beta = a^2\alpha^2 - b^2$, and hence that

$$u = e^{\beta t \pm \frac{x}{a}\sqrt{b^2+\beta}}, \quad u = e^{\pm\beta t i \pm \frac{x}{a}\sqrt{b^2 \pm \beta i}}, \quad u = e^{\pm\beta t i \pm \frac{x}{a\sqrt{2}}(p \pm q i)},$$

$$u = e^{\pm\frac{px}{a\sqrt{2}}} \sin\left(\beta t \pm \frac{qx}{a\sqrt{2}}\right), \quad \text{and} \quad u = e^{\pm\frac{px}{a\sqrt{2}}} \cos\left(\beta t \pm \frac{qx}{a\sqrt{2}}\right),$$

where

$$p = [\sqrt{\beta^2 + b^4} + b^2]^{\frac{1}{2}} \quad \text{and} \quad q = [\sqrt{\beta^2 + b^4} - b^2]^{\frac{1}{2}},$$

are solutions. Hence

$$u = \rho_m e^{-\frac{px}{a\sqrt{2}}} \sin\left(\beta t - \frac{qx}{a\sqrt{2}} + \lambda_m\right)$$

is a solution.

If $\beta = m\alpha$ this last result reduces to $u = \rho_m \sin(m\alpha t + \lambda_m)$ when $x = 0$ and by the reasoning of Art. 52 it must be the value u approaches as t increases if we have the same conditions as in the last part of Art. 51 Ex. 5.

53. The whole problem of the flow of heat is treated by Sir William Thomson (v. Math. and Phys. Papers, Vol. II), and other recent writers from a different and decidedly interesting point of view, which we shall briefly sketch in connection with the problem of *Linear Flow*.

Suppose we are dealing with a bar having a small cross-section and an adiathermanous surface, and take as our unit of heat the amount required to raise by a unit the temperature of a unit of length of the bar. If at a point of the bar a quantity Q of heat is suddenly generated the point is called an *instantaneous heat source* of strength Q.

If the heat instead of being suddenly generated is generated gradually and at a rate that would give Q units of heat per unit of time the point is called a *permanent heat source* of strength Q.

The temperature at any point of the bar at any time due to an instantaneous source of strength Q at the point $x = \lambda$ is easily found by the aid of formula (8) Art. 49 as follows:—

If a quantity of heat Q is suddenly generated along the portion of the bar from $x = \lambda$ to $x = \lambda + \Delta\lambda$, where $\Delta\lambda$ is any arbitrary length, the temperature of that portion will be suddenly raised to $\frac{Q}{\Delta\lambda}$, and we shall have by (8) Art. 49

$$u = \frac{Q}{2a\sqrt{\pi t}} \frac{1}{\Delta\lambda} \int\limits_{\lambda}^{\lambda+\Delta\lambda} e^{-\frac{(\lambda-x)^2}{4a^2t}} d\lambda \tag{1}$$

as the temperature of any point of the bar at any time t thereafter.

If now we write u equal to the limiting value approached by the second member of (1) as $\Delta\lambda$ is made to approach zero we get

$$u = \frac{Q}{2a\sqrt{\pi t}} e^{-\frac{(\lambda-x)^2}{4a^2t}} \tag{2}$$

as the solution for the case where we have an instantaneous source at the point $x = \lambda$.

It is to be observed that in (2) $u = 0$ when $t = 0$ and $u = \frac{Q}{2a\sqrt{\pi t}}$ when $x = \lambda$ and $t > 0$.

If we have several sources we have only to add the temperatures due to the separate sources.

Formula (8) Art. 49 may now be regarded as the solution for the case where we start with an instantaneous heat source of strength $f(\lambda)d\lambda$ in every element of length of the bar.

A source of strength $-Q$ is called a sink of strength Q; and (6) Art. 50 may be regarded as the solution for the case where we have at the start an instantaneous source of strength $f(\lambda)d\lambda$ in every element of the bar whose distance to the right of the origin is λ, and an instantaneous sink of strength $f(\lambda)d\lambda$ in every element of the bar whose distance to the left of the origin is λ.

If we have an instantaneous source at the origin (2) reduces to

$$u = \frac{Q}{2a\sqrt{\pi t}} e^{-\frac{x^2}{4a^2t}} \tag{3}$$

For a permanent source of constant strength Q at the origin (3) gives

$$u = \frac{Q}{2a\sqrt{\pi}} \int_0^t e^{-\frac{x^2}{4a^2(t-\tau)}} (t-\tau)^{-\frac{1}{2}} d\tau \tag{4}$$

and for a permanent source of variable strength $f(t)$

$$u = \frac{1}{2a\sqrt{\pi}} \int_0^t e^{-\frac{x^2}{4a^2(t-\tau)}} (t-\tau)^{-\frac{1}{2}} f(\tau) d\tau. \tag{5}$$

In (4) and (5) u obviously reduces to zero when $t = 0$ and $x > 0$, but its value when $x = 0$ is not easily determined. We can avoid the difficulty by introducing the conception of a *doublet.*

54. If a source and a sink of equal strength Q are made to approach each other while Q multiplied by their distance apart is kept equal to a constant P the limiting state of things is said to be due to a *doublet* of strength P whose axis is tangent to the line of approach and points from sink to source. A *doublet* of strength $-P$ differs from a doublet of strength P only in that its axis has the opposite direction.

Let us find the temperature due to an instantaneous doublet of strength P placed at the origin. For a source of strength Q at $x = \eta$ and an equal sink at $x = -\eta$ we have

$$u = \frac{Q}{2a\sqrt{\pi t}}\left(e^{-\frac{(\eta-x)^2}{4a^2t}} - e^{-\frac{(\eta+x)^2}{4a^2t}}\right),$$

or if $2\eta Q = P$,

$$\begin{aligned} u &= \frac{P}{4a\eta\sqrt{\pi t}} e^{-\frac{(\eta^2+x^2)}{4a^2t}} \left(e^{\frac{\eta x}{2a^2t}} - e^{-\frac{\eta x}{2a^2t}}\right) \\ &= \frac{P}{2a\eta\sqrt{\pi t}} e^{-\frac{(\eta^2+x^2)}{4a^2t}} \sinh \frac{\eta x}{2a^2t}. \end{aligned}$$

If η is made to approach zero

$$\text{limit}\left[\frac{1}{\eta}\sinh\frac{\eta x}{2a^2t}\right] = \frac{x}{2a^2t},$$

and

$$u = \frac{Px}{4a^3\sqrt{\pi t^3}} e^{-\frac{x^2}{4a^2t}} \tag{1}$$

is the solution for the temperature at any time and place due to an instantaneous doublet of strength P placed at the origin. For a doublet at any other point $x = \lambda$ we have

$$u = \frac{P(x-\lambda)}{4a^3\sqrt{\pi t^3}} e^{-\frac{(x-\lambda)^2}{4a^2t}}. \tag{2}$$

For a permanent doublet of constant strength P placed at the origin we have

$$u = \frac{Px}{4a^3\sqrt{\pi}} \int_0^t e^{-\frac{x^2}{4a^2(t-\tau)}} (t-\tau)^{-\frac{3}{2}} d\tau; \tag{3}$$

and for a permanent doublet of variable strength $f(t)$

$$u = \frac{x}{4a^3\sqrt{\pi}} \int_0^t e^{-\frac{x^2}{4a^2(t-\tau)}} (t-\tau)^{-\frac{3}{2}} f(\tau) d\tau, \tag{4}$$

or

$$u = \frac{1}{a^2\sqrt{\pi}} \int_{\frac{x}{2a\sqrt{t}}}^{\infty} e^{-\beta^2} f\left(t - \frac{x^2}{4a^2\beta^2}\right) d\beta \tag{5}$$

if $x > 0$, and

$$u = \frac{1}{a^2\sqrt{\pi}} \int_{\frac{x}{2a\sqrt{t}}}^{-\infty} e^{-\beta^2} f\left(t - \frac{x^2}{4a^2\beta^2}\right) d\beta \tag{6}$$

if $x < 0$, if we let $\beta = \dfrac{x}{2a\sqrt{t-\tau}}$.

From (5) and (6) we see readily that $u = 0$ when $t = 0$ and that $u = \dfrac{f(t)}{2a^2}$ when $x = 0$ if we approach the origin from the right and that $u = -\dfrac{f(t)}{2a^2}$ when $x = 0$ if we approach the origin from the left.

If the point $x = 0$ is kept at the constant temperature b and we are concerned only with positive values of x we can get from (5) the solution given in Art. 50 Ex. 4 by supposing a permanent doublet of strength $2a^2b$ placed at the origin.

To solve the problem treated in Art. 51 we have only to suppose a permanent doublet of strength $2a^2F(t)$ placed at $x = 0$ and from (5) we get at once (10) Art. 51.

EXAMPLE.

Show that if $D_t u = a^2 D_x^2 u - b^2 u$ and an instantaneous source of strength Q is placed at $x = \lambda$

$$u = \frac{Q}{2a\sqrt{\pi t}} e^{-b^2 t - \frac{(\lambda - x)^2}{4a^2 t}}$$ v. Art. 51, Ex. 5.

Show that if an instantaneous doublet of strength P is placed at the point $x = 0$

$$u = \frac{Px}{4a^3\sqrt{\pi t^3}} e^{-b^2 t - \frac{x^2}{4a^2 t}}.$$

If a permanent doublet of strength $f(t)$ is placed at $x = 0$

$$u = \frac{x}{4a^3\sqrt{\pi}}\int_0^t e^{-b^2(t-\tau)-\frac{x^2}{4a^2(t-\tau)}}(t-\tau)^{-\frac{3}{2}}f(\tau)d\tau$$

$$= \frac{1}{a^2\sqrt{\pi}}\int_{\frac{x}{2a\sqrt{t}}}^{\pm\infty} e^{-\beta^2-\frac{b^2x^2}{4a^2\beta^2}} f\left(t-\frac{x^2}{4a^2\beta^2}\right)d\beta,$$

whence $u = 0$ when $t = 0$ and $x > 0$ or $x < 0$ and $u = \pm\dfrac{f(t)}{2a^2}$ when $x = 0$.

Hence if we place at $x = 0$ a permanent doublet of strength $2a^2F(t)$ we get the solution given in Art. 51 Ex. 5 for the case where $u = F(t)$ when $x = 0$ and $u = 0$ when $t = 0$ provided we are concerned only with positive values of x.

If $F(t) = c$ this reduces to

$$u = \frac{2c}{\sqrt{\pi}}\int_{\frac{x}{2a\sqrt{t}}}^{\infty} e^{-\beta^2-\frac{b^2x^2}{4a^2\beta^2}}d\beta.$$

55. As another example of the use of Fourier's Integral we shall consider the transmission of a disturbance along a stretched elastic string.

Suppose we have a stretched elastic string so long that we need not consider what happens at its ends, that is so long that we may treat its length as infinite. Let the string be initially distorted into some given form and then released; to investigate its subsequent motion.

Let us take the position of equilibrium of the string as the axis of X and any given point as origin.

We have, then, to solve the differential equation

$$D_t^2 y = a^2 D_x^2 y \tag{1}$$

[v. (VIII) Art. 1] subject to the conditions

$$y = f(x) \quad \text{when} \quad t = 0 \tag{2}$$

$$D_t y = 0 \quad \text{"} \quad t = 0. \tag{3}$$

As in Art. 8 we find

$$y = \cos\alpha(x \pm at) \quad \text{and} \quad y = \sin\alpha(x \pm at)$$

as particular solutions of (1).

From these we must build up a value that will reduce to

$$f(x) = \frac{1}{\pi}\int_0^\infty d\alpha \int_{-\infty}^{\infty} f(\lambda)\cos\alpha(\lambda - x).d\lambda \tag{4}$$

when $t=0$ and will at the same time satisfy (3).

$$y = \cos\alpha\lambda\cos\alpha(x+at) + \sin\alpha\lambda\sin\alpha(x+at)$$

or
$$y = \cos\alpha(\lambda - x - at)$$

is a solution of (1).

Hence
$$y = \frac{1}{\pi}\int_0^\infty d\alpha \int_{-\infty}^\infty f(\lambda)\cos\alpha(\lambda - x - at).d\lambda \tag{5}$$

is also a solution of (1).

(5) reduces to $y = f(x)$ when $t=0$ but it gives

$$D_t y = \frac{a}{\pi}\int_0^\infty \alpha d\alpha \int_{-\infty}^\infty f(\lambda)\sin\alpha(\lambda - x).d\lambda$$

when $t=0$ and consequently does not satisfy equation (3).

If in forming (5) we use $\cos\alpha(x-at)$ and $\sin\alpha(x-at)$ instead of $\cos\alpha(x+at)$ and $\sin\alpha(x+at)$ we get

$$y = \frac{1}{\pi}\int_0^\infty d\alpha \int_{-\infty}^\infty f(\lambda)\cos\alpha(\lambda - x + at).d\lambda \tag{6}$$

which is a solution of (1), and reduces to $y = f(x)$ when $t=0$, but it gives

$$D_t y = -\frac{a}{\pi}\int_0^\infty \alpha d\alpha \int_{-\infty}^\infty f(\lambda)\sin\alpha(\lambda - x).d\lambda$$

when $t=0$ and does not satisfy (3).

If, however, we take one-half the sum of the values of y in (5) and (6) we get

$$y = \frac{1}{2}\left[\frac{1}{\pi}\int_0^\infty d\alpha \int_{-\infty}^\infty f(\lambda)\cos\alpha(\lambda - x - at).d\lambda + \frac{1}{\pi}\int_0^\infty d\alpha \int_{-\infty}^\infty f(\lambda)\cos\alpha(\lambda - x + at).d\lambda\right], \tag{7}$$

a solution of (1) which satisfies both (2) and (3), and is, therefore, our required solution.

This result can be very much simplified.

If we substitute $z = x + at$

$$\frac{1}{\pi}\int_0^\infty d\alpha \int_{-\infty}^\infty f(\lambda)\cos\alpha(\lambda - x - at).d\lambda$$

$$= \frac{1}{\pi}\int_0^\infty d\alpha \int_{-\infty}^\infty f(\lambda)\cos\alpha(\lambda - z).d\lambda = f(z) = f(x+at);$$

and in like manner we can show that

$$\frac{1}{\pi}\int_0^\infty d\alpha \int_{-\infty}^{\infty} f(\lambda)\cos\alpha(\lambda - x + at).d\lambda = f(x - at).$$

Hence our solution becomes

$$y = \frac{1}{2}[f(x + at) + f(x - at)]. \tag{8}$$

This result is of great importance in the theory of elastic strings and it shows that the initial disturbance splits into two equal waves which run along the string, one to the right and the other to the left, with a uniform velocity a, and that there is nothing like a periodic motion or vibration of any sort unless the ends of the string produce some effect.

56. If the string is not initially distorted but starts from its position of equilibrium with a given initial velocity impressed upon each point we have to solve the equation

$$D_t^2 y = a^2 D_x^2 y \tag{1}$$

subject to the conditions

$$y = 0 \quad \text{when} \quad t = 0 \tag{2}$$

$$D_t y = F(x) \quad \text{``} \quad t = 0. \tag{3}$$

We get by the process used in Art. 55

$$y = \frac{1}{2\pi a}\int_0^\infty d\alpha \int_{-\infty}^{\infty} F(\lambda)\left[\frac{\sin\alpha(\lambda - x + at)}{\alpha} - \frac{\sin\alpha(\lambda - x - at)}{\alpha}\right] d\lambda$$

$$= \frac{1}{2\pi a}\int_{-\infty}^{\infty} F(\lambda)d\lambda \int_0^\infty \left[\frac{\sin\alpha(\lambda - x + at)}{\alpha} - \frac{\sin\alpha(\lambda - x - at)}{\alpha}\right] d\alpha;$$

but
$$\int_0^\infty \frac{\sin\alpha(\lambda - x + at)}{\alpha}d\alpha - \int_0^\infty \frac{\sin\alpha(\lambda - x - at)}{\alpha}d\alpha = \pi$$

if $x - at < \lambda < x + at$, and is equal to zero for all other values of λ; since

$$\begin{aligned}\int_0^\infty \frac{\sin mx}{x}dx &= \frac{\pi}{2} \quad \text{if} \quad m > 0\\ &= -\frac{\pi}{2} \quad \text{if} \quad m < 0\\ &= 0 \quad \text{if} \quad m = 0.\end{aligned}$$

v. Int. Cal. Art. 92 (3).

Hence
$$y = \frac{1}{2a}\int_{x-at}^{x+at} F(\lambda)d\lambda \tag{4}$$

is our required solution.

EXAMPLES.

1. If the string is initially distorted and starts with initial velocity so that $y = f(x)$ and $D_t y = F(x)$ when $t = 0$

$$y = \frac{1}{2}[f(x+at) + f(x-at)] + \frac{1}{2a}\int\limits_{x-at}^{x+at} F(\lambda)d\lambda.$$

2. If the initial disturbance is caused by a blow, as from the hammer in a piano, which impresses upon all the points in a portion of the string of length c an equal transverse velocity b show that the front of the wave which will be seen to run to the left along the string will be a straight line having a slope equal to $\frac{b}{2a}$ and a length equal to $\frac{c}{2a}\sqrt{4a^2+b^2}$. Of course a wave having a front of the same length with a slope equal to $-\frac{b}{2a}$ will be seen to run to the right along the string, and the effect of the two waves will be to lift the string bodily and permanently to a distance $\frac{bc}{2a}$ above its original position.

57. We shall now take up a few examples of the use of *Fourier's Series.*

In the problem of Art. 7 let the temperature of the base of the plate be a given function of x, the other conditions remaining unchanged.

Since
$$f(x) = \sum_{m=1}^{m=\infty} (a_m \sin mx)$$

where
$$a_m = \frac{2}{\pi}\int\limits_0^{\pi} f(\alpha)\sin m\alpha.d\alpha$$

we have
$$u = \frac{2}{\pi}\sum_{m=1}^{m=\infty}\left[e^{-my}\sin mx\int\limits_0^{\pi} f(\alpha)\sin m\alpha.d\alpha\right]. \tag{1}$$

If the breadth of the plate is a instead of π

$$u = \frac{2}{a}\sum_{m=1}^{m=\infty}\left[e^{-\frac{m\pi y}{a}}\sin\frac{m\pi x}{a}\int\limits_0^{a} f(\lambda)\sin\frac{m\pi\lambda}{a}d\lambda\right]. \tag{2}$$

58. If the temperature of the base is unity and the breadth of the plate is π the solution is, as we have seen in Art. 7,

$$u = \frac{4}{\pi}\left[e^{-y}\sin x + \frac{1}{3}e^{-3y}\sin 3x + \frac{1}{5}e^{-5y}\sin 5x + \cdots\right]. \tag{1}$$

This series can be summed without difficulty. We have the development

$$\log(1+z) = \frac{z}{1} - \frac{z^2}{2} + \frac{z^3}{3} - \frac{z^4}{4} + \cdots$$

if the modulus of z is less than 1. Int. Cal. Art. 221 (4).

Hence $$\log(1-z) = -\frac{z}{1} - \frac{z^2}{2} - \frac{z^3}{3} - \frac{z^4}{4} - \cdots$$

if mod. $z < 1$.

and $$\frac{1}{2}[\log(1+z) - \log(1-z)] = \frac{z}{1} + \frac{z^3}{3} + \frac{z^5}{5} + \cdots \tag{2}$$

if mod. $z < 1$.

But

$$\begin{aligned}\log(1+z) &= \log[1 + r(\cos\phi + i\sin\phi)] \\ &= \frac{1}{2}\log[(1+r\cos\phi)^2 + (r\sin\phi)^2] + i\tan^{-1}\frac{r\sin\phi}{1+r\cos\phi} \\ &= \frac{1}{2}\log(1 + 2r\cos\phi + r^2) + i\tan^{-1}\frac{r\sin\phi}{1+r\cos\phi},\end{aligned}$$

and

$$\log(1-z) = \frac{1}{2}\log(1 - 2r\cos\phi + r^2) - i\tan^{-1}\frac{r\sin\phi}{1-r\cos\phi},$$

[Int. Cal. Art. 33 (2)],
and (2) becomes

$$\frac{1}{2}\left[\frac{1}{2}\log\frac{1+2r\cos\phi+r^2}{1-2r\cos\phi+r^2} + i\tan^{-1}\frac{2r\sin\phi}{1-r^2}\right] = \frac{r(\cos\phi + i\sin\phi)}{1} + \frac{r^3(\cos 3\phi + i\sin 3\phi)}{3} + \cdots \tag{3}$$

From (3) we get two equations

$$\frac{1}{4}\log\frac{1+2r\cos\phi+r^2}{1-2r\cos\phi+r^2} = \frac{r\cos\phi}{1} + \frac{r^3\cos 3\phi}{3} + \frac{r^5\cos 5\phi}{5} + \cdots \tag{4}$$

$$\frac{1}{2}\tan^{-1}\frac{2r\sin\phi}{1-r^2} = \frac{r\sin\phi}{1} + \frac{r^3\sin 3\phi}{3} + \frac{r^5\sin 5\phi}{5} + \cdots \tag{5}$$

both valid for all values of ϕ provided $r < 1$.
e^{-y} is less than 1 if y is positive.
Hence from (5)

$$\frac{e^{-y}\sin x}{1} + \frac{e^{-3y}\sin 3x}{3} + \frac{e^{-5y}\sin 5x}{5} + \cdots = \frac{1}{2}\tan^{-1}\frac{2e^{-y}\sin x}{1-e^{-2y}} = \frac{1}{2}\tan^{-1}\frac{2\sin x}{e^y - e^{-y}} = \frac{1}{2}\tan^{-1}\frac{\sin x}{\sinh y},$$

and (1) may be written

$$u = \frac{2}{\pi}\tan^{-1}\frac{\sin x}{\sinh y}. \tag{6}$$

If we replace r by e^{-y} and ϕ by x in

$$\log[1 + r(\cos\phi + i\sin\phi)]$$

it becomes

$$\log[1 + e^{-y}\cos x + ie^{-y}\sin x]$$

or

$$\log[1 + \cos z + i\sin z]$$

v. Int. Cal. Art. 35 (3) and (4)
a function of z as a whole; and

$$\log[1 - r(\cos\phi + i\sin\phi)]$$

becomes

$$\log(1 - \cos z - i\sin z);$$

hence by Int. Cal. Arts. 212 and 213,

$$\frac{1}{4}\log\frac{1 + 2e^{-y}\cos x + e^{-2y}}{1 - 2e^{-y}\cos x + e^{-2y}} \quad \text{and} \quad \frac{1}{2}\tan^{-1}\frac{2e^{-y}\sin x}{1 - e^{-2y}}$$

or

$$\frac{1}{4}\log\frac{\cosh y + \cos x}{\cosh y - \cos x} \quad \text{and} \quad \frac{1}{2}\tan^{-1}\frac{\sin x}{\sinh y}$$

are conjugate functions, and

$$u_1 = \frac{1}{\pi}\log\frac{\cosh y + \cos x}{\cosh y - \cos x} \tag{7}$$

is the solution for the problem where the isothermal lines are the lines of flow of the present problem and the lines of flow are the isothermal lines of the present problem.

For our problem, then, the isothermal lines are given by the equation

$$\frac{2}{\pi}\tan^{-1}\frac{\sin x}{\sinh y} = a$$

or

$$\frac{\sin x}{\sinh y} = \tan\frac{a\pi}{2} \tag{8}$$

and the lines of flow by

$$\frac{1}{\pi}\log\frac{\cosh y + \cos x}{\cosh y - \cos x} = b,$$

or

$$\frac{\cosh y + \cos x}{\cosh y - \cos x} = e^{\pi b}. \tag{9}$$

EXAMPLES.

1. If $D_x^2 u + D_y^2 u = 0$, and $u = 1$ when $y = 0$, and $u = 0$ when $x = 0$ and when $x = a$,

$$u = \frac{4}{\pi}\left[e^{-\frac{\pi y}{a}} \sin\frac{\pi x}{a} + \frac{1}{3}e^{-\frac{3\pi y}{a}} \sin\frac{3\pi x}{a} + \frac{1}{5}e^{-\frac{5\pi y}{a}} \sin\frac{5\pi x}{a} + \cdots\right]$$

$$= \frac{2}{\pi}\tan^{-1}\frac{\sin\frac{\pi x}{a}}{\sinh\frac{\pi y}{a}}.$$

2. If $u = \phi(x)$ when $y = 0$, $u = f(y)$ when $x = 0$, and $u = F(y)$ when $x = a$

$$u = \frac{2}{a}\sum_{m=1}^{m=\infty} e^{-\frac{m\pi y}{a}} \sin\frac{m\pi x}{a}\int_0^a \phi(\lambda)\sin\frac{m\pi\lambda}{a}d\lambda$$

$$+ \frac{1}{2a}\sin\frac{\pi x}{a}\int_0^\infty\left[\frac{1}{\cosh\frac{\pi}{a}(\lambda - y) - \cos\frac{\pi x}{a}} - \frac{1}{\cosh\frac{\pi}{a}(\lambda + y) - \cos\frac{\pi x}{a}}\right] f(\lambda)\,d\lambda$$

$$+ \frac{1}{2a}\sin\frac{\pi x}{a}\int_0^\infty\left[\frac{1}{\cosh\frac{\pi}{a}(\lambda - y) + \cos\frac{\pi x}{a}} - \frac{1}{\cosh\frac{\pi}{a}(\lambda + y) + \cos\frac{\pi x}{a}}\right] F(\lambda)\,d\lambda$$

v. Art. 48, Exs. 4, 5, and 6.

59. If three sides of a plane rectangular sheet of conducting material be kept at potential zero and the value of the potential function at every point of the fourth side be given; to find the value of this potential function at any point of the sheet.

To formulate:—

$$D_x^2 V + D_y^2 V = 0. \qquad (1)$$

$$V = 0 \quad \text{when} \quad x = 0. \qquad (2)$$

$$V = 0 \quad \text{"} \quad x = a. \qquad (3)$$

$$V = 0 \quad \text{"} \quad y = b. \qquad (4)$$

$$V = f(x) \quad \text{"} \quad y = 0. \qquad (5)$$

Working as in Art. 48 we get

$$\frac{\sinh\frac{m\pi}{a}(b - y)}{\sinh\frac{m\pi b}{a}}\sin\frac{m\pi x}{a}$$

as a value of V which satisfies equations (1), (2), (3), and (4) if m is an integer. Therefore

$$V = \frac{2}{a}\sum_{m=1}^{m=\infty}\left[\frac{\sinh\frac{m\pi}{a}(b - y)}{\sinh\frac{m\pi b}{a}}\sin\frac{m\pi x}{a}\int_0^a f(\lambda)\sin\frac{m\pi\lambda}{a}d\lambda\right] \qquad (6)$$

is our required solution.

EXAMPLES.

1. If $f(x) = 1$ Eq. (6) Art. 59 reduces to

$$V = \frac{4}{\pi}\left[\frac{\sinh\frac{\pi}{a}(b-y)}{\sinh\frac{\pi b}{a}}\sin\frac{\pi x}{a} + \frac{1}{3}\frac{\sinh\frac{3\pi}{a}(b-y)}{\sinh\frac{3\pi b}{a}}\sin\frac{3\pi x}{a} + \frac{1}{5}\frac{\sinh\frac{5\pi}{a}(b-y)}{\sinh\frac{5\pi b}{a}}\sin\frac{5\pi x}{a} + \cdots\right]$$

2. If $V = 0$ when $x = 0$, $V = 0$ when $x = a$, $V = 0$ when $y = 0$, and $V = F(x)$ when $y = b$, then

$$V = \frac{2}{a}\sum_{m=1}^{m=\infty}\left[\frac{\sinh\frac{m\pi y}{a}}{\sinh\frac{m\pi b}{a}}\sin\frac{m\pi x}{a}\int_0^a F(\lambda)\sin\frac{m\pi\lambda}{a}d\lambda\right].$$

3. If $F(x) = 1$ the answer of Ex. 2 reduces to

$$V = \frac{4}{\pi}\left[\frac{\sinh\frac{\pi y}{a}}{\sinh\frac{\pi b}{a}}\sin\frac{\pi x}{a} + \frac{1}{3}\frac{\sinh\frac{3\pi y}{a}}{\sinh\frac{3\pi b}{a}}\sin\frac{3\pi x}{a} + \frac{1}{5}\frac{\sinh\frac{5\pi y}{a}}{\sinh\frac{5\pi b}{a}}\sin\frac{5\pi x}{a} + \cdots\right].$$

4. If $V = 0$ when $x = 0$, $V = 0$ when $x = a$, $V = f(x)$ when $y = 0$, and $V = F(x)$ when $y = b$, then

$$V = \frac{2}{a}\sum_{m=1}^{m=\infty}\left[\sin\frac{m\pi x}{a}\left(\frac{\sinh\frac{m\pi}{a}(b-y)}{\sinh\frac{m\pi b}{a}}\int_0^a f(\lambda)\sin\frac{m\pi\lambda}{a}d\lambda + \frac{\sinh\frac{m\pi y}{a}}{\sinh\frac{m\pi b}{a}}\int_0^a F(\lambda)\sin\frac{m\pi\lambda}{a}d\lambda\right)\right].$$

5. If $f(x) = F(x)$ the answer of Ex. 4 reduces to

$$V = \frac{2}{a}\sum_{m=1}^{m=\infty}\left[\frac{\cosh\frac{m\pi}{a}\left(\frac{b}{2}-y\right)}{\cosh\frac{m\pi b}{2a}}\sin\frac{m\pi x}{a}\int_0^a f(\lambda)\sin\frac{m\pi\lambda}{a}d\lambda\right].$$

6. If $f(x) = F(x) = 1$ the answer of Ex. 5 reduces to

$$V = \frac{4}{\pi}\left[\frac{\cosh\frac{\pi}{a}\left(\frac{b}{2}-y\right)}{\cosh\frac{\pi b}{2a}}\sin\frac{\pi x}{a} + \frac{1}{3}\frac{\cosh\frac{3\pi}{a}\left(\frac{b}{2}-y\right)}{\cosh\frac{3\pi b}{2a}}\sin\frac{3\pi x}{a}\right.$$

$$\left. + \frac{1}{5}\frac{\cosh\frac{5\pi}{a}\left(\frac{b}{2}-y\right)}{\cosh\frac{5\pi b}{2a}}\sin\frac{5\pi x}{a} + \cdots\right].$$

7. If $V = f(x)$ when $y = 0$, $V = F(x)$ when $y = b$, $V = \phi(y)$ when $x = 0$, and $V = \chi(y)$ when $x = a$, then

$$V = \frac{2}{a}\sum_{m=1}^{m=\infty}\left[\sin\frac{m\pi x}{a}\left(\frac{\sinh\frac{m\pi}{a}(b-y)}{\sinh\frac{m\pi b}{a}}\int_0^a f(\lambda)\sin\frac{m\pi\lambda}{a}d\lambda\right.\right.$$

$$\left.\left. + \frac{\sinh\frac{m\pi y}{a}}{\sinh\frac{m\pi b}{a}}\int_0^a F(\lambda)\sin\frac{m\pi\lambda}{a}d\lambda\right)\right]$$

$$+ \frac{2}{b}\sum_{m=1}^{m=\infty}\left[\sin\frac{m\pi y}{b}\left(\frac{\sinh\frac{m\pi}{b}(a-x)}{\sinh\frac{m\pi a}{b}}\int_0^b \phi(\lambda)\sin\frac{m\pi\lambda}{b}d\lambda\right.\right.$$

$$\left.\left. + \frac{\sinh\frac{m\pi x}{b}}{\sinh\frac{m\pi a}{b}}\int_0^b \chi(\lambda)\sin\frac{m\pi\lambda}{b}d\lambda\right)\right].$$

8. If $f(x) = \phi(y) = 0$ and $F(x) = \chi(y) = 1$ the answer of Ex. 7 may be reduced to

$$V = \frac{2}{\pi}\left[\frac{\pi y}{2b} - \frac{\sinh\frac{\pi}{b}\left(\frac{a}{2}-x\right)}{\sinh\frac{\pi a}{2b}}\sin\frac{\pi y}{b} + \frac{1}{2}\frac{\cosh\frac{2\pi}{b}\left(\frac{a}{2}-x\right)}{\cosh\frac{2\pi a}{2b}}\sin\frac{2\pi y}{b}\right.$$

$$\left. - \frac{1}{3}\frac{\sinh\frac{3\pi}{b}\left(\frac{a}{2}-x\right)}{\sinh\frac{3\pi a}{2b}}\sin\frac{3\pi y}{b} + \frac{1}{4}\frac{\cosh\frac{4\pi}{b}\left(\frac{a}{2}-x\right)}{\cosh\frac{4\pi a}{2b}}\sin\frac{4\pi y}{b} - \cdots\right].$$

9. Find the temperature of the middle point of a thin square plate whose faces are impervious to heat; 1st, when three edges are kept at the temperature $0°$ and the fourth edge at the temperature $100°$; 2d, when two opposite edges are kept at the temperature $0°$ and the other two at the temperature $100°$; 3d, when two adjacent edges are kept at the temperature $0°$ and the other edges at the temperature $100°$. See examples 3, 6, and 8.

Ans., (1) $25°$; (2) $50°$; (3) $50°$.

60. Let us pass on to the consideration of the flow of heat in one dimension.

Suppose that we have an infinite solid with two parallel plane faces whose distance apart is c.

Take the origin in one face and the axis of X perpendicular to the faces. Let the initial temperature be any given function of x and let the two faces be kept at the constant temperature zero; to find the temperature at any point of the slab at any time.

We have to solve the equation

$$D_t u = a^2 D_x^2 u \tag{1}$$

subject to the conditions

$$u = 0 \quad \text{when} \quad x = 0 \tag{2}$$

$$u = 0 \quad \text{"} \quad x = c \tag{3}$$

$$u = f(x) \quad \text{"} \quad t = 0. \tag{4}$$

In Art. 49 we have found

$$u = e^{-a^2\alpha^2 t} \sin \alpha x$$

and

$$u = e^{-a^2\alpha^2 t} \cos \alpha x$$

as particular solutions of (1).

$u = e^{-a^2\alpha^2 t} \sin \alpha x$ satisfies (2) whatever value is given to α. It satisfies (3) if $\alpha = \dfrac{m\pi}{c}$ provided m is an integer. Let us try to build a value of u out of terms of the form $Ae^{-\frac{a^2m^2\pi^2t}{c^2}} \sin \dfrac{m\pi x}{c}$ which shall satisfy (4).

We have

$$f(x) = \frac{2}{c} \sum_{m=1}^{m=\infty} \left[\sin \frac{m\pi x}{c} \int_0^c f(\lambda) \sin \frac{m\pi\lambda}{c} d\lambda \right]. \tag{5}$$

$$u = \frac{2}{c} \sum_{m=1}^{m=\infty} \left[e^{-\frac{m^2a^2\pi^2t}{c^2}} \sin \frac{m\pi x}{c} \int_0^c f(\lambda) \sin \frac{m\pi\lambda}{c} d\lambda \right], \tag{6}$$

reduces to (5) when $t = 0$ and is our required solution.

EXAMPLES.

1. If $f(\lambda) = b$, a constant, (6) Art. 60 reduces to

$$u = \frac{4b}{\pi} \left[e^{-\frac{a^2\pi^2t}{c^2}} \sin \frac{\pi x}{c} + \frac{1}{3} e^{-\frac{9a^2\pi^2t}{c^2}} \sin \frac{3\pi x}{c} + \frac{1}{5} e^{-\frac{25\pi^2a^2t}{c^2}} \sin \frac{5\pi x}{c} + \cdots \right].$$

2. An iron slab 10 cm. thick is placed between and in contact with two other iron slabs each 10 cm. thick. The temperature of the middle slab is at

first 100° throughout, and of the outside slabs 0° throughout. The outer faces of the outside slabs are kept at the temperature 0°. Required the temperature of a point in the middle of the middle slab fifteen minutes after the slabs have been placed in contact. Given $a^2 = 0.185$ in C.G.S. units. *Ans.*, 10°.3.

3. Two iron slabs each 20 cm. thick one of which is at the temperature 0° and the other at the temperature 100° throughout, are placed together face to face, and their outer faces are kept at the temperature 0°. Find the temperature of a point in their common face and of points 10 cm. from the common face fifteen minutes after the slabs have been put together.

Ans., 22°.8; 15°.1; 17°.2.

4. One face of an iron slab 40 cm. thick is kept at the temperature 0° and the other face at the temperature 100° until the permanent state of temperatures is set up. Each face is then kept at the temperature 0°. Required the temperature of a point in the middle of the slab, and of points 10 cm. from the faces fifteen minutes after the cooling has begun. *Ans.*, 22°.8; 15°.6; 16°.7.

61. If the faces of the slab treated in Art. 60 instead of being kept at the temperature zero are rendered impervious to heat, the solution of the problem is easy.

In this case we have to solve the equation

$$D_t u = a^2 D_x^2 u$$

subject to the conditions

$$\begin{aligned} D_x u &= 0 \quad \text{when} \quad x = 0 \\ D_x u &= 0 \quad \quad \text{"} \quad\quad x = c \\ u &= f(x) \quad \text{"} \quad\quad t = 0. \end{aligned}$$

We have only to use the particular solution

$$u = e^{-a^2\alpha^2 t} \cos \alpha x$$

as we used

$$u = e^{-a^2\alpha^2 t} \sin \alpha x$$

in Art. 60. We get

$$u = \frac{2}{c}\left[\frac{1}{2}\int_0^c f(\lambda)d\lambda + \sum_{m=1}^{m=\infty}\left(e^{-\frac{m^2a^2\pi^2 t}{c^2}} \cos \frac{m\pi x}{c} \int_0^c f(\lambda)\cos\frac{m\pi\lambda}{c}d\lambda\right)\right]. \qquad (1)$$

EXAMPLES.

1. Solve example 2 Art. 60 supposing that the outer surfaces are blanketed after the slabs are placed together so that heat can neither enter nor escape. Find in addition the temperature of the outer surfaces fifteen minutes after the slabs are placed in contact. *Ans.*, 33°.3; 33°.3.

2. Solve example 3 Art. 60 on the hypothesis just stated, getting in addition the temperatures of points on the outer surfaces.
Ans., 50°; 33°.9; 66°.1; 27°.2; 72°.8.

3. Solve example 4 Art. 60 supposing that heat neither enters nor escapes at the outer surfaces after the permanent state of temperatures has been set up. Find also the temperatures of points in the outer surfaces.
Ans., 50°; 39°.7; 60°.3; 35°.5; 64°.5.

4. Show that if $u = 0$ when $x = 0$, $D_x u = 0$ when $x = c$, and $u = f(x)$ when $t = 0$,

$$u = \frac{2}{c}\sum_{m=0}^{m=\infty}\left(e^{-\frac{(2m+1)^2 a^2 \pi^2 t}{4c^2}} \sin\frac{(2m+1)\pi x}{2c}\int_0^c f(\lambda)\sin\frac{(2m+1)\pi\lambda}{2c}d\lambda\right).$$

Suggestion: Assume $u = 0$ when $x = 2c$ and $f(2c - x) = f(x)$, and see (6) Art. 60.

62. If the temperature of the right-hand face of the slab considered in Art. 60 is a constant γ instead of zero we have only to add to the second member of (6) Art. 60 a term u_1 which shall satisfy the conditions

$$D_t u_1 = a^2 D_x^2 u_1 \quad (1)$$
$$u_1 = 0 \quad \text{when} \quad x = 0 \quad (2)$$
$$u_1 = 0 \quad \text{"} \quad t = 0 \quad (3)$$
$$u_1 = \gamma \quad \text{"} \quad x = c. \quad (4)$$

$u_1 = \dfrac{\gamma x}{c}$ obviously satisfies (1), (2), and (4); to make it satisfy (3) as well we must add a term u_2 which shall be equal to zero when $x = 0$ and when $x = c$ and to $-\dfrac{\gamma x}{c}$ when $t = 0$, while always satisfying (1). It is given immediately by (6) Art. 60 and is

$$u_2 = -\frac{2\gamma}{c^2}\sum_{m=1}^{m=\infty}\left(e^{-\frac{m^2 a^2 \pi^2 t}{c^2}} \sin\frac{m\pi x}{c}\int_0^c \lambda\sin\frac{m\pi\lambda}{c}d\lambda\right). \quad (5)$$

$$\int_0^c \lambda\sin\frac{m\pi\lambda}{c}d\lambda = -\frac{c^2}{m\pi}\cos m\pi = (-1)^{m+1}\frac{c^2}{m\pi},$$

and
$$u_2 = \frac{2\gamma}{\pi}\sum_{m=1}^{m=\infty}\left(\frac{(-1)^m}{m}e^{-\frac{m^2a^2\pi^2t}{c^2}}\sin\frac{m\pi x}{c}\right). \tag{6}$$

Hence
$$u_1 = \gamma\left[\frac{x}{c} + \frac{2}{\pi}\sum_{m=1}^{m=\infty}\left(\frac{(-1)^m}{m}e^{-\frac{m^2a^2\pi^2t}{c^2}}\sin\frac{m\pi x}{c}\right)\right]. \tag{7}$$

If the left-hand face of the slab considered in Art. 60 is to be kept at a constant temperature β and the right-hand face at the temperature zero we can get the term u_3 which must be added to the second member of (6) Art. 60 by replacing γ by β and x by $c - x$ in (7). We then have

$$u_3 = \beta\left[\frac{c-x}{c} - \frac{2}{\pi}\sum_{m=1}^{m=\infty}\left(\frac{1}{m}e^{-\frac{m^2a^2\pi^2t}{c^2}}\sin\frac{m\pi x}{c}\right)\right]. \tag{8}$$

EXAMPLES.

1. Show that if $u = \beta$ when $x = 0$, $u = \gamma$ when $x = c$, and $u = f(x)$ when $t = 0$

$$u = \beta + (\gamma - \beta)\left[\frac{x}{c} + \frac{2}{\pi}\sum_{m=1}^{m=\infty}\left(\frac{(-1)^m}{m}e^{-\frac{m^2\pi^2a^2t}{c^2}}\sin\frac{m\pi x}{c}\right)\right]$$
$$+\frac{2}{c}\sum_{m=1}^{m=\infty}\left(e^{-\frac{m^2a^2\pi^2t}{c^2}}\sin\frac{m\pi x}{c}\int_0^c[f(\lambda)-\beta]\sin\frac{m\pi\lambda}{c}d\lambda\right).$$

2. Show that if $u = \beta$ when $x = 0$, $u = 0$ when $t = 0$, and $D_x u = 0$ when $x = c$

$$u = \beta\left[1 - \frac{4}{\pi}\sum_{m=0}^{m=\infty}\left(\frac{1}{2m+1}e^{-\frac{(2m+1)^2a^2\pi^2t}{4c^2}}\sin\frac{(2m+1)\pi x}{2c}\right)\right]$$
$$= \beta\left[1 - \frac{4}{\pi}\left(e^{-\frac{a^2\pi^2t}{4c^2}}\sin\frac{\pi x}{2c} + \frac{1}{3}e^{-\frac{9a^2\pi^2t}{4c^2}}\sin\frac{3\pi x}{2c} + \frac{1}{5}e^{-\frac{25a^2\pi^2t}{4c^2}}\sin\frac{5\pi x}{2c} + \cdots\right)\right].$$

63. If the temperature of the right-hand face of the slab just considered is a function of the time instead of a constant and the temperature of the left-hand face is zero the problem can be solved by a method nearly identical with that of Art. 51.

Let $\phi(x,t)$ be a function of x and t which shall be zero if t is less than zero and shall be equal to

$$\frac{x}{c} + \frac{2}{\pi}\sum_{m=1}^{m=\infty}\left(\frac{(-1)^m}{m}e^{-\frac{m^2a^2\pi^2t}{c^2}}\sin\frac{m\pi x}{c}\right)$$

[v. (7) Art. 62] if t is equal to or greater than zero. So that

$$\begin{aligned}
\phi(x,t) &= 0 \quad \text{if} \quad t<0 \\
\phi(x,t) &= 0 \quad \text{"} \quad t=0 \quad \text{unless} \quad x=c \\
\phi(x,t) &= 1 \quad \text{"} \quad t=0 \quad \text{and} \quad x=c \\
\phi(x,t) &= 1 \quad \text{"} \quad x=c \\
\phi(x,t) &= 0 \quad \text{"} \quad x=0.
\end{aligned}$$

Precisely as in Art. 51 we get

$$u = \underset{\tau \doteq 0}{\text{limit}} \sum_{k=0}^{k=n} \left[F(k\tau) \frac{[\phi(x,t-k\tau) - \phi(x,t-(k+1)\tau)]\tau}{\tau} \right] \tag{1}$$

as the required solution of our problem, n being as in Art. 51 the largest integer in $\frac{t}{\tau}$ where t is any given value of the time.

On our hypothesis the last term of (1), that is, $-F(n\tau)\phi[x,t-(n+1)\tau] = 0$; the next to the last term $F(n\tau)\phi(x,t-n\tau)$ has for its limiting value

$$F(t)\phi(x,0) = F(t)\left[\frac{x}{c} + \frac{2}{\pi}\sum_{m=1}^{m=\infty}\left(\frac{(-1)^m}{m}\sin\frac{m\pi x}{c}\right)\right],$$

while as in Art. 51 the limiting value of the rest of the sum is

$$-\int_0^t F(\lambda) D_\lambda \phi(x,t-\lambda)\, d\lambda.$$

$$D_\lambda \phi(x,t-\lambda) = \frac{2a^2\pi}{c^2}\sum_{m=1}^{m=\infty}\left[(-1)^m m e^{-\frac{m^2a^2\pi^2}{c^2}(t-\lambda)} \sin\frac{m\pi x}{c}\right].$$

Hence

$$\begin{aligned}
u = F(t)&\left[\frac{x}{c} + \frac{2}{\pi}\sum_{m=1}^{m=\infty}\left(\frac{(-1)^m}{m}\sin\frac{m\pi x}{c}\right)\right] \\
&- \frac{2a^2\pi}{c^2}\sum_{m=1}^{m=\infty}\left((-1)^m m \sin\frac{m\pi x}{c}\int_0^t F(\lambda) e^{-\frac{m^2a^2\pi^2}{c^2}(t-\lambda)} d\lambda\right),
\end{aligned}$$

$$\begin{aligned}
u = \frac{x}{c}F(t) + \frac{2}{\pi}\sum_{m=1}^{m=\infty}&\left[\frac{(-1)^m}{m}\sin\frac{m\pi x}{c}\left(F(t)\right.\right. \\
&\left.\left. - \frac{m^2a^2\pi^2}{c^2}\int_0^t F(\lambda) e^{-\frac{m^2a^2\pi^2}{c^2}(t-\lambda)} d\lambda\right)\right]. \quad (2)
\end{aligned}$$

If we substitute $\beta = \frac{m^2a^2\pi^2}{c^2}(t-\lambda)$ we get

$$u = \frac{x}{c}F(t) + \frac{2}{\pi}\sum_{m=1}^{m=\infty}\left[\frac{(-1)^m}{m}\sin\frac{m\pi x}{c}\left(F(t) - \int_0^{\frac{m^2a^2\pi^2t}{c^2}} e^{-\beta}F\left(t - \frac{\beta c^2}{m^2a^2\pi^2}\right)d\beta\right)\right]. \quad (3)$$

EXAMPLES.

1. If the temperature of the left-hand face is a function of t and the temperature of the right-hand face is zero and the initial temperature is zero

$$u = \left(1 - \frac{x}{c}\right)F(t) - \frac{2}{\pi}\sum_{m=1}^{m=\infty}\left[\frac{1}{m}\sin\frac{m\pi x}{c}\left(F(t) - \int_0^{\frac{m^2a^2\pi^2t}{c^2}} e^{-\beta}F\left(t - \frac{\beta c^2}{m^2a^2\pi^2}\right)d\beta\right)\right].$$

2. If the temperature of the left-hand face is a function of t, the initial temperature is zero, and the right-hand face is impervious to heat

$$u = F(t) - \frac{4}{\pi}\sum_{m=0}^{m=\infty}\left[\frac{1}{2m+1}\sin\frac{(2m+1)\pi x}{2c}\left(F(t) - \frac{(2m+1)^2a^2\pi^2}{4c^2}\int_0^t F(\lambda)e^{-\frac{(2m+1)^2a^2\pi^2}{4c^2}(t-\lambda)}d\lambda\right)\right].$$

3. If in Arts. 60–63 we are dealing with a bar of small cross-section and of length c and heat is radiating from the surface of the bar into air at the temperature zero so that $D_tu = a^2D_x^2u - b^2u$, show that: (a) the second members of (6) Art. 60 and (1) Art. 61 must be multiplied by e^{-b^2t}; (b) equation (7) Art. 62 becomes

$$u_1 = \gamma\left\{\frac{\sinh\frac{bx}{a}}{\sinh\frac{bc}{a}} + 2a^2\pi e^{-b^2t}\sum_{m=1}^{m=\infty}\left[(-1)^m\frac{m}{b^2c^2+m^2a^2\pi^2}e^{-\frac{m^2a^2\pi^2t}{c^2}}\sin\frac{m\pi x}{c}\right]\right\};$$

(c) equation (2) Art. 63 becomes

$$u = \frac{\sinh\frac{bx}{a}}{\sinh\frac{bc}{a}}F(t) + 2a^2\pi\sum_{m=1}^{m=\infty}\left\{\frac{(-1)^m m}{b^2c^2+m^2a^2\pi^2}\sin\frac{m\pi x}{c}\left[F(t) - \frac{b^2c^2+m^2a^2\pi^2}{c^2}\int_0^t e^{-\frac{b^2c^2+m^2a^2\pi^2}{c^2}(t-\lambda)}F(\lambda)d\lambda\right]\right\}.$$

64. The problem of the motion of a finite stretched elastic string of length l fastened at the ends and distorted at first into some given curve $y = f(x)$, and then allowed to swing, has been treated and partially solved in Art. 8.

The complete solution is easily seen to be

$$y = \frac{2}{l}\sum_{m=1}^{m=\infty} \sin\frac{m\pi x}{l}\cos\frac{m\pi at}{l}\int_0^l f(\lambda)\sin\frac{m\pi\lambda}{l}d\lambda. \tag{1}$$

The second member of (1) is a periodic function of t having the period $\frac{2l}{a}$. The motion, then, unlike that in the case of an infinite string (Art. 55) is a true vibration, a periodic motion. The period $\frac{2l}{a}$ is the time it takes a disturbance to travel twice the length of the string (v. Art. 55).

A careful examination of (1) will show that the actual motion is a good deal like that in the case considered in Art. 55. The original disturbance breaks up into two waves one of which runs to the right until it reaches the end of the string and is then reflected, and runs back to the left or the under side of the string, while the other wave runs to the left and is reflected at the left-hand end of the string and runs back to the right under the string and is again reflected, runs back to the left over the string and so on indefinitely.

If the curve into which the string is distorted at the start is of the form $y = b\sin\frac{m\pi x}{l}$ the solution is

$$y = b\sin\frac{m\pi x}{l}\cos\frac{m\pi at}{l}. \tag{2}$$

No matter what value t may have the curve is always of the form

$$y = A\sin\frac{m\pi x}{l};$$

that is, for different values of t we have a set of sine curves differing only in the amplitude and not at all in the period of the curve. In this case either the whole string if $m = 1$, or each mth of the string if m is not equal to one, rises and falls, and there is no apparent onward motion. When this is the case we are said to have a *steady* vibration.

If $m = 1$ we get steady motion of the string as a whole and if the vibration is rapid enough to give a musical note the note is said to be the pure fundamental note of the string. If $m = 2$ the vibration is twice as rapid as when $m = 1$, the middle point of the string does not move and is called a node, the two halves of the string are in opposite phases of vibration at any instant, and the note given is an octave higher than the fundamental note and is called its pure *first harmonic*.

If $m = 3$ the vibration is three times as rapid as in the first case, there are two nodes $x = \frac{l}{3}$ and $x = \frac{2l}{3}$, and the note is the pure *second harmonic* of the fundamental note.

For any value of m the vibration is m times as rapid as when $m = 1$, there are $m - 1$ nodes at the points $x = \frac{l}{m}, x = \frac{2l}{m}, \cdots x = \frac{m-1}{m}l$, and we get the $m - 1$st harmonic of the fundamental note.

It is clear from (1) that no matter what the original form of the string the resulting vibration can be regarded as a combination of steady vibrations each of which alone would give the fundamental note of the string or one of its harmonics, and that the complex note resulting is really a concord of the fundamental note and some of its harmonics.

A finely trained ear can often recognize in a complex note the fundamental note of the string and some of its harmonics and is capable of analyzing a complex note into its component pure notes precisely as Fourier's Theorem enables us to analyze the complex function representing the initial form of the string into the simpler sine-functions which must be combined to form it.

EXAMPLES.

1. Show that if a point whose distance from the end of a harp string is $\frac{1}{n}$th the length of the string is drawn aside by the player's finger to a distance b from its position of equilibrium and then released, the form of the vibrating string at any instant is given by the equation

$$y = \frac{2bn^2}{(n-1)\pi^2} \sum_{m=1}^{m=\infty} \left(\frac{1}{m^2} \sin \frac{m\pi}{n} \sin \frac{m\pi x}{l} \cos \frac{m\pi at}{l} \right).$$

Show from this that all the harmonics of the fundamental note of the string which correspond to forms of vibration having nodes at the point drawn aside by the finger will be wanting in the complex note actually sounded.

2. If a stretched string starts from its position of equilibrium, each of its points having a given initial velocity, so that we have

$$\begin{aligned} y &= 0 \quad \text{when} \quad t = 0 \\ D_t y &= F(x) \quad \text{"} \quad t = 0 \\ y &= 0 \quad \text{"} \quad x = 0 \\ y &= 0 \quad \text{"} \quad x = l, \end{aligned}$$

the solution of the problem of its vibration is easy and gives

$$y = \frac{2}{a\pi} \sum_{m=1}^{m=\infty} \left(\frac{1}{m} \sin \frac{m\pi x}{l} \sin \frac{m\pi at}{l} \int_0^l F(\lambda) \sin \frac{m\pi\lambda}{l} d\lambda \right).$$

3. Write down the solution for the case where the string is initially distorted and each point has a given initial velocity.

65. If we do not neglect the resistance of the air in the problem of the vibration of a stretched string the differential equation is rather more complicated and the solution is not so easily obtained. The equation is given as (IX) Art. 1.

Let us solve the problem for the case where there is no initial velocity.

Here we have
$$D_t^2 y + 2kD_t y = a^2 D_x^2 y. \tag{1}$$
$$y = 0 \quad \text{when} \quad x = 0 \tag{2}$$
$$y = 0 \quad \text{"} \quad x = l \tag{3}$$
$$y = f(x) \quad \text{"} \quad t = 0 \tag{4}$$
$$D_t y = 0 \quad \text{"} \quad t = 0. \tag{5}$$

We get particular solutions of (1) in the usual way. Assume $y = e^{\alpha x + \beta t}$ and substitute in (1). We have
$$\beta^2 + 2k\beta = a^2\alpha^2$$
as the only necessary relation between β and α. This gives
$$\beta = -k \pm \sqrt{a^2\alpha^2 + k^2}.$$
Hence
$$y = e^{\alpha x - kt \pm t\sqrt{a^2\alpha^2+k^2}} \tag{6}$$
is a solution of (1) no matter what the value of α.

To throw it into Trigonometric form replace α by αi, and since in actual problems k, which is proportional to the resistance, is very small, take -1 out as a factor of the radical. We have
$$y = e^{-kt}e^{(\alpha x \pm t\sqrt{a^2\alpha^2-k^2})i}.$$

Since α may be positive or negative we can get
$$y = e^{-kt}\sin(\alpha x \pm t\sqrt{a^2\alpha^2 - k^2})$$
and
$$y = e^{-kt}\cos(\alpha x \pm t\sqrt{a^2\alpha^2 - k^2})$$
as solutions of (1), or by combining these
$$y = e^{-kt}\sin\alpha x \cos t\sqrt{a^2\alpha^2 - k^2} \tag{7}$$
$$y = e^{-kt}\sin\alpha x \sin t\sqrt{a^2\alpha^2 - k^2} \tag{8}$$
$$y = e^{-kt}\cos\alpha x \cos t\sqrt{a^2\alpha^2 - k^2} \tag{9}$$
$$y = e^{-kt}\cos\alpha x \sin t\sqrt{a^2\alpha^2 - k^2} \tag{10}$$

(7) and (8) satisfy (1) and (2) for all values of α. They satisfy (3) if $\alpha = \dfrac{m\pi}{l}$. Let us see if out of them we cannot build up a value that will satisfy (4) and

(5) as well.

$$f(x) = \frac{2}{l}\sum_{m=1}^{m=\infty}\left(\sin\frac{m\pi x}{l}\int_0^l f(\lambda)\sin\frac{m\pi\lambda}{l}d\lambda\right). \tag{11}$$

$$y = \frac{2}{l}e^{-kt}\sum_{m=1}^{m=\infty}\left(\sin\frac{m\pi x}{l}\cos t\sqrt{\frac{m^2\pi^2a^2}{l^2} - k^2}.\int_0^l f(\lambda)\sin\frac{m\pi\lambda}{l}d\lambda\right) \tag{12}$$

reduces to (11) when $t = 0$ and therefore satisfies (4).

$$D_t y = -\frac{2}{le^{kt}}\sum_{m=1}^{m=\infty}\left(\sqrt{\frac{m^2\pi^2a^2}{l^2} - k^2}.\sin\frac{m\pi x}{l}\sin t\sqrt{\frac{m^2\pi^2a^2}{l^2} - k^2}\right.$$
$$\left. .\int_0^l f(\lambda)\sin\frac{m\pi\lambda}{l}d\lambda\right)$$
$$-\frac{2k}{le^{kt}}\sum_{m=1}^{m=\infty}\left(\sin\frac{m\pi x}{l}\cos t\sqrt{\frac{m^2\pi^2a^2}{l^2} - k^2}.\int_0^l f(\lambda)\sin\frac{m\pi\lambda}{l}d\lambda\right). \tag{13}$$

When $t = 0$ the first line of the second member of (13) vanishes but the second line reduces to

$$-\frac{2k}{l}\sum_{m=1}^{m=\infty}\left(\sin\frac{m\pi x}{l}\int_0^l f(\lambda)\sin\frac{m\pi\lambda}{l}d\lambda\right).$$

We must, then, introduce into (12) an additional term which shall equal zero when $t = 0$ and whose derivative with respect to t shall cancel the term above when $t = 0$.

This is easily seen to be

$$\frac{2k}{l}e^{-kt}\sum_{m=1}^{m=\infty}\frac{1}{\sqrt{\frac{m^2\pi^2a^2}{l^2} - k^2}}\sin\frac{m\pi x}{l}\sin t\sqrt{\frac{m^2\pi^2a^2}{l^2} - k^2}.\int_0^l f(\lambda)\sin\frac{m\pi\lambda}{l}d\lambda.$$

Hence our complete solution is

$$y = \frac{2}{l}e^{-kt}\sum_{m=1}^{m=\infty}\left[\left(\cos t\sqrt{\frac{m^2\pi^2a^2}{l^2} - k^2}\right.\right.$$
$$\left.\left. + \frac{k}{\sqrt{\frac{m^2\pi^2a^2}{l^2} - k^2}}\sin t\sqrt{\frac{m^2\pi^2a^2}{l^2} - k^2}\right)\sin\frac{m\pi x}{l}\int_0^l f(\lambda)\sin\frac{m\pi\lambda}{l}d\lambda\right]. \tag{14}$$

Here the fact that e^{-kt}, which decreases rapidly as t increases, is a factor of the whole second member shows that the amplitude of the vibration rapidly decreases.

Comparing this solution with that given in Art. 64 for the case where there is no resistance we see that the period of any given term

$$A\sin\frac{m\pi x}{l}\cos t\sqrt{\frac{m^2\pi^2a^2}{l^2}-k^2},$$

is greater than that of the corresponding term $A_1\sin\frac{m\pi x}{l}\cos\frac{m\pi at}{l}$ in Art. 64. In other words the effect of the resistance of the air is to flatten somewhat each component part of the note given by the string. More than this since the periods of the different terms of (14) are no longer exact submultiples of the period of the first term, the component notes are no longer in perfect harmony with the fundamental note of the string, and the ideal perfect harmony between the fundamental note and its harmonics is not quite realized in any actual case.

When k is very small, as in the case of a fine string, the departure from perfect harmony is very slight; but in the case of a coarse string or worse still of an elastic ribbon, where the resistance of the air is considerable, the unmusical character of the sound is very noticeable.

EXAMPLES.

1. Solve Ex. 1 Art. 64 allowing for the resistance of the air.

2. Solve Ex. 2 Art. 64 allowing for the resistance of the air;

$$y=\frac{2}{l}e^{-kt}\sum_{m=1}^{m=\infty}\left(\frac{1}{\sqrt{\frac{m^2\pi^2a^2}{l^2}-k^2}}\sin\frac{m\pi x}{l}\sin t\sqrt{\frac{m^2\pi^2a^2}{l^2}-k^2}\cdot\int_0^l F(\lambda)\sin\frac{m\pi\lambda}{l}d\lambda\right).$$

3. Find a particular solution of (1) Art. 65 on the assumption that it is of the form $y=T.X$, where T is a function of t alone and X a function of x alone.

66. We pass on now to a couple of problems that require the modification and extension of Fourier's Theorem, *the cooling of a sphere in air*, and the *vibration of a stretched rectangular membrane*, but as an introduction to the former we shall first consider the following very simple problem; to find the temperature of any point of a sphere whose initial temperature is any given function of r the distance of the point from the centre, and whose surface is kept at the constant temperature b.

Here we are to solve

$$D_t(ru)=a^2D_r^2(ru), \tag{1}$$

see [v] Art. 1, subject to the conditions

$$u=f(r)\text{ when }\quad t=0 \tag{2}$$

$$u=b\qquad\text{"}\qquad r=c \tag{3}$$

if c is the radius.

Let $v = ru$, then our equations become

$$D_t v = a^2 D_r^2 v \tag{4}$$
$$v = rf(r) \text{ when } t = 0 \tag{5}$$
$$v = bc \quad \text{"} \quad r = c \tag{6}$$
$$v = 0 \quad \text{"} \quad r = 0. \tag{7}$$

Our problem is now precisely that of Art. 62 and we have as our solution

$$\begin{aligned} ru = & \frac{2}{c}\sum_{m=1}^{m=\infty}\left(e^{-\frac{m^2a^2\pi^2}{c^2}t}\sin\frac{m\pi r}{c}\int_0^c \lambda f(\lambda)\sin\frac{m\pi\lambda}{c}d\lambda\right) \\ & + b\left[r + \frac{2c}{\pi}\sum_{m=1}^{m=\infty}\left(\frac{(-1)^m}{m}e^{-\frac{m^2a^2\pi^2}{c^2}t}\sin\frac{m\pi r}{c}\right)\right]. \end{aligned} \tag{8}$$

EXAMPLES.

1. If $f(r) = b$ (8) Art. 66 reduces to $u = b$ and there is no change of temperature.
2. If the initial temperature is constant and equal to β

$$\begin{aligned} u = b + \frac{2c}{\pi r}(\beta - b)\Bigg[& e^{-\frac{a^2\pi^2}{c^2}t}\sin\frac{\pi r}{c} - \frac{1}{2}e^{-\frac{4a^2\pi^2}{c^2}t}\sin\frac{2\pi r}{c} \\ & + \frac{1}{3}e^{-\frac{9a^2\pi^2}{c^2}t}\sin\frac{3\pi r}{c} - \cdots\Bigg]. \end{aligned}$$

3. An iron sphere 40 cm. in diameter is heated to the temperature 100° centigrade throughout; its surface is then kept at the constant temperature 0°. Find the temperature of a point 10 cm. from the centre, and find the temperature of the centre, 15 minutes after cooling has begun. Given $a^2 = 0.185$ in C.G.S. units. *Ans.*, 2°.1; 3°.3.

67. If instead of having the temperature of the surface of the sphere constant, the sphere is placed in air which is kept at the constant temperature zero, the problem is much more complicated. For in this case the surface temperature can no longer be simply expressed but is given by a new differential equation

$$D_r u + hu = 0 \quad \text{when} \quad r = c, \tag{1}$$

where h is an experimental constant depending upon what is called the surface conductivity of the sphere.

Our equations, then, are

$$D_t(ru) = a^2 D_r^2(ru) \tag{2}$$
$$u = f(r) \quad \text{when} \quad t = 0 \tag{3}$$
$$D_r u + hu = 0 \quad \text{when} \quad r = c. \tag{4}$$

As in Art. 66 let $v = ru$; then we have

$$D_t v = a^2 D_r^2 v \tag{5}$$

$$v = rf(r) \quad \text{when} \quad t = 0 \tag{6}$$

$$v = 0 \quad \text{"} \quad r = 0 \tag{7}$$

$$D_r v + \left(h - \frac{1}{c}\right) v = 0 \quad \text{when} \quad r = c. \tag{8}$$

$v = e^{-a^2\alpha^2 t} \cos \alpha r$ and $v = e^{-a^2\alpha^2 t} \sin \alpha r$ have already been found as particular solutions of (5) (see Art. 60).

$$v = e^{-a^2\alpha^2 t} \sin \alpha r \tag{9}$$

satisfies (7) for all values of α.

Substitute this value of v in (8) and we have

$$\alpha c \cos \alpha c + (hc - 1) \sin \alpha c = 0. \tag{10}$$

If α_k is a value of α which is a root of the transcendental equation (10)

$$v = e^{-a^2\alpha_k^2 t} \sin \alpha_k r \tag{11}$$

will satisfy (5), (7), and (8).

It remains to see whether out of terms of the form given in (11) we can build up a value of v which will satisfy (6).

When $t = 0$ the second member of (11) reduces to $\sin \alpha_k r$. If then we can express $rf(r)$ as a sum of terms of the form $b_k \sin \alpha_k r$ where α_k is a root of (10)

$$v = \sum b_k e^{-a^2\alpha_k^2 t} \sin \alpha_k r \tag{12}$$

will satisfy all of the equations (5), (6), (7), and (8), and will be the required solution.

Here, then, we have a new problem analogous to that of developing in a Fourier's Series, but rather more complicated, namely, to develop any function of x in a series of the form $\sum a_m \sin \alpha_m x$ where α_m is a root of the equation (10); or if we call $ac = \phi$ and $hc - 1 = p$, where $a_m = \dfrac{\phi_m}{c}$, ϕ_m being a root of the equation

$$\phi \cos \phi + p \sin \phi = 0 \tag{13}$$

or more simply of

$$\phi + p \tan \phi = 0; \tag{14}$$

remembering that the series and the function must be equal for all values of x between zero and c.

If ϕ_m is a root of (14) $-\phi_m$ is also a root.

Since $\sin \frac{\phi_m}{c} x = -\sin\left(-\frac{\phi_m}{c} x\right)$ the terms of the required development which correspond to negative roots may be combined with those corresponding to positive roots, and therefore we need consider only positive roots.

$\phi = 0$ is a root of (14) but as $\sin 0 = 0$ there will be no corresponding term in the development. If we construct the curve

$$y = -\frac{1}{p}x \tag{15}$$

and the curve

$$y = \tan x \tag{16}$$

the abscissas of their points of intersection are values of x which satisfy $\frac{x}{p} + \tan x = 0$, that is, are roots of equation (14). It is easy to see that there will always be an infinite number of real positive roots, one for each of the branches of the periodic curve $y = \tan x$ which lie to the right of the origin. The numerical values of these roots can be obtained by an easy computation. The construction suggested above shows that as m increases ϕ_m will rapidly approach the value $(2m-1)\frac{\pi}{2}$ if p is positive or if p is negative and numerically less than unity, and $(2m+1)\frac{\pi}{2}$ if p is negative and numerically greater than unity.

There exist, then, an infinite number of positive real roots of $\phi + p\tan\phi = 0$ and consequently of

$$\alpha c \cos \alpha c + (hc - 1)\sin \alpha c = 0.$$

68. The development called for in the last article can be obtained very easily from a simpler one which we shall now consider, namely, to develop $f(x)$ into a series of the form

$$f(x) = a_1 \sin \phi_1 x + a_2 \sin \phi_2 x + a_3 \sin \phi_3 x + \cdots \tag{1}$$

where ϕ_1, ϕ_2, $\phi_3 \cdots$ are roots of the equation

$$\phi \cos \phi + p \sin \phi = 0, \tag{2}$$

the development to hold good for all values of x between $x = 0$ and $x = 1$.

Let us proceed as in Arts. 24 and 27. Call $\frac{1}{n+1} = \Delta x$ and form n equations by substituting for x in turn in the equation

$$f(x) = a_1 \sin \phi_1 x + a_2 \sin \phi_2 x + a_3 \sin \phi_3 x + \cdots + a_n \sin \phi_n x \tag{3}$$

the values Δx, $2\Delta x$, $3\Delta x$, $\cdots n\Delta x$; this being equivalent to making the values of the sum and the function coincide for the n values of x substituted.

To determine any coefficient a_m multiply the first equation by $\Delta x.\sin(\phi_m \Delta x)$, the second by $\Delta x.\sin(2\phi_m \Delta x)$, the third by $\Delta x.\sin(3\phi_m \Delta x)$, and so on, the nth equation by $\Delta x.\sin(n\phi_m \Delta x)$; add the equations and compute the limiting values of the terms of the resulting equation as n is indefinitely increased. This as in Art. 24 is seen to be equivalent to multiplying (3) by $\sin \phi_m x.dx$ and integrating between the limits $x = 0$ and $x = 1$.

The first member of the resulting equation is

$$\int_0^1 f(x)\sin \phi_m x.dx;$$

The coefficient of a_k is

$$\int_0^1 \sin \phi_k x \sin \phi_m x.dx,$$

and of a_m is

$$\int_0^1 \sin^2 \phi_m x.dx.$$

$$\begin{aligned}\int_0^1 \sin \phi_k x \sin \phi_m x.dx &= \frac{1}{2}\int_0^1 [\cos(\phi_k - \phi_m)x - \cos(\phi_k + \phi_m)x]dx \\ &= \frac{1}{2}\left[\frac{\sin(\phi_k - \phi_m)}{\phi_k - \phi_m} - \frac{\sin(\phi_k + \phi_m)}{\phi_k + \phi_m}\right] \\ &= -\frac{\phi_k \cos \phi_k \sin \phi_m - \phi_m \sin \phi_k \cos \phi_m}{\phi_k^2 - \phi_m^2} \qquad (4)\end{aligned}$$

But $$\phi_k \cos \phi_k + p \sin \phi_k = 0$$
and $$\phi_m \cos \phi_m + p \sin \phi_m = 0 \qquad \text{by (2).}$$

Hence the numerator of the second member of (4) is zero, and the coefficient of a_k vanishes if k is not equal to m.

$$\int_0^1 \sin^2 \phi_m x.dx = \frac{1}{2\phi_m}[\phi_m - \sin \phi_m \cos \phi_m] = \frac{1}{2}\left[1 - \frac{\sin 2\phi_m}{2\phi_m}\right]. \qquad (5)$$

Therefore $$a_m = \frac{2}{1 - \dfrac{\sin 2\phi_m}{2\phi_m}} \int_0^1 f(x)\sin \phi_m x.dx. \qquad (6)$$

The coefficient of the integral in (6) can be transformed as follows so as not to involve trigonometric functions.

$$\phi_m \cos\phi_m + p\sin\phi_m = 0, \qquad \text{by (2)}$$

$$\phi_m \cos^2\phi_m + \frac{p}{2}\sin 2\phi_m = 0,$$

$$\frac{\sin 2\phi_m}{2\phi_m} = -\frac{\cos^2\phi_m}{p}. \tag{7}$$

$$\phi_m^2 \cos^2\phi_m = p^2\sin^2\phi_m,$$

$$(\phi_m^2 + p^2)\cos^2\phi_m = p^2,$$

$$\frac{\cos^2\phi_m}{p} = \frac{p}{\phi_m^2 + p^2}. \tag{8}$$

Hence by (7) and (8)

$$1 - \frac{\sin 2\phi_m}{2\phi_m} = \frac{\phi_m^2 + p(p+1)}{\phi_m^2 + p^2},$$

and
$$a_m = \frac{2(\phi_m^2 + p^2)}{\phi_m^2 + p(p+1)}\int_0^1 f(\alpha)\sin\phi_m\alpha.d\alpha. \tag{9}$$

Therefore our required development is

$$f(x) = \sum_{m=1}^{m=\infty}\left(\frac{2(\phi_m^2 + p^2)}{\phi_m^2 + p(p+1)}\sin\phi_m x\int_0^1 f(\alpha)\sin\phi_m\alpha.d\alpha\right). \tag{10}$$

From (10) it easily follows that for values of x between 0 and c

$$f(x) = a_1\sin a_1x + a_2\sin a_2x + a_3\sin a_3x + \cdots \tag{11}$$

where
$$a_m = \frac{2}{c}\cdot\frac{\alpha_m^2c^2 + p^2}{\alpha_m^2c^2 + p(p+1)}\int_0^c f(\lambda)\sin\alpha_m\lambda.d\lambda, \tag{12}$$

and α_m is a root of the equation

$$\alpha c\cos\alpha c + p\sin\alpha c = 0. \tag{13}$$

It is to be observed that if p is infinite (13) reduces to $\sin\alpha c = 0$, α_m becomes $\frac{m\pi}{c}$ and (11) and (12) give our regulation Fourier sine series (v. Art. 31), and therefore the ordinary Fourier development in sine series is merely a special case of the problem just solved.

Moreover since the Fourier method of determining the coefficients of such a series requires that

$$\int_0^c \sin\alpha_m x \sin\alpha_n x.dx = 0,$$

that is that
$$\frac{\sin(\alpha_m-\alpha_n)c}{\alpha_m-\alpha_n} - \frac{\sin(\alpha_m+\alpha_n)c}{\alpha_m+\alpha_n} = 0$$

or reducing, that
$$\frac{\alpha_m c\cos\alpha_m c}{\sin\alpha_m c} = \frac{\alpha_n c\cos\alpha_n c}{\sin\alpha_n c},$$

or that α_m and α_n should be roots of the equation

$$\frac{\alpha c\cos\alpha c}{\sin\alpha c} = p$$

where p is some constant, it follows that we have obtained in (11) the most general sine development that can be obtained by Fourier's method.

EXAMPLES.

1. Show that the solution of the problem of Art. 67 is

$$ru = \sum_{m=1}^{m=\infty} b_m e^{-a^2\alpha_m^2 t}\sin\alpha_m r,$$

where
$$b_m = \frac{2}{c}.\frac{\alpha_m^2 c^2+(hc-1)^2}{\alpha_m^2 c^2+hc(hc-1)}\int_0^c \lambda f(\lambda)\sin\alpha_m\lambda.d\lambda$$

and α_m is a root of

$$\alpha c\cos\alpha c + (hc-1)\sin\alpha c = 0.$$

2. If the initial temperature of the sphere is constant and equal to β

$$ru = \sum_{m=1}^{m=\infty} b_m e^{-a^2\alpha_m^2 t}\sin\alpha_m r$$

where
$$b_m = 2\beta h.\frac{\alpha_m^2 c^2+(hc-1)^2}{\alpha_m^2 c^2+hc(hc-1)}.\frac{\sin\alpha_m c}{\alpha_m^2}$$
$$= \frac{2\beta hc}{\alpha_m}.\frac{[\alpha_m^2 c^2+(hc-1)^2]^{\frac{1}{2}}}{\alpha_m^2 c^2+hc(hc-1)}.$$

3. If the temperature of the air is a constant γ instead of zero the surface equation of condition is

$$D_r u + h(u-\gamma) = 0 \quad \text{when} \quad r = c.$$

The substitution of $u_1 = u - \gamma$, however, brings the problem under Ex. 1 and we get

$$r(u-\gamma) = \sum_{m=1}^{m=\infty} b_m e^{-a^2\alpha_m^2 t} \sin \alpha_m r$$

where $$b_m = \frac{2}{c} \cdot \frac{\alpha_m^2 c^2 + (hc-1)^2}{\alpha_m^2 c^2 + hc(hc-1)} \int_0^c \lambda[f(\lambda) - \gamma] \sin \alpha_m \lambda . d\lambda.$$

4. An iron sphere 40 cm. in diameter is heated to the temperature 100° centigrade throughout; it is then allowed to cool in air which is kept at the constant temperature 0°. Find the temperature at the centre; at a point 10 cm. from the centre; and at the surface; 15 minutes after cooling has begun. Given $a^2 = 0.185$ and $h = \frac{1}{800}$ in C.G.S. units. (v. Ex. 3, Art. 66.)

Ans., 97°.67; 97°.36; 96°.46.

5. Show that if in the slab considered in Art. 60 one face is exposed to air at the temperature zero, so that we have $D_t u = a^2 D_x^2 u$, $u = 0$ when $x = 0$, $u = f(x)$ when $t = 0$, and $D_x u + hu = 0$ when $x = c$, then

$$u = \sum_{m=1}^{m=\infty} a_m e^{-a^2\alpha_m^2 t} \sin \alpha_m x$$

where $$a_m = 2 \frac{\alpha_m^2 + h^2}{\alpha_m^2 c + h(hc+1)} \int_0^c f(\lambda) \sin \alpha_m \lambda . d\lambda,$$

α_m being a root of $\alpha c \cos \alpha c + hc \sin \alpha c = 0$.

6. If in the problem of Art. 57 heat escapes from one side of the plate into air at the temperature zero so that we have $D_x^2 u + D_y^2 u = 0$, $u = 0$ when $x = 0$, $u = f(x)$ when $y = 0$, and $D_x u + hu = 0$ when $x = a$, then

$$u = \sum_{m=1}^{m=\infty} a_m e^{-\alpha_m y} \sin \alpha_m x$$

where $$a_m = 2 \frac{\alpha_m^2 + h^2}{\alpha_m^2 a + h(ha+1)} \int_0^a f(\lambda) \sin \alpha_m \lambda . d\lambda,$$

α_m being a root of $\alpha a \cos \alpha a + ha \sin \alpha a = 0$.

7. If in the problem of Art. 59 there is leakage at one side of the sheet so that we have $D_x^2 V + D_y^2 V = 0$, $V = 0$ when $x = 0$, $V = 0$ when $y = b$, $V = f(x)$ when $y = 0$, and $D_x V + hV = 0$ when $x = a$, then

$$V = \sum_{m=1}^{m=\infty} a_m \frac{\sinh \alpha_m (b-y)}{\sinh \alpha_m b} \sin \alpha_m x,$$

where a_m has the value given in Ex. 6.

69. If we have an infinite solid with one plane face which is exposed to air at the temperatures $U = F(t)$ and heat can flow only at right angles to this face, we can solve the problem readily for the case where the initial temperatures are zero. We have

$$D_t u = a^2 D_x^2 u$$

subject to the conditions

$$u = 0 \quad \text{when} \quad t = 0$$

and
$$D_x u + h(U - u) = 0 \quad \text{when} \quad x = 0.$$

Let
$$v = u - \frac{1}{h} D_x u. \tag{1}$$

Then v will satisfy the equation

$$D_t v = a^2 D_x^2 v,$$

and we shall also have $v = U$ when $x = 0$.

Since $U = F(t)$
$$v = \frac{2}{\sqrt{\pi}} \int_{\frac{x}{2a\sqrt{t}}}^{\infty} e^{-\beta^2} F\left(t - \frac{x^2}{4a^2\beta^2}\right) d\beta \tag{2}$$

by Art. 51 (10).

$$D_x u - hu = -hv \qquad \text{by (1).}$$

Hence
$$ue^{-hx} = -h \int e^{-hx} v dx + C;$$

v. Int. Cal. § 4, page 314.

Determining C by the fact that $ue^{-hx} = 0$ when $x = \infty$ we have

$$u = he^{hx} \int_x^{\infty} e^{-hx} v dx. \tag{3}$$

Substituting the value of v from (2) we have

$$u = \frac{2he^{hx}}{\sqrt{\pi}} \int_x^{\infty} e^{-hx} dx \int_{\frac{x}{2a\sqrt{t}}}^{\infty} e^{-\beta^2} F\left(t - \frac{x^2}{4a^2\beta^2}\right) d\beta, \tag{4}$$

as our required solution.

For an extension of this method to the flow of heat in two and three dimensions and for the interpretation of the results by the aid of the theory of *Images*, see E. W. Hobson, Proc. Lond. Math. Soc., Vol. XIX.

EXAMPLES.

1. If the temperature of the air is a periodic function of the time, say $\rho_m \sin(m\alpha t + \lambda_m)$ and we care only for the limiting value of u as t increases, show that this value is

$$\frac{h\rho_m e^{-\frac{x}{a}\frac{\sqrt{m\alpha}}{2}}}{\left(h+\frac{1}{a}\sqrt{\frac{m\alpha}{2}}\right)^2+\frac{m\alpha}{2a^2}}\left[\left(h+\frac{1}{a}\sqrt{\frac{m\alpha}{2}}\right)\sin\left(m\alpha t-\frac{x}{a}\sqrt{\frac{m\alpha}{2}}+\lambda_m\right)\right.$$
$$\left.-\frac{1}{a}\sqrt{\frac{m\alpha}{2}}\cos\left(m\alpha t-\frac{x}{a}\sqrt{\frac{m\alpha}{2}}+\lambda_m\right)\right].$$

v. Art. 52 and Art. 51 Ex. 4.

Note that $$\int e^{ax}\sin bx.dx=\frac{e^{ax}(a\sin bx-b\cos bx)}{a^2+b^2}$$

and $$\int e^{ax}\cos bx.dx=\frac{e^{ax}(a\cos bx+b\sin bx)}{a^2+b^2}$$

v. Int. Cal. Table of Int. (235) and (236).

2. If $D_x^2V + D_y^2V = 0$, $V = 0$ when $y = 0$ and $D_xV + h[F(y) - V] = 0$ when $x = 0$ show that

$$V=\frac{he^{hx}}{\pi}\int_x^\infty e^{-hx}dx\int_0^\infty F(\lambda)d\lambda\left[\frac{x}{x^2+(\lambda-y)^2}-\frac{x}{x^2+(\lambda+y)^2}\right];$$

v. Art. 47 Ex. 1.

70. The solution for an instantaneous heat source of strength Q at the point $x = \lambda$ if heat escapes at the origin into air at the temperature zero, so that $D_xu - hu = 0$ when $x = 0$, can be obtained by the aid of Art. 53.

Let $u = u_1 + u_2$ where u_1 is the temperature that would be due to the given source if we had no boundary at the origin, so that

$$u_1=\frac{Q}{2a\sqrt{\pi t}}e^{-\frac{(\lambda-x)^2}{4a^2t}}. \qquad \text{[Art. 53 (2)]}$$

$$D_xu-hu=D_xu_1-hu_1+D_xu_2-hu_2=0 \quad \text{when} \quad x=0.$$

Therefore $$D_xu_2-hu_2=-(D_xu_1-hu_1) \qquad (1)$$

when $x = 0$.

But $$-(D_xu_1-hu_1)=-\frac{Q}{2a\sqrt{\pi t}}\left(\frac{\lambda-x}{2a^2t}-h\right)e^{-\frac{(\lambda-x)^2}{4a^2t}}$$
$$=-\frac{Q}{2a\sqrt{\pi t}}\left(\frac{\lambda}{2a^2t}-h\right)e^{-\frac{\lambda^2}{4a^2t}}$$

when $x = 0$.

This is easily seen to be the value to which

$$-\frac{Q}{2a\sqrt{\pi t}}\left(\frac{\lambda+x}{2a^2t}-h\right)e^{-\frac{(\lambda+x)^2}{4a^2t}}$$

reduces when $x=0$, and this last expression is

$$(D_x+h)\frac{Q}{2a\sqrt{\pi t}}e^{-\frac{(\lambda+x)^2}{4a^2t}}$$

and therefore satisfies the equation

$$D_tu = a^2D_x^2u; \tag{2}$$

since $\dfrac{Q}{2a\sqrt{\pi t}}e^{-\frac{(\lambda+x)^2}{4a^2t}}$ is the temperature due to a source at $x=-\lambda$.

If, then, we determine u_2 from the condition that

$$D_xu_2 - hu_2 = -\frac{Q}{2a\sqrt{\pi t}}\left(\frac{\lambda+x}{2a^2t}-h\right)e^{-\frac{(\lambda+x)^2}{4a^2t}} \tag{3}$$

taking care not to introduce any arbitrary constant or arbitrary function of t in our integration, u_2 will satisfy equation (2) and condition (1).

Integrating (3) [v. Int. Cal. § 4, page 314] and determining the constants of integration suitably we get

$$u_2 = \frac{Q}{2a\sqrt{\pi t}}\left[e^{-\frac{(\lambda+x)^2}{4a^2t}} - 2he^{hx}\int_x^\infty e^{-hx-\frac{(\lambda+x)^2}{4a^2t}}dx\right]. \tag{4}$$

Therefore the solution of our problem is

$$u = \frac{Q}{2a\sqrt{\pi t}}\left[e^{-\frac{(\lambda-x)^2}{4a^2t}} + e^{-\frac{(\lambda+x)^2}{4a^2t}} - 2he^{hx}\int_x^\infty e^{-hx-\frac{(\lambda+x)^2}{4a^2t}}dx\right]. \tag{5}$$

If we replace Q by $f(\lambda)d\lambda$ and integrate from 0 to ∞ we get as the solution for the case where $u=f(x)$ when $t=0$ and $x>0$, and $D_xu-hu=0$ when $x=0$

$$u = \frac{1}{2a\sqrt{\pi t}}\int_0^\infty f(\lambda)d\lambda\left[e^{-\frac{(\lambda-x)^2}{4a^2t}} + e^{-\frac{(\lambda+x)^2}{4a^2t}} - 2he^{hx}\int_x^\infty e^{-hx-\frac{(\lambda+x)^2}{4a^2t}}dx\right]. \tag{6}$$

For an interpretation of this result by the theory of Images and the extension of the method to the conduction of heat in n dimensions see G. H. Bryan, Proc. Lond. Math. Soc., Vol. XXII.

EXAMPLE.

Show that if $u = f(x)$ when $t = 0$ and $D_x u + h[F(t) - u] = 0$ when $x = 0$ we must take u equal to the sum of the second members of (6) Art. 70 and of (4) Art. 69.

71. As another problem requiring a slight extension of Fourier's Theorem let us consider the vibration of a rectangular stretched elastic membrane fastened at the edges, that is of a rectangular drumhead.

If two of the sides are taken as axes and the plane of equilibrium of the membrane as the plane of XY the equation for the motion of the membrane is

$$D_t^2 z = c^2(D_x^2 z + D_y^2 z) \tag{1}$$

see [x] Art. 1.

Let the membrane be distorted at the start into some given form $z = f(x, y)$ and then allowed to swing. Our equations of conditions are then

$$z = 0 \quad \text{when} \quad x = 0 \tag{2}$$
$$z = 0 \quad \text{"} \quad x = a \tag{3}$$
$$z = 0 \quad \text{"} \quad y = 0 \tag{4}$$
$$z = 0 \quad \text{"} \quad y = b \tag{5}$$
$$z = f(x, y) \quad \text{"} \quad t = 0 \tag{6}$$
$$D_t z = 0 \quad \text{"} \quad t = 0. \tag{7}$$

We can get a particular solution of (1) by our usual device. Assume

$$z = e^{\alpha x + \beta y + \gamma t}$$

and substitute in (1). We get $\gamma^2 = c^2(\alpha^2 + \beta^2)$ as the only relation that need hold between α, β, and γ, in order that $z = e^{\alpha x + \beta y + \gamma t}$ may be a solution. This gives

$$\gamma = \pm\, c\sqrt{\alpha^2 + \beta^2}.$$

Therefore

$$z = e^{\alpha x + \beta y \pm ct\sqrt{\alpha^2 + \beta^2}}$$

is a solution of (1) no matter what values are given to α and β.

Replace α and β by αi and βi and we have

$$z = e^{(\alpha x + \beta y \pm ct\sqrt{\alpha^2 + \beta^2})i}$$

as a solution, and from this we get

$$z = \sin(\alpha x + \beta y \pm ct\sqrt{\alpha^2 + \beta^2}) \tag{8}$$

and

$$z = \cos(\alpha x + \beta y \pm ct\sqrt{\alpha^2 + \beta^2}) \tag{9}$$

as particular solutions of (1), α and β being unrestricted.

From (8) and (9) we can get solutions of the following forms

$$\left.\begin{aligned}
z &= \sin\alpha x \sin\beta y \sin ct\sqrt{\alpha^2+\beta^2}\\
z &= \sin\alpha x \sin\beta y \cos ct\sqrt{\alpha^2+\beta^2}\\
z &= \sin\alpha x \cos\beta y \sin ct\sqrt{\alpha^2+\beta^2}\\
z &= \sin\alpha x \cos\beta y \cos ct\sqrt{\alpha^2+\beta^2}\\
z &= \cos\alpha x \sin\beta y \sin ct\sqrt{\alpha^2+\beta^2}\\
z &= \cos\alpha x \sin\beta y \cos ct\sqrt{\alpha^2+\beta^2}\\
z &= \cos\alpha x \cos\beta y \sin ct\sqrt{\alpha^2+\beta^2}\\
z &= \cos\alpha x \cos\beta y \cos ct\sqrt{\alpha^2+\beta^2},
\end{aligned}\right\} \qquad (10)$$

each of which will satisfy equation (1). The second of these will satisfy also (2), (4) and (7) whatever values be taken for α and β. It will satisfy (3) and (5) if α and β are equal $\frac{m\pi}{a}$ and $\frac{n\pi}{b}$ respectively.

If, then, we can so combine terms of the form

$$\sin\frac{m\pi x}{a}\sin\frac{n\pi y}{b}\cos c\pi t\sqrt{\frac{m^2}{a^2}+\frac{n^2}{b^2}}$$

as to satisfy (6) our problem will be completely solved.

This can be done if we can express $f(x,y)$ as a sum of terms of the form $A\sin\frac{m\pi x}{a}\sin\frac{n\pi y}{b}$, the sum and the function being equal when x lies between 0 and a and y between 0 and b.

$f(x,y)$ can be expressed in terms of $\sin\frac{m\pi x}{a}$ by Fourier's Theorem if we regard y as constant. We have

$$f(x,y)=\sum_{m=1}^{m=\infty}a_m\sin\frac{m\pi x}{a} \qquad (11)$$

where
$$a_m=\frac{2}{a}\int_0^a f(\lambda,y)\sin\frac{m\pi\lambda}{a}d\lambda. \qquad (12)$$

$f(\lambda,y)$ in (12) is a function of y and may be developed by Fourier's Theorem.

We have
$$f(\lambda,y)=\sum_{n=1}^{n=\infty}b_n\sin\frac{n\pi y}{b} \qquad (13)$$

where
$$b_n=\frac{2}{b}\int_0^b f(\lambda,\mu)\sin\frac{n\pi\mu}{b}d\mu. \qquad (14)$$

Substituting for $f(\lambda,y)$ in (12) the value just obtained we have

$$a_m=\frac{2}{a}\frac{2}{b}\sum_{n=1}^{n=\infty}\left(\int_0^a d\lambda\int_0^b f(\lambda,\mu)\sin\frac{m\pi\lambda}{a}\sin\frac{n\pi\mu}{b}d\mu\right)\sin\frac{n\pi y}{b}$$

and

$$f(x,y) = \frac{4}{ab}\sum_{m=1}^{m=\infty}\sum_{n=1}^{n=\infty}\left(\sin\frac{m\pi x}{a}\sin\frac{n\pi y}{b}\int_0^a d\lambda\int_0^b f(\lambda,\mu)\sin\frac{m\pi\lambda}{a}\sin\frac{n\pi\mu}{b}d\mu\right). \tag{15}$$

Hence
$$z = \sum_{m=1}^{m=\infty}\sum_{n=1}^{n=\infty}\left(A_{m,n}\sin\frac{m\pi x}{a}\sin\frac{n\pi y}{b}\cos c\pi t\sqrt{\frac{m^2}{a^2}+\frac{n^2}{b^2}}\right), \tag{16}$$

where
$$A_{m,n} = \frac{4}{ab}\int_0^a d\lambda\int_0^b f(\lambda,\mu)\sin\frac{m\pi\lambda}{a}\sin\frac{n\pi\mu}{b}d\mu. \tag{17}$$

is our required solution.

EXAMPLES.

1. Show that if the membrane starts from its position of equilibrium but with a given initial velocity impressed upon each point so that $z = 0$ when $t = 0$ and $D_t z = F(x,y)$ when $t = 0$ the solution is

$$z = \frac{1}{c\pi}\sum_{m=1}^{m=\infty}\sum_{n=1}^{n=\infty}\left(A_{m,n}\frac{1}{\sqrt{\frac{m^2}{a^2}+\frac{n^2}{b^2}}}\sin\frac{m\pi x}{a}\sin\frac{n\pi y}{b}\sin c\pi t\sqrt{\frac{m^2}{a^2}+\frac{n^2}{b^2}}\right)$$

where
$$A_{m,n} = \frac{4}{ab}\int_0^a d\lambda\int_0^b F(\lambda,\mu)\sin\frac{m\pi\lambda}{a}\sin\frac{n\pi\mu}{b}d\mu.$$

2. If there is both initial distortion and initial velocity

$$z = \frac{4}{ab}\sum_{m=1}^{m=\infty}\sum_{n=1}^{n=\infty}\sin\frac{m\pi x}{a}\sin\frac{n\pi y}{b}\left[A_{m,n}\cos c\pi t\sqrt{\frac{m^2}{a^2}+\frac{n^2}{b^2}}\right.$$
$$\left.+B_{m,n}\sin c\pi t\sqrt{\frac{m^2}{a^2}+\frac{n^2}{b^2}}\right]$$

where
$$A_{m,n} = \int_0^a d\lambda\int_0^b f(\lambda,\mu)\sin\frac{m\pi\lambda}{a}\sin\frac{n\pi\mu}{b}d\mu,$$

and
$$B_{m,n} = \frac{1}{c\pi\sqrt{\frac{m^2}{a^2}+\frac{n^2}{b^2}}}\int_0^a d\lambda\int_0^b F(\lambda,\mu)\sin\frac{m\pi\lambda}{a}\sin\frac{n\pi\mu}{b}d\mu.$$

3. Obtain a particular solution of (1) Art. 71 by assuming $z = T.X.Y$ where T is a function of t alone, X of x alone, and Y of y alone.

72. A number of interesting conclusions can be drawn from the results of Art. 71 and Exs. 1 and 2.

(a) No one of the three values of z is in general a periodic function of t, and consequently a vibrating rectangular membrane will not in general give a musical note.

(b) A stretched rectangular membrane can be made to give a musical note by starting the vibration properly. For if the initial circumstances are such that the solution reduces to a single term, as will be the case if the initial distortion in the problem of Art. 71 be such that $f(x,y) = A_{m,n}\sin\frac{m\pi x}{a}\sin\frac{n\pi y}{b}$, or the initial velocity in Ex. 1 be such that $F(x,y) = B_{m,n}\sin\frac{m\pi x}{a}\sin\frac{n\pi y}{b}$, or the initial distortion and initial velocity in Ex. 2 be the values just given, then the vibration will be periodic and will have the period

$$T = \frac{2}{c\sqrt{\frac{m^2}{a^2}+\frac{n^2}{b^2}}}. \tag{1}$$

Since T is a function of m and n and m and n are any whole numbers, the same membrane is capable of giving a great variety of musical notes of different pitches. If m and n are both unity we get the lowest note the membrane can give, which is called its fundamental note. Its period

$$T_1 = \frac{2}{c\sqrt{\frac{1}{a^2}+\frac{1}{b^2}}} = \frac{2ab}{c\sqrt{a^2+b^2}} \tag{2}$$

If m and n are both equal to k we get

$$T_k = \frac{2ab}{kc\sqrt{a^2+b^2}}; \tag{3}$$

therefore the membrane can be made to give any harmonic of its fundamental note.

More than this, since as we have seen

$$T_{m,n} = \frac{2}{c\sqrt{\frac{m^2}{a^2}+\frac{n^2}{b^2}}}$$

is the period of any note the membrane can give, and since if m and n are replaced by mk and nk we get

$$T_{mk,nk} = \frac{2}{ck\sqrt{\frac{m^2}{a^2}+\frac{n^2}{b^2}}}$$

the membrane can sound all the harmonics of any note which it can give.

(c) In the case considered above, where the solution reduces to the single term

$$z = \sin\frac{m\pi x}{a}\sin\frac{n\pi y}{b}\left[A_{m,n}\cos c\pi t\sqrt{\frac{m^2}{a^2}+\frac{n^2}{b^2}} + B_{m,n}\sin c\pi t\sqrt{\frac{m^2}{a^2}+\frac{n^2}{b^2}}\right],$$

if $x = \frac{a}{m}$, or $\frac{2a}{m}$, or $\frac{3a}{m}\cdots$ or $\frac{(m-1)a}{m}$, $z = 0$ for all values of t, and the lines $x = \frac{a}{m}$, $x = \frac{2a}{m}$, $\cdots$ $x = \frac{(m-1)a}{m}$ remain at rest during the whole vibration and are nodes. The same thing is true of the lines

$$y = \frac{b}{n},\ y = \frac{2b}{n},\ y = \frac{3b}{n},\ \cdots\ y = \frac{(n-1)b}{n}.$$

73. If the membrane is square it may have much more complicated nodes than if the length and breadth are unequal, as in this case the period of any term of the general solution reduces to

$$T = \frac{2a}{c\sqrt{m^2+n^2}} \tag{1}$$

and there will in general be two terms having the same period, and a musical note of the pitch corresponding to that period may be produced by initial circumstances that bring in both terms. Thus

$$\begin{aligned} z = {} & \sin\frac{m\pi x}{a}\sin\frac{n\pi y}{a}\left[A_{m,n}\cos\frac{c\pi t}{a}\sqrt{m^2+n^2} + B_{m,n}\sin\frac{c\pi t}{a}\sqrt{m^2+n^2}\right] \\ & + \sin\frac{n\pi x}{a}\sin\frac{m\pi y}{a}\left[A_{n,m}\cos\frac{c\pi t}{a}\sqrt{m^2+n^2} + B_{n,m}\sin\frac{c\pi t}{a}\sqrt{m^2+n^2}\right] \end{aligned}$$

is a form of vibration that will give a musical note. Let us write this

$$\begin{aligned} z = {} & \cos\frac{c\pi t}{a}\sqrt{m^2+n^2}\left[A\sin\frac{m\pi x}{a}\sin\frac{n\pi y}{a} + B\sin\frac{n\pi x}{a}\sin\frac{m\pi y}{a}\right] \\ & + \sin\frac{c\pi t}{a}\sqrt{m^2+n^2}\left[C\sin\frac{m\pi x}{a}\sin\frac{n\pi y}{a} + D\sin\frac{n\pi x}{a}\sin\frac{m\pi y}{a}\right] \end{aligned} \tag{2}$$

and in studying the forms of musical vibration of which the membrane is capable we may take A, B, C, and D at pleasure. Consider the simple case where $A = C$ and $B = D$; then (2) reduces to

$$\begin{aligned} z = {} & \left(A\sin\frac{m\pi x}{a}\sin\frac{n\pi y}{a} + B\sin\frac{n\pi x}{a}\sin\frac{m\pi y}{a}\right)\left(\cos\frac{c\pi t}{a}\sqrt{m^2+n^2}\right. \\ & \left. + \sin\frac{c\pi t}{a}\sqrt{m^2+n^2}\right). \end{aligned} \tag{3}$$

Values of x and y that will reduce the first parenthesis in (3) to zero will correspond to points of the membrane remaining motionless during the vibration.

Let us consider a few cases at length.

(a) If $m = 1$ and $n = 1$, the first parenthesis in (3) becomes

$$(A+B)\sin\frac{\pi x}{a}\sin\frac{\pi y}{a},$$

which is equal to zero only when $x = 0$ or $y = 0$, or $x = a$ or $y = a$, that is, for the four edges of the membrane. If, then, the membrane is sounding its fundamental note it has no nodes.

(b) If $m = 1$ and $n = 2$, we have

$$A\sin\frac{\pi x}{a}\sin\frac{2\pi y}{a} + B\sin\frac{2\pi x}{a}\sin\frac{\pi y}{a} = 0$$

to give the nodes.

Let $B = 0$, then $\sin\frac{\pi x}{a}\sin\frac{2\pi y}{a} = 0$, which is satisfied by $y = \frac{a}{2}$; and in addition to the edges the line $y = \frac{a}{2}$ is at rest and is a node.

If $A = 0$ $\quad x = \frac{a}{2}$ is a node.

If $A = B$

$$\sin\frac{\pi x}{a}\sin\frac{2\pi y}{a} + \sin\frac{2\pi x}{a}\sin\frac{\pi y}{a} = 0$$

$$2\sin\frac{\pi x}{a}\sin\frac{\pi y}{a}\cos\frac{\pi y}{a} + 2\sin\frac{\pi x}{a}\cos\frac{\pi x}{a}\sin\frac{\pi y}{a} = 0$$

$$\sin\frac{\pi x}{a}\sin\frac{\pi y}{a}\left(\cos\frac{\pi y}{a} + \cos\frac{\pi x}{a}\right) = 0.$$

The first factor gives the four edges of the membrane. The second written equal to zero gives

$$\cos\frac{\pi y}{a} = -\cos\frac{\pi x}{a} = \cos\left(\pi - \frac{\pi x}{a}\right)$$

$$\frac{\pi y}{a} = \pi - \frac{\pi x}{a}$$

$$x + y = a,$$

which is a diagonal of the square.

If $B = -A$

$$\sin\frac{\pi x}{a}\sin\frac{2\pi y}{a} - \sin\frac{2\pi x}{a}\sin\frac{\pi y}{a} = 0$$

$$\cos\frac{\pi y}{a} = \cos\frac{\pi x}{a}$$

$$x - y = 0,$$

which is the other diagonal of the square.

Other relations between A and B will give Trigonometric curves of the form

$$\cos\frac{\pi y}{a} = -\frac{B}{A}\cos\frac{\pi x}{a}$$

which are easily constructed and which obviously all agree in passing through the middle point of the square.

We give the figures for a few of the cases

$A=0$

$B=0$

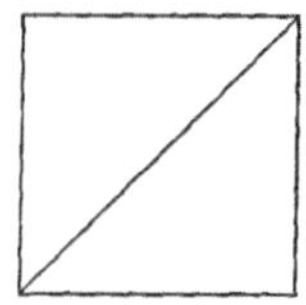

$A=-B$

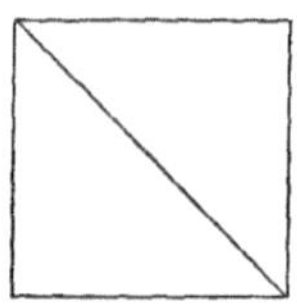

$A=B$

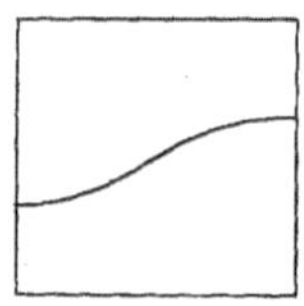

$A=2B$

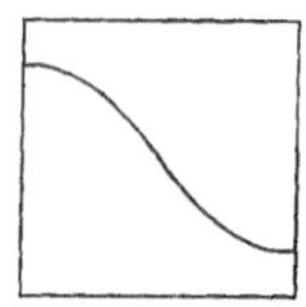

$A=\frac{1}{2}B$

(*c*) If $m = n = 2$ we have

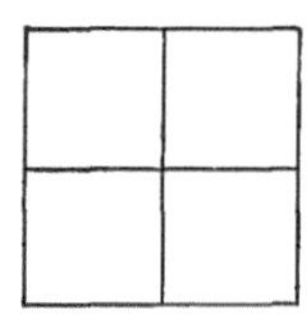

$$(A+B)\sin\frac{2\pi x}{a}\sin\frac{2\pi y}{a} = 0$$

to give the nodes, which are merely the lines

$$x = \frac{a}{2}, \quad \text{and} \quad y = \frac{a}{2}.$$

This form gives the octave of the fundamental note.

(*d*) If $m = 1$ and $n = 3$ we have

$$A\sin\frac{\pi x}{a}\sin\frac{3\pi y}{a} + B\sin\frac{3\pi x}{a}\sin\frac{\pi y}{a} = 0$$

to give the nodes.

If $A = 0$ we get $\quad x = \frac{a}{3}$ and $x = \frac{2a}{3} \quad$ (1)

If $B = 0$ we get $\quad y = \frac{a}{3}$ and $y = \frac{2a}{3} \quad$ (2)

If $A = -B$ we get

$$\sin\frac{\pi x}{a}\sin\frac{3\pi y}{a} - \sin\frac{3\pi x}{a}\sin\frac{\pi y}{a} = 0$$

$$\sin\frac{\pi x}{a}\sin\frac{\pi y}{a}\left[4\cos^2\frac{\pi y}{a} - 1 - 4\cos^2\frac{\pi x}{a} + 1\right] = 0$$

$$\cos^2\frac{\pi y}{a} - \cos^2\frac{\pi x}{a} = 0$$

$$\left(\cos\frac{\pi y}{a} - \cos\frac{\pi x}{a}\right)\left(\cos\frac{\pi y}{a} + \cos\frac{\pi x}{a}\right) = 0$$

or $$x - y = 0 \quad \text{and} \quad x + y = a. \tag{3}$$

If $A = B$ we get $$\cos^2\frac{\pi y}{a} + \cos^2\frac{\pi x}{a} = \frac{1}{2}$$

or $$\cos\frac{2\pi y}{a} + \cos\frac{2\pi x}{a} = -1, \tag{4}$$

a Trigonometric curve easily constructed.

For other relations between A and B we get more complicated Trigonometric curves coming under the general form

$$A\cos\frac{2\pi y}{a} + B\cos\frac{2\pi x}{a} = -\frac{A+B}{2} \tag{5}$$

which all agree in containing the points

$$\left(\frac{a}{3},\frac{a}{3}\right),\left(\frac{a}{3},\frac{2a}{3}\right),\left(\frac{2a}{3},\frac{a}{3}\right), \text{ and } \left(\frac{2a}{3},\frac{2a}{3}\right).$$

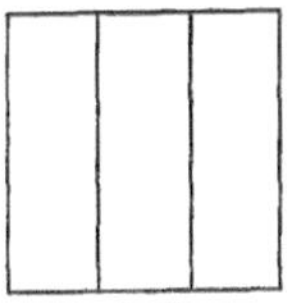
$A=0$

$B=0$

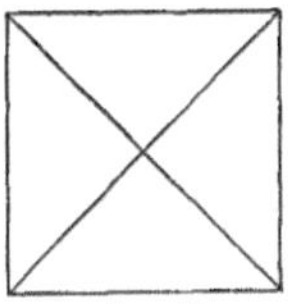
$A=-B$

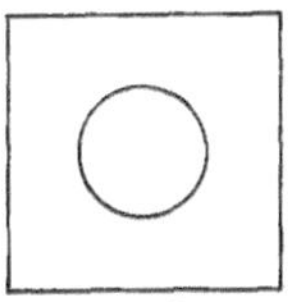
$A=B$

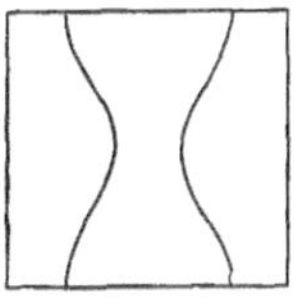
$B=-2A$

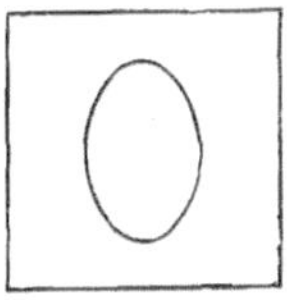
$B=2A$

MISCELLANEOUS PROBLEMS.

I. *Logarithmic Potential. Polar Coördinates.*

1. Show that $D_x^2V + D_y^2V = 0$ becomes

$$D_r^2V + \frac{1}{r}D_rV + \frac{1}{r^2}D_\phi^2V = 0$$

if we transform to Polar Coördinates.

2. If in
$$D_r^2V + \frac{1}{r}D_rV + \frac{1}{r^2}D_\phi^2V = 0 \qquad (1)$$
we let $V = R.\Phi$ we get

$$\left.\begin{array}{l}\Phi = A\cos\alpha\phi + B\sin\alpha\phi\\ R = A_1r^\alpha + B_1r^{-\alpha}\end{array}\right\} \quad \text{or} \quad \left.\begin{array}{l}\Phi = Ae^{\alpha\phi} + Be^{-\alpha\phi}\\ R = A_1\cos(\alpha\log r) + B_1\sin(\alpha\log r);\end{array}\right\}$$

whence

$V = r^\alpha\cos\alpha\phi$	$V = e^{\alpha\phi}\cos(\alpha\log r)$	$V = \cosh\alpha\phi\cos(\alpha\log r)$
$V = r^\alpha\sin\alpha\phi$	$V = e^{\alpha\phi}\sin(\alpha\log r)$	$V = \cosh\alpha\phi\sin(\alpha\log r)$
$V = \dfrac{1}{r^\alpha}\cos\alpha\phi$	$V = e^{-\alpha\phi}\cos(\alpha\log r)$	$V = \sinh\alpha\phi\cos(\alpha\log r)$
$V = \dfrac{1}{r^\alpha}\sin\alpha\phi$	$V = e^{-\alpha\phi}\sin(\alpha\log r)$	$V = \sinh\alpha\phi\sin(\alpha\log r)$

are particular solutions of (1).

3. Show that if V satisfies (1) Ex. 2 and $V = f(\phi)$ when $r = a$

$$V = \frac{1}{2}b_0 + \sum_{m=1}^{m=\infty}\left(\frac{r}{a}\right)^m(b_m\cos m\phi + a_m\sin m\phi) \quad \text{for} \quad r < a$$

and
$$V = \frac{1}{2}b_0 + \sum_{m=1}^{m=\infty}\left(\frac{a}{r}\right)^m(b_m\cos m\phi + a_m\sin m\phi) \quad \text{for} \quad r > a,$$

where $b_m = \dfrac{1}{\pi}\displaystyle\int_{-\pi}^{\pi} f(\phi)\cos m\phi.d\phi$ and $a_m = \dfrac{1}{\pi}\displaystyle\int_{-\pi}^{\pi} f(\phi)\sin m\phi.d\phi$

4. Show that if V satisfies (1) Ex. 2 and $V = f(r)$ when $\phi = 0$ and $r > 0$

$$V = \frac{1}{\pi}\int_{-\infty}^{\infty} f(e^\lambda)d\lambda\int_0^\infty \frac{\cosh\alpha(\pi-\phi)}{\cosh\alpha\pi}\cos\alpha(\lambda - \log r).d\alpha$$

$$= \frac{1}{\pi}\sin\frac{\phi}{2}\int_{-\infty}^{\infty} f(e^\lambda)\frac{\cosh\frac{1}{2}(\lambda-\log r)}{\cosh(\lambda-\log r)-\cos\phi}d\lambda.$$

5. If $V = 1$ when $\phi = 0$ and $0 < r < 1$, and $V = 0$ when $\phi = 0$ and $r > 1$

$$V = \frac{1}{\pi}\left\{\frac{\pi}{2} - \tan^{-1}\left[\frac{\sinh\dfrac{\log r}{2}}{\sin\dfrac{\phi}{2}}\right]\right\} = \frac{1}{\pi}\left[\frac{\pi}{2} - \tan^{-1}\left(\frac{r-1}{2\sqrt{r}.\sin\dfrac{\phi}{2}}\right)\right].$$

6. If $V = f(r)$ when $\phi = 0$ and $V = 0$ when $\phi = \beta$

$$V = \frac{1}{\pi}\int_{-\infty}^{\infty} f(e^\lambda)d\lambda \int_0^\infty \frac{\sinh(\beta-\phi)\alpha}{\sinh\beta\alpha}\cos\alpha(\lambda - \log r).d\alpha$$

$$= \frac{1}{2\beta}\sin\frac{\pi\phi}{\beta}\int_{-\infty}^{\infty}\frac{f(e^\lambda)d\lambda}{\cosh\dfrac{\pi}{\beta}(\lambda-\log r) - \cos\dfrac{\pi}{\beta}\phi},$$

if $0 < \phi < \beta$.

7. If $V = 0$ when $\phi = 0$ and $V = F(r)$ when $\phi = \beta$

$$V = \frac{1}{\pi}\int_{-\infty}^{\infty} F(e^\lambda)d\lambda \int_0^\infty \frac{\sinh\phi\alpha}{\sinh\beta\alpha}\cos\alpha(\lambda - \log r).d\alpha$$

$$= \frac{1}{2\beta}\sin\frac{\pi\phi}{\beta}\int_{-\infty}^{\infty}\frac{F(e^\lambda)d\lambda}{\cosh\dfrac{\pi}{\beta}(\lambda-\log r) + \cos\dfrac{\pi}{\beta}\phi}.$$

8. If $V = \chi(r)$ when $\phi = 0$ and $r < a$, $V = 0$ when $\phi = \beta$, and $V = 0$ when $r = a$

$$V = \frac{1}{2\beta}\sin\frac{\pi\phi}{\beta}\int_{-\infty}^{0}\chi(ae^\lambda)\left[\frac{d\lambda}{\cosh\dfrac{\pi}{\beta}\left(\lambda - \log\dfrac{r}{a}\right) - \cos\dfrac{\pi\phi}{\beta}} - \frac{d\lambda}{\cosh\dfrac{\pi}{\beta}\left(\lambda + \log\dfrac{r}{a}\right) - \cos\dfrac{\pi\phi}{\beta}}\right].$$

9. If $V = 0$ when $r = 1$, $V = 1$ when $\phi = 0$, $V = 0$ when $\phi = \dfrac{\pi}{2}$

$$V = \frac{2}{\pi}\tan^{-1}\left[\frac{1-r^2}{1+r^2}\operatorname{ctn}\phi\right].$$

10. If $V = 0$ when $r = 1$, $V = 1$ when $\phi = 0$, $V = 1$ when $\phi = \dfrac{\pi}{2}$

$$V = \frac{2}{\pi}\tan^{-1}\left[\frac{1-r^4}{2r^2\sin 2\phi}\right].$$

11. If $V = f(\phi)$ when $r = a$, $V = 0$ when $\phi = 0$, and $V = 0$ when $\phi = \beta$

$$V = \sum_{m=1}^{m=\infty} a_m \left(\frac{r}{a}\right)^{\frac{m\pi}{\beta}} \sin \frac{m\pi\phi}{\beta} \quad \text{if} \quad r < a$$

$$V = \sum_{m=1}^{m=\infty} a_m \left(\frac{a}{r}\right)^{\frac{m\pi}{\beta}} \sin \frac{m\pi\phi}{\beta} \quad \text{if} \quad r > a$$

where
$$a_m = \frac{2}{\beta}\int_0^\beta f(\phi) \sin \frac{m\pi\phi}{\beta} d\phi \quad \text{and} \quad 0 < \phi < \beta.$$

12. If $V = f(\phi)$ when $r = a$, $V = 0$ when $r = b$, $V = 0$ when $\phi = 0$, and $V = 0$ when $\phi = \beta$, then if $a < r < b$ and $0 < \phi < \beta$

$$V = \sum_{m=1}^{m=\infty} \left\{ \frac{a^{\frac{m\pi}{\beta}} b^{\frac{m\pi}{\beta}}}{a^{\frac{2m\pi}{\beta}} - b^{\frac{2m\pi}{\beta}}} \left[\left(\frac{r}{b}\right)^{\frac{m\pi}{\beta}} - \left(\frac{b}{r}\right)^{\frac{m\pi}{\beta}} \right] a_m \sin \frac{m\pi\phi}{\beta} \right\}$$

where
$$a_m = \frac{2}{\beta}\int_0^\beta f(\phi) \sin \frac{m\pi\phi}{\beta} d\phi.$$

13. If $V = F(\phi)$ when $r = b$, $V = 0$ when $r = a$, $V = 0$ when $\phi = 0$, and $V = 0$ when $\phi = \beta$, then if $a < r < b$ and $0 < \phi < \beta$

$$V = \sum_{m=1}^{m=\infty} \left\{ \frac{a^{\frac{m\pi}{\beta}} b^{\frac{m\pi}{\beta}}}{b^{\frac{2m\pi}{\beta}} - a^{\frac{2m\pi}{\beta}}} \left[\left(\frac{r}{a}\right)^{\frac{m\pi}{\beta}} - \left(\frac{a}{r}\right)^{\frac{m\pi}{\beta}} \right] a_m \sin \frac{m\pi\phi}{\beta} \right\}$$

where
$$a_m = \frac{2}{\beta}\int_0^\beta F(\phi) \sin \frac{m\pi\phi}{\beta} d\phi.$$

14. If $V = \chi(r)$ when $\phi = 0$, $V = 0$ when $\phi = \beta$, $V = 0$ when $r = a$, and $V = 0$ when $r = b$, then if $a < r < b$ and $0 < \phi < \beta$

$$V = \sum_{m=1}^{m=\infty} \left\{ a_m \frac{\sinh \dfrac{m\pi(\beta - \phi)}{\log b - \log a}}{\sinh \dfrac{m\pi\beta}{\log b - \log a}} \sin \frac{m\pi(\log r - \log a)}{\log b - \log a} \right\}$$

where
$$a_m = \frac{2}{\log b - \log a} \int_0^{\log \frac{b}{a}} \chi(ae^x) \sin \frac{m\pi x}{\log b - \log a} dx.$$

15. If $V = \psi(r)$ when $\phi = \beta$, $V = 0$ when $\phi = 0$, $V = 0$ when $r = a$, and $V = 0$ when $r = b$, then if $a < r < b$ and $0 < \phi < \beta$

$$V = \sum_{m=1}^{m=\infty} \left\{ a_m \frac{\sinh \dfrac{m\pi\phi}{\log b - \log a}}{\sinh \dfrac{m\pi\beta}{\log b - \log a}} \sin \frac{m\pi(\log r - \log a)}{\log b - \log a} \right\}$$

where $$a_m = \frac{2}{\log b - \log a}\int_0^{\log\frac{b}{a}} \psi(ae^x)\sin\frac{m\pi x}{\log b - \log a}dx.$$

II. *Potential Function in Space.*

1. Show that

$$f(x,y) = \frac{1}{\pi^2}\int_0^\infty d\alpha\int_0^\infty d\beta\int_0^\infty d\lambda\int_0^\infty f(\lambda,\mu)\cos\alpha(\lambda-x)\cos\beta(\mu-y).d\mu,$$

for all values of x and y.

2. Find particular solutions of $D_x^2V + D_y^2V + D_z^2V = 0$ in the forms

$$V = e^{\pm z\sqrt{\alpha^2+\beta^2}}\cos(\alpha x \pm \beta y)$$
$$V = e^{\pm z\sqrt{\alpha^2+\beta^2}}\sin(\alpha x \pm \beta y)$$
$$V = \sinh z\sqrt{\alpha^2+\beta^2}.\sin(\alpha x \pm \beta y)$$
$$V = \cosh z\sqrt{\alpha^2+\beta^2}.\sin(\alpha x \pm \beta y)$$

&c.

3. Given $D_x^2V + D_y^2V + D_z^2V = 0$, and $V = f(x,y)$ when $z = 0$, solve for positive values of z.

Result: $$V = \frac{1}{2\pi}\int_{-\infty}^\infty d\lambda\int_{-\infty}^\infty \frac{zf(\lambda,\mu)d\mu}{[z^2+(\lambda-x)^2+(\mu-y)^2]^{\frac{3}{2}}}.$$

4. Confirm the result of the last example by showing that if $f(x,y)$ is independent of y

$$V = \frac{1}{\pi}\int_{-\infty}^\infty \frac{zf(\lambda,\mu)d\lambda}{z^2+(\lambda-x)^2} \qquad \text{(v. Ex. 3 Art. 45).}$$

5. If $D_x^2V + D_y^2V + D_z^2V = 0$, and $V = 1$ when $z = 0$ for all points within the rectangle bounded by the lines $x = a$, $x = -a$, $y = b$, and $y = -b$; and $V = 0$ when $z = 0$ for all points outside of this rectangle, then

$$2\pi V =$$

$$\frac{b-y}{\sqrt{(b-y)^2}}\left\{\frac{\pi}{2} + \frac{1}{2}\sin^{-1}\frac{(a-x)^2(b-y)^2 - z^2[(a-x)^2+(b-y)^2+z^2]}{(a-x)^2(b-y)^2 + z^2[(a-x)^2+(b-y)^2+z^2]}\right.$$
$$\left. + \frac{1}{2}\sin^{-1}\frac{(a+x)^2(b-y)^2 - z^2[(a+x)^2+(b-y)^2+z^2]}{(a+x)^2(b-y)^2 + z^2[(a+x)^2+(b-y)^2+z^2]}\right\}$$
$$+\frac{b+y}{\sqrt{(b+y)^2}}\left\{\frac{\pi}{2} + \frac{1}{2}\sin^{-1}\frac{(a-x)^2(b+y)^2 - z^2[(a-x)^2+(b+y)^2+z^2]}{(a-x)^2(b+y)^2 + z^2[(a-x)^2+(b+y)^2+z^2]}\right.$$
$$\left. + \frac{1}{2}\sin^{-1}\frac{(a+x)^2(b+y)^2 - z^2[(a+x)^2+(b+y)^2+z^2]}{(a+x)^2(b+y)^2 + z^2[(a+x)^2+(b+y)^2+z^2]}\right\}$$

if $-a < x < a$, and

$$4\pi V = \frac{b-y}{\sqrt{(b-y)^2}}\left\{\sin^{-1}\frac{(a-x)^2(b-y)^2 - z^2[(a-x)^2+(b-y)^2+z^2]}{(a-x)^2(b-y)^2 + z^2[(a-x)^2+(b-y)^2+z^2]}\right.$$
$$\left.-\sin^{-1}\frac{(a+x)^2(b-y)^2 - z^2[(a+x)^2+(b-y)^2+z^2]}{(a+x)^2(b-y)^2 + z^2[(a+x)^2+(b-y)^2+z^2]}\right\}$$
$$+\frac{b+y}{\sqrt{(b+y)^2}}\left\{\sin^{-1}\frac{(a-x)^2(b+y)^2 - z^2[(a-x)^2+(b+y)^2+z^2]}{(a-x)^2(b+y)^2 + z^2[(a-x)^2+(b+y)^2+z^2]}\right.$$
$$\left.-\sin^{-1}\frac{(a+x)^2(b+y)^2 - z^2[(a+x)^2+(b+y)^2+z^2]}{(a+x)^2(b+y)^2 + z^2[(a+x)^2+(b+y)^2+z^2]}\right\};$$

if $x < -a$ or $x > a$.

6. If the value of the potential function V is given at every point of the base of an infinite rectangular prism and if the sides of the prism are at potential zero the value of V at any point within the prism is

$$V = \frac{4}{ab}\sum_{m=1}^{m=\infty}\sum_{n=1}^{n=\infty} e^{-\pi z\sqrt{\frac{m^2}{a^2}+\frac{n^2}{b^2}}}\sin\frac{m\pi x}{a}\sin\frac{n\pi y}{b}$$
$$\int_0^a d\lambda\int_0^b f(\lambda,\,\mu)\sin\frac{m\pi\lambda}{a}\sin\frac{n\pi\mu}{b}d\mu.$$

If $V = 1$ on the base of the prism this reduces to

$$V = \frac{16}{\pi^2}\sum_{m=0}^{m=\infty}\sum_{n=0}^{n=\infty} e^{-\pi z\sqrt{\frac{(2m+1)^2}{a^2}+\frac{(2n+1)^2}{b^2}}}\frac{\sin\frac{(2m+1)\pi x}{a}\sin\frac{(2n+1)\pi y}{b}}{(2m+1)(2n+1)}.$$

7. If the value of the potential function on five faces of a rectangular parallelopiped, whose length, breadth, and height are a, b, and c, is zero, and if the value of V is given for every point of the sixth face, then for any point within the parallelopiped

$$V = \sum_{m=1}^{m=\infty}\sum_{n=1}^{n=\infty} A_{m,n}\frac{\sinh\pi(c-z)\sqrt{\frac{m^2}{a^2}+\frac{n^2}{b^2}}}{\sinh\pi c\sqrt{\frac{m^2}{a^2}+\frac{n^2}{b^2}}}\sin\frac{m\pi x}{a}\sin\frac{n\pi y}{b}$$

where
$$A_{m,n} = \frac{4}{ab}\int_0^a d\lambda\int_0^b f(\lambda,\,\mu)\sin\frac{m\pi\lambda}{a}\sin\frac{n\pi\mu}{b}d\mu.$$

8. If the value of the potential function is given on two opposite faces of a rectangular parallelopiped and is zero on the four remaining faces, then within

the parallelopiped

$$V=\sum_{m=1}^{m=\infty}\sum_{n=1}^{n=\infty}A_{m,n}\frac{\sinh\pi(c-z)\sqrt{\frac{m^2}{a^2}+\frac{n^2}{b^2}}}{\sinh\pi c\sqrt{\frac{m^2}{a^2}+\frac{n^2}{b^2}}}\sin\frac{m\pi x}{a}\sin\frac{n\pi y}{b}$$

$$+\sum_{m=1}^{m=\infty}\sum_{n=1}^{n=\infty}B_{m,n}\frac{\sinh\pi z\sqrt{\frac{m^2}{a^2}+\frac{n^2}{b^2}}}{\sinh\pi c\sqrt{\frac{m^2}{a^2}+\frac{n^2}{b^2}}}\sin\frac{m\pi x}{a}\sin\frac{n\pi y}{b}$$

where $$A_{m,n}=\frac{4}{ab}\int_0^a d\lambda\int_0^b f(\lambda,\,\mu)\sin\frac{m\pi\lambda}{a}\sin\frac{n\pi\mu}{b}d\mu$$

and $$B_{m,n}=\frac{4}{ab}\int_0^a d\lambda\int_0^b F(\lambda,\,\mu)\sin\frac{m\pi\lambda}{a}\sin\frac{n\pi\mu}{b}d\mu.$$

9. If the value of the potential function is given at every point on the surface of a rectangular parallelopiped, what is its value at any point within the parallelopiped?

III. *Conduction of Heat in a Plane.*

1. Find particular solutions of $D_t u=a^2(D_x^2u+D_y^2u)$ of the forms

$$u=e^{-a^2(\alpha^2+\beta^2)t}\sin(\alpha x\pm\beta y)$$
$$u=e^{-a^2(\alpha^2+\beta^2)t}\cos(\alpha x\pm\beta y).$$

2. Given the initial temperature of every point in a thin plane plate, find the temperature of any point at any time,

$$u=\frac{1}{4a^2\pi t}\int_{-\infty}^{\infty}d\lambda\int_{-\infty}^{\infty}e^{-\frac{(\lambda-x)^2+(\mu-y)^2}{4a^2t}}f(\lambda,\,\mu)d\mu$$
$$=\frac{1}{\pi}\int_{-\infty}^{\infty}e^{-\beta^2}d\beta\int_{-\infty}^{\infty}e^{-\gamma^2}f(x+2a\sqrt{t}.\beta,\,y+2a\sqrt{t}.\gamma)d\gamma.$$

3. For an instantaneous *source* of strength Q at (λ,μ)

$$u=\frac{Q}{4\pi a^2t}e^{-\frac{(\lambda-x)^2+(\mu-y)^2}{4a^2t}}$$ v. Art. 53.

For an instantaneous *doublet* of strength P at $(0,\mu)$ with its axis perpendicular to the axis of Y

$$u=\frac{Px}{8\pi a^4t^2}e^{-\frac{x^2+(\mu-y)^2}{4a^2t}}$$ v. Art. 54.

For a permanent doublet of strength P at $(0, \mu)$ with its axis perpendicular to the axis of Y

$$u = \frac{P}{2\pi a^2} \frac{x}{x^2 + (\mu - y)^2} e^{-\frac{x^2+(\mu-y)^2}{4a^2 t}}.$$

If the strength of the doublet were $Pd\mu$ and the heat were uniformly generated and absorbed along the element $d\mu$ of the axis of Y beginning at $(0, \mu)$ we should have

$$u = \frac{P}{2\pi a^2} e^{-\frac{x^2+(\mu-y)^2}{4a^2 t}} \frac{x d\mu}{x^2 + (\mu - y)^2} = \frac{P}{2\pi a^2} e^{-\frac{x^2+(\mu-y)^2}{4a^2 t}} d\tan^{-1}\frac{\mu - y}{x},$$

and since $d\tan^{-1}\dfrac{\mu - y}{x}$ is the angle ARA', where A and A' are the points $(0, \mu)$ and $(0, \mu{+}d\mu)$ and R is the point (x, y), $u = 0$ when $x = 0$ unless $\mu < y < \mu{+}d\mu$, in which case $u = \dfrac{P}{2a^2}$ if x approaches zero from the positive side; and $u = 0$ when $t = 0$ except in the element $d\mu$. If then $u = 0$ when $t = 0$ and $u = f(y)$ when $x = 0$ we have only to suppose a doublet of strength $2a^2 f(x)dx$ placed in each element of the axis of Y and then to integrate; we get

$$u = \frac{1}{\pi}\int_{-\infty}^{\infty} e^{-\frac{x^2+(\mu-y)^2}{4a^2 t}} \frac{x f(\mu)}{x^2 + (\mu - y)^2} d\mu.$$

For a permanent doublet of strength $F(t)$ at $(0, \mu)$ we have

$$u = \frac{x}{8\pi a^4}\int_0^t e^{-\frac{x^2+(\mu-y)^2}{4a^2(t-\tau)}} (t-\tau)^{-2} F(\tau) d\tau.$$

$$= \frac{1}{2\pi a^2}\left[\frac{xF(0)}{x^2 + (\mu - y)^2} e^{-\frac{x^2+(\mu-y)^2}{4a^2 t}} + \int_0^t \frac{xF'(\tau)}{x^2 + (\mu - y)^2} e^{-\frac{x^2+(\mu-y)^2}{4a^2(t-\tau)}} d\tau\right].$$

From the reasoning above this must be zero when $t = 0$ except at the point $(0, \mu)$, must be $2a^2F(t)$ at the point $(0, \mu)$, and 0 at every other point of the axis of Y when t is not zero.

Hence if $u = 0$ when $t = 0$ and $u = F(y, t)$ when $x = 0$

$$u = \frac{1}{\pi}\int_{-\infty}^{\infty} \frac{xF(\mu, 0)}{x^2 + (\mu - y)^2} e^{-\frac{x^2+(\mu-y)^2}{4a^2 t}} d\mu + \frac{1}{\pi}\int_{-\infty}^{\infty} d\mu \int_0^t \frac{xD_\tau F(\mu, \tau)}{x^2 + (\mu - y)^2} e^{-\frac{x^2+(\mu-y)^2}{4a^2(t-\tau)}} d\tau.$$

For an extension of this solution by the method of images to the case where there are other rectilinear boundaries and for its application to the corresponding problems in the flow of heat in three dimensions see E. W. Hobson in Vol. XIX. Proc. Lond. Math. Soc.

4. If the perimeter of a thin plane rectangular plate is kept at the temperature zero and the initial temperatures of all points of the plate are given, then

for any point of the plate

$$u = \frac{4}{bc}\sum_{m=1}^{m=\infty}\sum_{n=1}^{n=\infty} e^{-a^2\pi^2\left(\frac{m^2}{b^2}+\frac{n^2}{c^2}\right)t}\sin\frac{m\pi x}{b}\sin\frac{n\pi y}{c}$$

$$\int_0^b d\lambda\int_0^c f(\lambda,\mu)\sin\frac{m\pi\lambda}{b}\sin\frac{n\pi\mu}{c}d\mu.$$

if b is the length and c the breadth of the plate.

5. A large mass of iron at the temperature 0° contains an iron core in the shape of a long prism 40 cm. square. The core is removed and heated to the temperature of 100° throughout and then replaced. Find the temperature of a point in the axis of the core fifteen minutes afterward. Given $a^2 = .185$ in C.G.S. units. *Ans.*, 52°.9.

6. If the prism described in Ex. 5 after being heated to 100° has its lateral faces kept for 15 minutes at the temperature 0° find the temperature of a point in its axis. *Ans.*, 20°.8.

IV. *Conduction of Heat in Space.*

1. Show that

$$\frac{1}{\pi^3}\int_0^\infty d\alpha\int_0^\infty d\beta\int_0^\infty d\gamma\int_{-\infty}^\infty d\lambda\int_{-\infty}^\infty d\mu\int_{-\infty}^\infty f(\lambda,\mu,\nu)$$

$$\cos\alpha(\lambda-x)\cos\beta(\mu-y)\cos\gamma(\nu-z).d\nu = f(x,y,z)$$

for all values of x, y, and z.

2. Show that

$$f(x,y,z) = \sum_{m=1}^{m=\infty}\sum_{n=1}^{n=\infty}\sum_{p=1}^{p=\infty} A_{m,n,p}\sin\frac{m\pi x}{a}\sin\frac{n\pi y}{b}\sin\frac{p\pi z}{c}$$

where $$A_{m,n,p} = \frac{8}{abc}\int_0^a d\lambda\int_0^b d\mu\int_0^c f(\lambda,\mu,\nu)\sin\frac{m\pi\lambda}{a}\sin\frac{n\pi\mu}{b}\sin\frac{p\pi\nu}{c}d\nu,$$

for $0 < x < a$, $0 < y < b$, $0 < z < c$.

3. Obtain particular solutions of $D_t u = a^2(D_x^2 u + D_y^2 u + D_z^2 u)$ of the forms

$$u = e^{-a^2(\alpha^2+\beta^2+\gamma^2)t}\sin(\alpha x \pm \beta y \pm \gamma z).$$

$$u = e^{-a^2(\alpha^2+\beta^2+\gamma^2)t}\cos(\alpha x \pm \beta y \pm \gamma z).$$

4. Given the initial temperature of every point in an infinite homogeneous solid find the temperature of any point at any time.

$$u = \frac{1}{8a^3(\pi t)^{\frac{3}{2}}} \int\limits_{-\infty}^{\infty} d\lambda \int\limits_{-\infty}^{\infty} d\mu \int\limits_{-\infty}^{\infty} e^{-\frac{(\lambda-x)^2+(\mu-y)^2+(\nu-z)}{4a^2t}} f(\lambda,\mu,\nu)d\nu$$

$$= \frac{1}{\pi^{\frac{3}{2}}} \int\limits_{-\infty}^{\infty} e^{-\beta^2} d\beta \int\limits_{-\infty}^{\infty} e^{-\gamma^2} d\gamma \int\limits_{-\infty}^{\infty} e^{-\delta^2} f(x + 2a\sqrt{t}.\beta,\ y + 2a\sqrt{t}.\gamma,\ z + 2a\sqrt{t}.\delta)d\delta.$$

5. If the surface of a rectangular parallelopiped is kept at the temperature zero and the initial temperatures of all points of the parallelopiped are given, then for any point of the parallelopiped

$$u = \sum_{m=1}^{m=\infty} \sum_{n=1}^{n=\infty} \sum_{p=1}^{p=\infty} A_{m,n,p} e^{-a^2\pi^2\left(\frac{m^2}{b^2}+\frac{n^2}{c^2}+\frac{p^2}{d^2}\right)t} \sin\frac{m\pi x}{b} \sin\frac{n\pi y}{c} \sin\frac{p\pi z}{d}$$

where $$A_{m,n,p} = \frac{8}{bcd} \int\limits_0^b d\lambda \int\limits_0^c d\mu \int\limits_0^d f(\lambda,\ \mu,\ \nu) \sin\frac{m\pi\lambda}{b} \sin\frac{n\pi\mu}{c} \sin\frac{p\pi\nu}{d} d\nu.$$

6. An iron cube 40 cm. on an edge is heated to the uniform temperature of 100° Centigrade and then tightly enclosed in a large iron mass which is at the uniform temperature of 0°. Find the temperature of the centre of the cube fifteen minutes afterwards. *Ans.*, 38°.4.

7. An iron cube 40 cm. on an edge is heated to the uniform temperature of 100° and then its surface is kept for fifteen minutes at the temperature 0°. Required the temperature of its centre. *Ans.*, 9°.5.

CHAPTER V.[1]

ZONAL HARMONICS.

74. In Art. 16 we obtained

$$z = Ap_m(x) + Bq_m(x) \tag{1}$$

[v. (6) Art. 16] as the general solution of Legendre's Equation

$$(1 - x^2)\frac{d^2 z}{dx^2} - 2x\frac{dz}{dx} + m(m+1)z = 0, \tag{2}$$

m being wholly unrestricted in value and x lying between -1 and 1; where

$$\begin{aligned} p_m(x) = 1 - \frac{m(m+1)}{2!}x^2 + \frac{m(m-2)(m+1)(m+3)}{4!}x^4 \\ - \frac{m(m-2)(m-4)(m+1)(m+3)(m+5)}{6!}x^6 + \cdots \end{aligned} \tag{3}$$

and

$$\begin{aligned} q_m(x) = x - \frac{(m-1)(m+2)}{3!}x^3 + \frac{(m-1)(m-3)(m+2)(m+4)}{5!}x^5 \\ - \frac{(m-1)(m-3)(m-5)(m+2)(m+4)(m+6)}{7!}x^7 + \cdots; \end{aligned} \tag{4}$$

and we found

$$\left.\begin{aligned} V &= r^m p_m(\cos\theta) \\ V &= \frac{1}{r^{m+1}} p_m(\cos\theta) \\ V &= r^m q_m(\cos\theta) \\ V &= \frac{1}{r^{m+1}} q_m(\cos\theta), \end{aligned}\right\} \tag{5}$$

m being unrestricted in value, as particular solutions of the special form assumed by Laplace's Equation in spherical coördinates when V is independent of ϕ; that is, of the equation

$$rD_r^2(rV) + \frac{1}{\sin\theta}D_\theta(\sin\theta D_\theta V) = 0. \tag{6}$$

For the important case where m is a positive integer we found

$$z = AP_m(x) + BQ_m(x) \tag{7}$$

[1] Before reading this chapter the student is advised to re-read carefully articles 9, 10, 13(c), 15, 16, and 18(c).

[v. (10) Art. 16] as the general solution of Legendre's Equation (2), whence

$$\left.\begin{aligned} V &= r^m P_m(\cos\theta) \\ V &= \frac{1}{r^{m+1}} P_m(\cos\theta) \\ V &= r^m Q_m(\cos\theta) \\ V &= \frac{1}{r^{m+1}} Q_m(\cos\theta) \end{aligned}\right\} \tag{8}$$

are particular solutions of (6) if m is a positive integer.

$$P_m(x) = \frac{(2m-1)(2m-3)\cdots 1}{m!}\left[x^m - \frac{m(m-1)}{2(2m-1)}x^{m-2} + \frac{m(m-1)(m-2)(m-3)}{2.4.(2m-1)(2m-3)}x^{m-4} - \cdots\right] \tag{9}$$

[v. (8) Art. 16] and is a finite sum terminating with the term which involves x if m is odd and with the term involving x^0 if m is even.

It is called a *Surface Zonal Harmonic*, or a *Legendre's Coefficient*, or more briefly a *Legendrian.*

$$Q_m(x) = \frac{m!}{(2m+1)(2m-1)\cdots 1}\left[\frac{1}{x^{m+1}} + \frac{(m+1)(m+2)}{2.(2m+3)}\frac{1}{x^{m+3}} + \frac{(m+1)(m+2)(m+3)(m+4)}{2.4.(2m+3)(2m+5)}\frac{1}{x^{m+5}} + \cdots\right] \tag{10}$$

if $x < -1$ or $x > 1$. [v. (9) Art. 16.]

It is called a *Surface Zonal Harmonic* of the *second kind.*

$$\begin{aligned} Q_m(x) &= (-1)^{\frac{m+1}{2}} \frac{2^{m-1}\left[\Gamma\left(\frac{m+1}{2}\right)\right]^2}{\Gamma(m+1)} p_m(x) \\ &= (-1)^{\frac{m+1}{2}} \frac{2.4.6.\cdots(m-1)}{3.5.7.\cdots m} p_m(x) \end{aligned} \tag{11}$$

[v. (13) Art. 16] if m is odd and $-1 < x < 1$.

$$\begin{aligned} Q_m(x) &= (-1)^{\frac{m}{2}} \frac{2^{m}\left[\Gamma\left(\frac{m+1}{2}\right)\right]^2}{\Gamma(m+1)} q_m(x) \\ &= (-1)^{\frac{m}{2}} \frac{2.4.6.\cdots m}{1.3.5.\cdots(m-1)} q_m(x) \end{aligned} \tag{12}$$

[v. (14) Art. 16] if m is even and $-1 < x < 1$.

In most of the work that immediately follows we shall regard x in $P_m(x)$as equal to $\cos\theta$ and therefore as lying between -1 and 1.[2]

[2]English writers on Spherical Harmonics generally use μ in place of x for $\cos\theta$. We shall follow them, however, only when we should thereby avoid confusion.

75. In Article 9 the undetermined coefficient a_m of x^m in $P_m(x)$ was arbitrarily written in the form $\dfrac{(2m-1)(2m-3)\cdots 1}{m!}$ for reasons which shall now be given.

In Articles 9 and 16 $z = P_m(x)$ was obtained as a particular solution of Legendre's Equation

$$(1-x^2)\frac{d^2z}{dx^2} - 2x\frac{dz}{dx} + m(m+1)z = 0 \tag{1}$$

by the device of assuming that z could be expressed as a sum or a series of terms of the form $a_n x^n$ and then determining the coefficients. We can, however, obtain a particular solution of Legendre's Equation by an entirely different method.

The potential function due to a unit of mass concentrated at a given point (x_1, y_1, z_1) is

$$V = \frac{1}{\sqrt{(x-x_1)^2 + (y-y_1)^2 + (z-z_1)^2}} \tag{2}$$

and this must be a particular solution of Laplace's Equation

$$D_x^2 V + D_y^2 V + D_z^2 V = 0, \tag{3}$$

as is easily verified by direct substitution.

If we transform (2) to spherical coördinates using the formulas of transformation

$$\begin{aligned} x &= r\cos\theta \\ y &= r\sin\theta\cos\phi \\ z &= r\sin\theta\sin\phi \end{aligned}$$

we get

$$V = \frac{1}{\sqrt{r^2 - 2rr_1[\cos\theta\cos\theta_1 + \sin\theta\sin\theta_1\cos(\phi-\phi_1)] + r_1^2}} \tag{4}$$

as a solution of Laplace's Equation in Spherical Coördinates

$$rD_r^2(rV) + \frac{1}{\sin\theta}D_\theta(\sin\theta D_\theta V) + \frac{1}{\sin^2\theta}D_\phi{}^2 V = 0 \quad \text{[XIII] Art. 1.}$$

If the given point (x_1, y_1, z_1) is taken on the axis of X, as it must be that (4) may be independent of ϕ, $\theta_1 = 0$, and

$$V = \frac{1}{\sqrt{r^2 - 2rr_1\cos\theta + r_1^2}} \tag{5}$$

is a solution of

$$rD_r^2(rV) + \frac{1}{\sin\theta}D_\theta(\sin\theta D_\theta V) = 0. \tag{6}$$

Equation (5) may be written

$$V = \frac{1}{r}\frac{1}{\sqrt{1 - 2\frac{r_1}{r}\cos\theta + \frac{r_1^2}{r^2}}} \tag{7}$$

or

$$V = \frac{1}{r_1}\frac{1}{\sqrt{1 - 2\frac{r}{r_1}\cos\theta + \frac{r^2}{r_1^2}}}. \tag{8}$$

$\sqrt{1 - 2z\cos\theta + z^2}$ is finite and continuous for all values real or complex of z. It is double-valued but the two branches of the function are distinct except for the values of z which make $1 - 2z\cos\theta + z^2 = 0$ namely $z = \cos\theta + i\sin\theta$ and $z = \cos\theta - i\sin\theta$, both of which have the modulus unity and which are *critical* values.

$\frac{1}{\sqrt{1 - 2z\cos\theta + z^2}}$ is finite and continuous except for the values of $z = \cos\theta - i\sin\theta$ and $z = \cos\theta + i\sin\theta$ for which it becomes infinite; it is double-valued but has as critical values only these values of z. It is then *holomorphic* within a circle described with the origin as centre and the radius unity, and can be developed into a power series which will be convergent for all values of z having moduli less than one. (Int. Cal. Arts. 207, 212, 214, 220.)

If then $r > r_1$ $\frac{1}{\sqrt{1 - \frac{2r_1}{r}\cos\theta + \frac{r_1^2}{r^2}}}$ can be developed into a convergent series involving whole powers of $\frac{r_1}{r}$.

Let $\sum p_m \frac{r_1^m}{r^m}$ be this series, p_m, of course, being a function of $\cos\theta$. Then

$$V = \frac{1}{r}\sum p_m \frac{r_1^m}{r^m}$$

[v. (7)] is a solution of (6). Substitute this value of V in (6) and we get

$$\sum\left[\frac{r_1^m}{r^{m+1}}m(m+1)p_m + \frac{r_1^m}{r^{m+1}}\frac{1}{\sin\theta}\frac{d}{d\theta}\left(\sin\theta\frac{dp_m}{d\theta}\right)\right] = 0.$$

As this must hold whatever the value of r provided $r > r_1$ the coefficient of each power of r must be zero, and hence the equation

$$\frac{1}{\sin\theta}\frac{d}{d\theta}\left(\sin\theta\frac{dp_m}{d\theta}\right) + m(m+1)p_m = 0 \tag{9}$$

must be true.

But as we have seen in Art. 9 the substitution of $x = \cos\theta$ in (9) reduces it to

$$(1 - x^2)\frac{d^2p_m}{dx^2} - 2x\frac{dp_m}{dx} + m(m+1)p_m = 0,$$

and therefore $$z = p_m$$

is a solution of Legendre's Equation (1).

If $r < r_1$ $\dfrac{1}{\sqrt{1 - \dfrac{2r}{r_1}\cos\theta + \dfrac{r^2}{r_1^2}}}$ can be developed into a convergent series involving whole powers of $\dfrac{r^2}{r_1^2}$.

Let $\sum p_m \dfrac{r^m}{r_1^m}$ be this series. Then

$$V = \frac{1}{r_1}\sum p_m \frac{r^m}{r_1^m}$$

(v. 8) is a solution of (6); substituting in (6) we get

$$\sum\left[\frac{r^m}{r_1^{m+1}} m(m+1)p_m + \frac{r^m}{r_1^{m+1}}\frac{1}{\sin\theta}\frac{d}{d\theta}\left(\sin\theta\frac{dp_m}{d\theta}\right)\right] = 0,$$

whence it follows as before that

$$z = p_m$$

is a solution of Legendre's Equation.

But p_m is the coefficient of the mth power of $\dfrac{r}{r_1}$ in the development of $\left(1 - 2\dfrac{r}{r_1}\cos\theta + \dfrac{r^2}{r_1^2}\right)^{-\frac{1}{2}}$ according to powers of $\dfrac{r}{r_1}$, or of the mth power of $\dfrac{r_1}{r}$ in the development of $\left(1 - 2\dfrac{r_1}{r}\cos\theta + \dfrac{r_1^2}{r^2}\right)^{-\frac{1}{2}}$ according to powers of $\dfrac{r_1}{r}$, or more briefly it is the coefficient of the mth power of z in the development of $(1 - 2xz + z^2)^{-\frac{1}{2}}$ according to powers of z, x standing for $\cos\theta$.

$$(1 - 2xz + z^2)^{-\frac{1}{2}} = [1 - z(2x - z)]^{-\frac{1}{2}}$$

and can be developed by the Binomial Theorem; the coefficient of z^m is easily picked out and is

$$\frac{(2m-1)(2m-3)\cdots 1}{m!}\left[x^m - \frac{m(m-1)}{2(2m-1)}x^{m-2} + \frac{m(m-1)(m-2)(m-3)}{2.4.(2m-1)(2m-3)}x^{m-4} - \cdots\right].$$

But this is precisely $P_m(x)$. [v. Art. 74 (9)]

Hence $P_m(x)$ is equal to the coefficient of the mth power of z in the development of $[1 - 2xz + z^2]^{-\frac{1}{2}}$ into a power series, the modulus of z being less than unity.

76. If $x = 1$ $P_m(x) = 1$. For if $x = 1$ $(1 - 2xz + z^2)^{-\frac{1}{2}}$ reduces to $(1 - 2z + z^2)^{-\frac{1}{2}}$ that is to $(1 - z)^{-1}$, which develops into

$$1 + z + z^2 + z^3 + z^4 + \cdots,$$

and the coefficient of each power of z is unity. Therefore

$$P_m(1) = 1. \tag{1}$$

We have seen that if m is even $P_m(x)$ contains only even powers of x and terminates with the term involving x^0, that is with the constant term.

The value of this constant term can be picked out from the formula for $P_m(x)$ [v. Art. 74 (9)]. It is $(-1)^{\frac{m}{2}} \dfrac{1.3.5.\cdots(m-1)}{2.4.6.\cdots m}$; or it can be found as follows:—It is clearly the value $P_m(x)$ assumes when $x = 0$; it is, then, the coefficient of z^m in the development of $(1 + z^2)^{-\frac{1}{2}}$; but

$$(1 + z^2)^{-\frac{1}{2}} = 1 - \frac{1}{2}z^2 + \frac{1.3}{2.4}z^4 - \frac{1.3.5}{2.4.6}z^6 + \frac{1.3.5.7}{2.4.6.8}z^8 - \cdots$$

and the coefficient of z^m, m being an even number, is $(-1)^{\frac{m}{2}} \dfrac{1.3.5\cdots(m-1)}{2.4.6\cdots m}$.

If m is odd $P_m(x)$ contains only odd powers of x and terminates with the term involving x to the first power. The coefficient of this term can be picked out from (9) Art. 74 and is $(-1)^{\frac{m-1}{2}} \dfrac{3.5.7.\cdots m}{2.4.6.\cdots(m-1)}$; or it can be found as follows:—It is clearly the value assumed by $\dfrac{dP_m(x)}{dx}$ when $x = 0$.

It is, then, the coefficient of z^m in the development of $\dfrac{z}{(1+z^2)^{\frac{3}{2}}}$.

$$\frac{z}{(1+z^2)^{\frac{3}{2}}} = z - \frac{3}{2}z^3 + \frac{3.5}{2.4}z^5 - \frac{3.5.7}{2.4.6}z^7 + \cdots$$

and the coefficient of z^m in this development is $(-1)^{\frac{m-1}{2}} \dfrac{3.5.7\cdots m}{2.4.6\cdots(m-1)}$, m being an odd number.

77. To recapitulate:

$$\begin{aligned} P_m(x) = \frac{1.3.5\cdots(2m-1)}{m!}\Bigg[x^m &- \frac{m(m-1)}{2(2m-1)}x^{m-2} \\ &+ \frac{m(m-1)(m-2)(m-3)}{2.4.(2m-1)(2m-3)}x^{m-4} \\ &- \frac{m(m-1)(m-2)(m-3)(m-4)(m-5)}{2.4.6.(2m-1)(2m-3)(2m-5)}x^{m-6} + \cdots\Bigg], \end{aligned} \tag{1}$$

m being a positive integer, is a *Surface Zonal Harmonic* or *Legendrian* of the mth order. It is a finite sum terminating with the first power of x if m is odd, and with the zeroth power of x if m is even.

$P_m(x)$ is the coefficient of the mth power of z in the development of $(1 - 2xz + z^2)^{-\frac{1}{2}}$ into a power series. Hence if $z < 1$

$$(1 - 2xz + z^2)^{-\frac{1}{2}} = P_0(x) + P_1(x).z + P_2(x).z^2 + P_3(x).z^3 + P_4(x).z^4 + P_5(x).z^5 + \cdots + P_m(x).z^m + \cdots . \quad (2)$$

Whence

$$\left.\begin{aligned} \frac{1}{\sqrt{r^2 - 2rr_1\cos\theta + r_1^2}} &= \frac{1}{r}\left[P_0(\cos\theta) + \frac{r_1}{r}P_1(\cos\theta) + \frac{r_1^2}{r^2}P_2(\cos\theta) + \cdots \right. \\ &\qquad \left. + \frac{r_1^m}{r^m}P_m(\cos\theta) + \cdots\right] \quad \text{if} \quad r > r_1 \\ &= \frac{1}{r_1}\left[P_0(\cos\theta) + \frac{r}{r_1}P_1(\cos\theta) + \frac{r^2}{r_1^2}P_2(\cos\theta) + \cdots \right. \\ &\qquad \left. + \frac{r^m}{r_1^m}P_m(\cos\theta) + \cdots\right] \quad \text{if} \quad r < r_1. \end{aligned}\right\} \quad (3)$$

$$z = P_m(x)$$

is a solution of Legendre's Equation

$$(1 - x^2)\frac{d^2z}{dx^2} - 2x\frac{dz}{dx} + m(m+1)z = 0$$

when m is a positive integer.

$$V = r^m P_m(\cos\theta)$$

and

$$V = \frac{1}{r^{m+1}}P_m(\cos\theta)$$

are solutions of the form of Laplace's Equation in Spherical Coördinates which is independent of ϕ, namely

$$rD_r^2(rV) + \frac{1}{\sin(\theta)}D_\theta(\sin\theta D_\theta V) = 0. \quad (4)$$

$$P_m(1) = 1. \quad (5)$$

$$P_{2m}(-x) = P_{2m}(x). \quad (6)$$

$$P_{2m+1}(-x) = -P_{2m+1}(x). \quad (7)$$

$$P_{2m+1}(0) = 0. \quad (8)$$

$$P_{2m}(0) = (-1)^m\frac{1.3.5.\cdots(2m-1)}{2.4.6.\cdots 2m}. \quad (9)$$

$$\left[\frac{dP_{2m+1}(x)}{dx}\right]_{x=0} = (-1)^m\frac{3.5.7.\cdots(2m+1)}{2.4.6.\cdots 2m}. \quad (10)$$

For convenience of reference we write out a few Zonal Harmonics. They are obtained by substituting successive integers for m in formula (1).

$$\left.\begin{aligned}
P_0(x) &= 1\\
P_1(x) &= x\\
P_2(x) &= \frac{1}{2}(3x^2 - 1)\\
P_3(x) &= \frac{1}{2}(5x^3 - 3x)\\
P_4(x) &= \frac{1}{8}(35x^4 - 30x^2 + 3)\\
P_5(x) &= \frac{1}{8}(63x^5 - 70x^3 + 15x)\\
P_6(x) &= \frac{1}{16}(231x^6 - 315x^4 + 105x^2 - 5)\\
P_7(x) &= \frac{1}{16}(429x^7 - 693x^5 + 315x^3 - 35x)\\
P_8(x) &= \frac{1}{128}(6435x^8 - 12012x^6 + 6930x^4 - 1260x^2 + 35).
\end{aligned}\right\} \tag{11}$$

Any Surface Zonal Harmonic may be obtained from the two of next lower orders by the aid of the formula

$$(n+1)P_{n+1}(x) - (2n+1)xP_n(x) + nP_{n-1}(x) = 0 \tag{12}$$

which is easily obtained and is convenient when the numerical value of x is given.

Differentiate (2) with respect to z and we get

$$\frac{-(z-x)}{(1-2xz+z^2)^{\frac{3}{2}}} = P_1(x) + 2P_2(x).z + 3P_3(x).z^2 + \cdots$$

whence

$$\frac{-(z-x)}{(1-2xz+z^2)^{\frac{1}{2}}} = (1-2xz+z^2)(P_1(x) + 2P_2(x).z + 3P_3(x).z^2 + \cdots).$$

Hence by (2)

$$\begin{aligned}
(1-2xz+z^2)&(P_1(x) + 2P_2(x).z + 3P_3(x).z^2 + \cdots)\\
&+ (z-x)(P_0(x) + P_1(x).z + P_2(x).z^2 + \cdots) = 0
\end{aligned} \tag{13}$$

(13) is identically true, hence the coefficient of each power of z must vanish. Picking out the coefficient of z^n and writing it equal to zero we have formula (12) above.[3]

[3]For tables of Surface Zonal Harmonics v. Appendix Tables I and II.

78. We are now able to solve completely the problem considered in Art. 9. We were to find a solution of the differential equation

$$rD_r^2(rV) + \frac{1}{\sin\theta} D_\theta(\sin\theta D_\theta V) = 0 \tag{1}$$

subject to the condition

$$V = \frac{M}{(c^2+r^2)^{\frac{1}{2}}} \quad \text{when} \quad \theta = 0. \tag{2}$$

We know (v. Art. 77) that

$$V = r^m P_m(\cos\theta)$$

and

$$V = \frac{1}{r^{m+1}} P_m(\cos\theta)$$

are solutions of (1).

For values of $r < c$

$$\frac{M}{(c^2+r^2)^{\frac{1}{2}}} = \frac{M}{c}\left[1 - \frac{1}{2}\frac{r^2}{c^2} + \frac{1.3}{2.4}\frac{r^4}{c^4} - \frac{1.3.5}{2.4.6}\frac{r^6}{c^6} + \cdots\right]. \tag{3}$$

Therefore for values of $r < c$

$$V = \frac{M}{c}\left[P_0(\cos\theta) - \frac{1}{2}\frac{r^2}{c^2}P_2(\cos\theta) + \frac{1.3}{2.4}\frac{r^4}{c^4}P_4(\cos\theta) - \frac{1.3.5}{2.4.6}\frac{r^6}{c^6}P_6(\cos\theta) + \cdots\right] \tag{4}$$

is our required solution; because each term satisfies equation (1), and therefore the whole value satisfies (1), and when $\theta = 0$

$$P_m(\cos\theta) = P_m(1) = 1$$

[v. (5) Art. 77], and hence (4) reduces to (3) and (2) is satisfied.

For values of $r > c$

$$\frac{M}{(c^2+r^2)^{\frac{1}{2}}} = \frac{M}{r}\left[1 - \frac{1}{2}\frac{c^2}{r^2} + \frac{1.3}{2.4}\frac{c^4}{r^4} - \frac{1.3.5}{2.4.6}\frac{c^6}{r^6} + \cdots\right] \tag{5}$$

$$= M\left[\frac{1}{r} - \frac{1}{2}\frac{c^2}{r^3} + \frac{1.3}{2.4}\frac{c^4}{r^5} - \frac{1.3.5}{2.4.6}\frac{c^6}{r^7} + \cdots\right].$$

Therefore for values of $r > c$

$$V = \frac{M}{c}\left[\frac{c}{r}P_0(\cos\theta) - \frac{1}{2}\frac{c^3}{r^3}P_2(\cos\theta) + \frac{1.3}{2.4}\frac{c^5}{r^5}P_4(\cos\theta) - \frac{1.3.5}{2.4.6}\frac{c^7}{r^7}P_6(\cos\theta) + \cdots\right] \tag{6}$$

is our required solution. For it satisfies (1) and reduces to (2) when $\theta = 0$.

79. As another example let us suppose a conductor in the form of a thin circular disc charged with electricity, and let it be required to find the value of the potential function at any point in space.

If the magnitude of the charge is M and the radius of the plate is a the surface density at a point of the plate at a distance r from the centre is

$$\sigma = \frac{M}{4a\pi\sqrt{a^2 - r^2}}$$

and all points of the conductor are at the potential $\frac{\pi M}{2a}$. (v. Peirce's Newtonian Potential Function, § 61.)

The value of the potential function at a point in the axis of the plate at the distance x from the plate is easily seen to be

$$V = \frac{M}{a}\int_0^a \frac{rdr}{\sqrt{(a^2 - r^2)(x^2 + r^2)}}$$

$$= \frac{M}{2a}\cos^{-1}\frac{x^2 - a^2}{x^2 + a^2}.$$

$$\frac{d}{dx}\left(\frac{M}{2a}\cos^{-1}\frac{x^2 - a^2}{x^2 + a^2}\right) = -\frac{M}{a^2 + x^2}$$

$$= -\frac{M}{a^2}\left[1 - \frac{x^2}{a^2} + \frac{x^4}{a^4} - \frac{x^6}{a^6} + \cdots\right]$$

if $x < a$,

$$= -\frac{M}{x^2}\left[1 - \frac{a^2}{x^2} + \frac{a^4}{x^4} - \frac{a^6}{x^6} + \cdots\right]$$

if $x > a$.

Integrating and then determining the arbitrary constant we have

$$\frac{M}{2a}\cos^{-1}\frac{x^2 - a^2}{x^2 + a^2} = \frac{M}{a}\left[\frac{\pi}{2} - \frac{x}{a} + \frac{x^3}{3a^3} - \frac{x^5}{5a^5} + \frac{x^7}{7a^7} - \cdots\right]$$

if $x < a$,

$$= \frac{M}{a}\left[\frac{a}{x} - \frac{a^3}{3x^3} + \frac{a^5}{5x^5} - \frac{a^7}{7x^7} + \cdots\right]$$

if $x > a$.

We have, then, to solve the equation

$$rD_r^2(rV) + \frac{1}{\sin\theta}D_\theta(\sin\theta D_\theta V) = 0$$

subject to the conditions

$$V = \frac{M}{a}\left[\frac{\pi}{2} - \frac{r}{a} + \frac{r^3}{3a^3} - \frac{r^5}{5a^5} + \frac{r^7}{7a^7} - \cdots\right]$$

when $\theta = 0$ and $r < a$

and
$$V = \frac{M}{a}\left[\frac{a}{r} - \frac{a^3}{3r^3} + \frac{a^5}{5r^5} - \frac{a^7}{7r^7} + \cdots\right]$$

when $\theta = 0$ and $r > a$.

The required solution is easily seen to be

$$V = \frac{M}{a}\left[\frac{\pi}{2} - \frac{r}{a}P_1(\cos\theta) + \frac{1}{3}\frac{r^3}{a^3}P_3(\cos\theta) - \frac{1}{5}\frac{r^5}{a^5}P_5(\cos\theta) + \cdots\right]$$

if $r < a$ and $\theta < \frac{\pi}{2}$,

and
$$V = \frac{M}{a}\left[\frac{a}{r} - \frac{1}{3}\frac{a^3}{r^3}P_2(\cos\theta) + \frac{1}{5}\frac{a^5}{r^5}P_4(\cos\theta) - \frac{1}{7}\frac{a^7}{r^7}P_6(\cos\theta) + \cdots\right]$$

if $r > a$.

EXAMPLES.

1. Given that if a charge M of electricity is placed on an ellipsoidal conductor the surface density at any point P of the conductor is equal to $\frac{Mp}{4\pi abc}$, where p is the distance from the centre of the conductor to the tangent plane at P (v. Peirce, New. Pot. Func. § 61); find the value of the potential function at any external point when the conductor is the oblate spheroid generated by the rotation of the ellipse $\frac{x^2}{a^2} + \frac{y^2}{b^2} = 1$ about its minor axis.

Ans. (1) If the point is on the axis of revolution

$$V = \frac{M}{2\sqrt{a^2 - b^2}}\left[\sin^{-1}\left(\frac{bx + a^2 - b^2}{a\sqrt{x^2 + a^2 - b^2}}\right) - \sin^{-1}\left(\frac{bx - a^2 + b^2}{a\sqrt{x^2 + a^2 - b^2}}\right)\right]$$

x being the distance from the centre.

(2) If the point is on the surface of the spheroid

$$V = \frac{M}{2\sqrt{a^2 - b^2}}\left[\frac{\pi}{2} - \sin^{-1}\left(\frac{2b^2 - a^2}{a^2}\right)\right] = \frac{M}{\sqrt{a^2 - b^2}}\left[\frac{\pi}{2} - \tan^{-1}\left(\frac{b}{\sqrt{a^2 - b^2}}\right)\right].$$

(3) If the distance r of the point from the centre is less than $\sqrt{a^2-b^2}$ and $\theta < \frac{\pi}{2}$

$$V = \frac{M}{\sqrt{a^2-b^2}}\left[\frac{\pi}{2} - \frac{r}{(a^2-b^2)^{\frac{1}{2}}}P_1(\cos\theta) + \frac{r^3}{3(a^2-b^2)^{\frac{3}{2}}}P_3(\cos\theta) - \frac{r^5}{5(a^2-b^2)^{\frac{5}{2}}}P_5(\cos\theta) + \cdots\right].$$

(4) If the distance r of the point from the centre is greater than $\sqrt{a^2-b^2}$

$$V = \frac{M}{\sqrt{a^2-b^2}}\left[\frac{(a^2-b^2)^{\frac{1}{2}}}{r} - \frac{(a^2-b^2)^{\frac{3}{2}}}{3r^3}P_2(\cos\theta) + \frac{(a^2-b^2)^{\frac{5}{2}}}{5r^5}P_4(\cos\theta) - \frac{(a^2-b^2)^{\frac{7}{2}}}{7r^7}P_6(\cos\theta) + \cdots\right].$$

2. If the conductor is the prolate spheroid generated by the rotation of the ellipse $\frac{x^2}{a^2} + \frac{y^2}{b^2} = 1$ about its major axis, show that if the point is an external point and is on the axis at a distance x from the centre,

$$V = \frac{M}{2\sqrt{a^2-b^2}}\log\frac{x+\sqrt{a^2-b^2}}{x-\sqrt{a^2-b^2}}.$$

If the point is not on the axis and $r > \sqrt{a^2-b^2}$

$$V = \frac{M}{\sqrt{a^2-b^2}}\left[\frac{(a^2-b^2)^{\frac{1}{2}}}{r} + \frac{(a^2-b^2)^{\frac{3}{2}}}{3r^3}P_2(\cos\theta) + \frac{(a^2-b^2)^{\frac{5}{2}}}{5r^5}P_4(\cos\theta) + \frac{(a^2-b^2)^{\frac{7}{2}}}{7r^7}P_6(\cos\theta) + \cdots\right].$$

80. As a third example we will find the value of the potential function due to a thin homogeneous circular disc, of density ρ, thickness k, and radius a.

The value of V at a point in the axis of the disc at a distance x from its centre is readily found and proves to be

$$V_0 = 2\pi\rho k(\sqrt{x^2+a^2} - x) = \frac{2M}{a^2}[\sqrt{x^2+a^2} - x].$$

If $x > a$

$$\sqrt{x^2+a^2} = x\left(1+\frac{a^2}{x^2}\right)^{\frac{1}{2}}$$
$$= x\left[1 + \frac{1}{2}\frac{a^2}{x^2} - \frac{1.1}{2.4}\frac{a^4}{x^4} + \frac{1.1.3}{2.4.6}\frac{a^6}{x^6} - \frac{1.1.3.5}{2.4.6.8}\frac{a^8}{x^8} + \cdots\right]$$

and
$$V_0 = \frac{2M}{a}\left[\frac{1}{2}\frac{a}{x} - \frac{1.1}{2.4}\frac{a^3}{x^3} + \frac{1.1.3}{2.4.6}\frac{a^5}{x^5} - \frac{1.1.3.5}{2.4.6.8}\frac{a^7}{x^7} + \cdots\right].$$

If $x < a$

$$\sqrt{x^2+a^2} = a\left(1+\frac{x^2}{a^2}\right)^{\frac{1}{2}}$$

$$= a\left[1+\frac{1}{2}\frac{x^2}{a^2} - \frac{1.1}{2.4}\frac{x^4}{a^4} + \frac{1.1.3}{2.4.6}\frac{x^6}{a^6} + \cdots\right]$$

and $$V_0 = \frac{2M}{a}\left[1 - \frac{x}{a} + \frac{1}{2}\frac{x^2}{a^2} - \frac{1.1}{2.4}\frac{x^4}{a^4} + \frac{1.1.3}{2.4.6}\frac{x^6}{a^6} - \frac{1.1.3.5}{2.4.6.8}\frac{x^8}{a^8} + \cdots\right].$$

Hence the solution for any external point is

$$V = \frac{2M}{a}\left[\frac{1}{2}\frac{a}{r} - \frac{1.1}{2.4}\frac{a^3}{r^3}P_2(\cos\theta) + \frac{1.1.3}{2.4.6}\frac{a^5}{r^5}P_4(\cos\theta) - \frac{1.1.3.5}{2.4.6.8}\frac{a^7}{r^7}P_6(\cos\theta) + \cdots\right]$$

if $r > a$, and

$$V = \frac{2M}{a}\left[1 - \frac{r}{a}P_1(\cos\theta) + \frac{1}{2}\frac{r^2}{a^2}P_2(\cos\theta) - \frac{1.1}{2.4}\frac{r^4}{a^4}P_4(\cos\theta) + \frac{1.1.3}{2.4.6}\frac{r^6}{a^6}P_6(\cos\theta) - \cdots\right]$$

if $r < a$ and $\theta < \frac{\pi}{2}$.

EXAMPLES.

1. The potential function due to a homogeneous hemisphere whose axis is taken as the polar axis, is

$$V = \frac{M}{a}\left[\frac{a}{r} + \frac{3.1}{2.4}\frac{a^2}{r^2}P_1(\cos\theta) - \frac{3.1.1}{2.4.6}\frac{a^4}{r^4}P_3(\cos\theta) + \frac{3.1.1.3}{2.4.6.8}\frac{a^6}{r^6}P_5(\cos\theta) - \cdots\right]$$

if $r > a$, and is

$$V = \frac{M}{a}\left[\frac{3}{2} + \frac{3}{2}\frac{r}{a}P_1(\cos\theta) + \frac{r^2}{a^2}P_2(\cos\theta) + \frac{3.1}{2.4}\frac{r^3}{a^3}P_3(\cos\theta) - \frac{3.1.1}{2.4.6}\frac{r^5}{a^5}P_5(\cos\theta) + \cdots\right]$$

if $r < a$ and $\theta > \frac{\pi}{2}$.

2. The potential function due to a solid sphere whose density is proportional to the distance from a diametral plane is, at an external point,

$$V = \frac{8}{15}\frac{M}{a}\left[\frac{5.3}{2.4}\frac{a}{r} + \frac{5.3.1}{2.4.6}\frac{a^3}{r^3}P_2(\cos\theta) - \frac{5.3.1.1}{2.4.6.8}\frac{a^5}{r^5}P_4(\cos\theta) + \frac{5.3.1.1.3}{2.4.6.8.10}\frac{a^7}{r^7}P_6(\cos\theta) - \cdots\right].$$

3. The potential function due to the homogeneous oblate spheroid generated by the rotation of $\frac{x^2}{a^2} + \frac{y^2}{b^2} = 1$ about its minor axis is, at an external point,

$$V = \frac{3}{2}\frac{M}{(a^2-b^2)}\left[\frac{x^2+a^2-b^2}{2(a^2-b^2)^{\frac{1}{2}}}\left(\sin^{-1}\frac{(a^2-b^2+bx)}{a\sqrt{x^2+a^2-b^2}} + \sin^{-1}\frac{(a^2-b^2-bx)}{a\sqrt{x^2+a^2-b^2}}\right) - x\right]$$

if the point is on the axis of the spheroid at a distance x from its centre.

$$V = \frac{3M}{(a^2-b^2)^{\frac{1}{2}}}\left[\frac{1}{1.3}\frac{(a^2-b^2)^{\frac{1}{2}}}{r} - \frac{1}{3.5}\frac{(a^2-b^2)^{\frac{3}{2}}}{r^3}P_2(\cos\theta) + \frac{1}{5.7}\frac{(a^2-b^2)^{\frac{5}{2}}}{r^5}P_4(\cos\theta) - \cdots\right]$$

if $r > (a^2-b^2)^{\frac{1}{2}}$, and

$$V = \frac{3M}{(a^2-b^2)^{\frac{1}{2}}}\left[\frac{\pi}{4} - \frac{r}{(a^2-b^2)^{\frac{1}{2}}}P_1(\cos\theta) + \frac{\pi}{4}\frac{r^2}{(a^2-b^2)}P_2(\cos\theta) - \frac{1}{1.3}\frac{r^3}{(a^2-b^2)^{\frac{3}{2}}}P_3(\cos\theta) + \frac{1}{3.5}\frac{r^5}{(a^2-b^2)^{\frac{5}{2}}}P_5(\cos\theta) - \cdots\right]$$

if $r < (a^2-b^2)^{\frac{1}{2}}$ and $\theta < \frac{\pi}{2}$.

4. The potential function due to the homogeneous prolate spheroid generated by the rotation of $\frac{x^2}{a^2} + \frac{y^2}{b^2} = 1$ about its major axis is, at an external point,

$$V = \frac{3M}{(a^2-b^2)^{\frac{1}{2}}}\left[\frac{1}{1.3}\frac{(a^2-b^2)^{\frac{1}{2}}}{r} + \frac{1}{3.5}\frac{(a^2-b^2)^{\frac{3}{2}}}{r^3}P_2(\cos\theta) + \frac{1}{5.7}\frac{(a^2-b^2)^{\frac{5}{2}}}{r^5}P_4(\cos\theta) + \cdots\right]$$

if $r > (a^2-b^2)^{\frac{1}{2}}$.

81. The method employed in the last three articles may be stated in general as follows:—Whenever in a problem involving the solving of the special form of Laplace's Equation

$$rD_r^2(rV) + \frac{1}{\sin\theta} D_\theta(\sin\theta D_\theta V) = 0,$$

the value of V is given or can be found for all points on the axis of X and this value can be expressed as a sum or a series involving only whole powers positive or negative of the radius vector of the point, the solution for a point not on the axis can be obtained by multiplying each term by the appropriate Zonal Harmonic, subject only to the condition that the result if a series must be convergent.

It will be shown in the next article that $P_m(\cos\theta)$ is never greater than one nor less than minus one. Hence the series in question will be convergent for all values of r for which the original series was *absolutely convergent.*

82. In addition to the form given in (1) Art. 77 for $P_m(x)$ other forms are often useful.

It ought to be possible to develop $P_m(\cos\theta)$, which may be regarded as a function of θ, into a Fourier's Series, and such a development may be obtained, though with much labor, by the methods of Chapter II.

The development in terms of cosines of multiples of θ may be obtained much more easily by the following device.

We have seen in Art. 75 that $P_m(\cos\theta)$ is the coefficient of the mth power of z in the development of $(1 - 2z\cos\theta + z^2)^{-\frac{1}{2}}$ in a power series, and that if mod $z < 1$ $(1 - 2z \cos\theta + z^2)^{-\frac{1}{2}}$ can be developed into such a series. We know by the Theory of Functions that only one such series exists, so that the method by which we may choose to obtain the development will not affect the result.

$$\begin{aligned}(1 - 2z\cos\theta + z^2)^{-\frac{1}{2}} &= (1 - z(e^{\theta i} + e^{-\theta i}) + z^2)^{-\frac{1}{2}} \\ &= (1 - ze^{\theta i})^{-\frac{1}{2}}(1 - ze^{-\theta i})^{-\frac{1}{2}}.\end{aligned}$$

$(1 - ze^{\theta i})^{-\frac{1}{2}}$ may be developed into an absolutely convergent series if mod $z < 1$, by the Binomial Theorem. We have

$$(1 - ze^{\theta i})^{-\frac{1}{2}} = 1 + \frac{1}{2}ze^{\theta i} + \frac{1.3}{2.4}z^2e^{2\theta i} + \frac{1.3.5}{2.4.6}z^3e^{3\theta i} + \frac{1.3.5.7}{2.4.6.8}z^4e^{4\theta i} + \cdots$$

$$\begin{aligned}(1 - ze^{-\theta i})^{-\frac{1}{2}} = 1 + \frac{1}{2}ze^{-\theta i} &+ \frac{1.3}{2.4}z^2e^{-2\theta i} \\ &+ \frac{1.3.5}{2.4.6}z^3e^{-3\theta i} + \frac{1.3.5.7}{2.4.6.8}z^4e^{-4\theta i} + \cdots\end{aligned}$$

The product of these series will give a development for $(1 - 2z\cos\theta + z^2)^{-\frac{1}{2}}$ in power series. The coefficient of z^m is easily picked out, and must be equal to

$P_m(\cos\theta)$. We thus get

$$P_m(\cos\theta) = \frac{1.3.5.\cdots(2m-1)}{2.4.6.\cdots 2m}\Big[e^{m\theta i} + e^{-m\theta i}$$
$$+\frac{1}{2}\cdot\frac{2m}{2m-1}(e^{(m-2)\theta i} + e^{-(m-2)\theta i})$$
$$+\frac{1.3}{2.4}\cdot\frac{2m(2m-2)}{(2m-1)(2m-3)}(e^{(m-4)\theta i} + e^{-(m-4)\theta i}) + \cdots\Big]$$

$$P_m(\cos\theta) = \frac{1.3.5\cdots(2m-1)}{2.4.6.\cdots 2m}\Big[2\cos m\theta + 2\frac{1.m}{1.(2m-1)}\cos(m-2)\theta$$
$$+2\frac{1.3\,m(m-1)}{1.2(2m-1)(2m-3)}\cos(m-4)\theta$$
$$+2\frac{1.3.5}{1.2.3}\frac{m(m-1)(m-2)}{(2m-1)(2m-3)(2m-5)}\cos(m-6)\theta + \cdots\Big]. \quad (1)$$

If m is odd the development runs down to $\cos\theta$; if m is even to $\cos(0)$, but in that case the coefficient of $\cos(0)$, that is, the constant term, will not contain the factor 2 which is common to all the other terms, but will be simply $\left[\frac{1.3.5\cdots(m-1)}{2.4.6.\cdots m}\right]^2$.

We write out the values of $P_m(\cos\theta)$ for a few values of m

$$\left.\begin{aligned}
P_0(\cos\theta) &= 1\\
P_1(\cos\theta) &= \cos\theta\\
P_2(\cos\theta) &= \frac{1}{4}(3\cos 2\theta + 1)\\
P_3(\cos\theta) &= \frac{1}{8}(5\cos 3\theta + 3\cos\theta)\\
P_4(\cos\theta) &= \frac{1}{64}(35\cos 4\theta + 20\cos 2\theta + 9)\\
P_5(\cos\theta) &= \frac{1}{128}[63\cos 5\theta + 35\cos 3\theta + 30\cos\theta]\\
P_6(\cos\theta) &= \frac{1}{512}[231\cos 6\theta + 126\cos 4\theta + 105\cos 2\theta + 50]\\
P_7(\cos\theta) &= \frac{1}{1024}[429\cos 7\theta + 231\cos 5\theta + 189\cos 3\theta + 175\cos\theta]\\
P_8(\cos\theta) &= \frac{1}{16384}[6435\cos 8\theta + 3432\cos 6\theta + 2772\cos 4\theta\\
&\qquad + 2520\cos 2\theta + 1225].
\end{aligned}\right\} \quad (2)$$

Since all the coefficients in the second member of (1) are positive, and since each cosine has unity for its maximum value it is clear that $P_m(\cos\theta)$ has its maximum value when $\theta = 0$; but we have shown in Art. 76 that $P_m(1) = 1$. Therefore $P_m(\cos\theta)$ is never greater than unity if θ is real. It is also easily seen from (1) that $P_m(\cos\theta)$ can never be less than -1.

83. $P_m(x)$ can be very simply expressed as a derivative. We have

$$P_m(x) = \frac{(2m-1)(2m-3)\cdots 1}{m!}\left[x^m - \frac{m(m-1)}{2.(2m-1)}x^{m-2} + \frac{m(m-1)(m-2)(m-3)}{2.4.(2m-1)(2m-3)}x^{m-4} - \cdots\right]$$

$$\int_0^x P_m(x)dx = \frac{(2m-1)(2m-3)\cdots 1}{(m+1)!}\left[x^{m+1} - \frac{(m+1)m}{2.(2m-1)}x^{m-1} + \frac{(m+1)m(m-1)(m-2)}{2.4.(2m-1)(2m-3)}x^{m-3} - \cdots\right]$$

$$\int_0^x {}^2 P_m(x)dx^2 = \int_0^x dx \int_0^x P_m(x)dx$$

$$= \frac{(2m-1)(2m-3)\cdots 1}{(m+2)!}\left[x^{m+2} - \frac{(m+2)(m+1)}{2.(2m-1)}x^{m} + \frac{(m+2)(m+1)m(m-1)}{2.4.(2m-1)(2m-3)}x^{m-2} - \cdots\right]$$

$$\int_0^x {}^m P_m(x)dx^m = \frac{(2m-1)(2m-3)\cdots 1}{(2m)!}\left[x^{2m} - \frac{2m(2m-1)}{2(2m-1)}x^{2m-2} + \frac{2m(2m-1)(2m-2)(2m-3)}{2.4.(2m-1)(2m-3)}x^{2m-4} - \cdots\right]$$

$$= \frac{(2m-1)(2m-3)\cdots 1}{(2m)!}\left[x^{2m} - mx^{2m-2} + \frac{m(m-1)}{2!}x^{2m-4} - \frac{m(m-1)(m-2)}{3!}x^{2m-6} + \cdots\right].$$

The quantity in brackets obviously differs from $(x^2-1)^m$ by terms involving lower powers of x than the mth.

Hence
$$P_m(x) = \frac{1.3.5\cdots(2m-1)}{(2m)!}\frac{d^m}{dx^m}(x^2-1)^m,$$

or
$$P_m(x) = \frac{1}{2^m m!}\frac{d^m}{dx^m}(x^2-1)^m. \tag{1}$$

This important formula is entirely general and holds not merely when $x = \cos\theta$, but for all values of x.

84. The last result is so important that it is worth while to confirm it by obtaining it directly from Legendre's Equation

$$(1-x^2)\frac{d^2z}{dx^2} - 2x\frac{dz}{dx} + m(m+1)z = 0 \tag{1}$$

v. (1) Art. 75.

Let us differentiate (1) with respect to x a few times representing $\frac{dz}{dx}$ by z', $\frac{d^2z}{dx^2}$ by z'', $\frac{d^3z}{dz^3}$ by z''', &c. We get

$$(1-x^2)\frac{d^2z'}{dx^2} - 2.2x\frac{dz'}{dx} + [m(m+1)-2]z' = 0,$$
$$(1-x^2)\frac{d^2z''}{dx^2} - 2.3x\frac{dz''}{dx} + [m(m+1)-2(1+2)]z'' = 0,$$
$$(1-x^2)\frac{d^2z'''}{dx^2} - 2.4x\frac{dz'''}{dx} + [m(m+1)-2(1+2+3)]z''' = 0,$$

and in general

$$(1-x^2)\frac{d^2z^{(n)}}{dx^2} - 2(n+1)x\frac{dz^{(n)}}{dx} + [m(m+1)-2(1+2+3+\cdots+n)]z^{(n)} = 0$$

or $$(1-x^2)\frac{d^2z^{(n)}}{dx^2} - 2(n+1)x\frac{dz^{(n)}}{dx} + m(m+1)-n(n+1)]z^{(n)} = 0. \quad (2)$$

Following the analogy of these steps it is easy to write equations that will differentiate into (1).

Let $\frac{dz_1}{dx} = z$, $\frac{d^2z_2}{dx^2} = z$, $\frac{d^3z_3}{dx^3} = z$, &c. Then

$$(1-x^2)\frac{d^2z_1}{dx^2} + m(m+1)z_1 = 0$$

will differentiate into (1),

$$(1-x^2)\frac{d^2z_2}{dx^2} + 2.1x\frac{dz_2}{dx} + [m(m+1)-2.1]z_2 = 0$$

if differentiated twice will give (1),

$$(1-x^2)\frac{d^2z_3}{dx^2} + 2.2x\frac{dz_3}{dx} + [m(m+1)-2(1+2)]z_3 = 0$$

if differentiated three times will give (1), and in general

$$(1-x^2)\frac{d^2z_n}{dx^2} + 2(n-1)x\frac{dz_n}{dx} + [m(m+1)-n(n-1)]z_n = 0 \quad (3)$$

if differentiated n times with respect to x will give (1).

If $n = m+1$ (3) reduces to

$$(1-x^2)\frac{d^2z_{m+1}}{dx^2} + 2mx\frac{dz_{m+1}}{dx} = 0, \quad (4)$$

and the $(m+1)$st derivative with respect to x of any function of x which satisfies (4) will be a solution of (1). (4) can be written

$$(1-x^2)\frac{dz_m}{dx} + 2mxz_m = 0$$

and can be readily solved by separating the variables and integrating. v. Int. Cal. (1) page 314. It gives

$$z_m = C(x^2 - 1)^m.$$

Hence
$$z = \frac{d^m z_m}{dx^m} = C\frac{d^m (x^2 - 1)^m}{dx^m} \tag{5}$$

is a solution of Legendre's Equation (1) and agrees with the value of $P_m(x)$ obtained in Art. 83.

85. The equations obtained in Art. 84 are so curious and so simply related that it is worth while to consider them a little more fully.

We have seen that

$$(1 - x^2)\frac{d^2 z}{dx^2} + 2mx\frac{dz}{dx} = 0 \tag{1}$$

differentiates into

$$(1 - x^2)\frac{d^2 z}{dx^2} + 2(m - 1)x\frac{dz}{dx} + 2mz = 0; \tag{2}$$

that if we differentiate (2) m times we get Legendre's Equation

$$(1 - x^2)\frac{d^2 z}{dx^2} - 2x\frac{dz}{dx} + m(m + 1)z = 0; \tag{3}$$

that if we differentiate (2) $2m$ times we get

$$(1 - x^2)\frac{d^2 z}{dx^2} - 2(m + 1)x\frac{dz}{dx} = 0; \tag{4}$$

that if we differentiate (2) $m - n$ times we have

$$(1 - x^2)\frac{d^2 z}{dx^2} + 2(n - 1)x\frac{dz}{dx} + [m(m + 1) - n(n - 1)]z = 0; \tag{5}$$

and that if we differentiate (2) $m + n$ times we have

$$(1 - x^2)\frac{d^2 z}{dx^2} - 2(n + 1)x\frac{dz}{dx} + [m(m + 1) - n(n + 1)]z = 0. \tag{6}$$

By the aid of (1) we found in the last article a particular solution of (2), namely

$$z = (x^2 - 1)^m.$$

If we substitute in (2) $z = u(x^2 - 1)^m$ following the method illustrated fully in Art. 18, we get as the general solution of (2)

$$z = A(x^2 - 1)^m + B(x^2 - 1)^m \int \frac{dx}{(x^2 - 1)^{m+1}}, \tag{7}$$

A and B being arbitrary constants.

$\int \frac{dx}{(x^2-1)^{m+1}}$ is easily written out [v. formula (42) page 6. Table of Integrals. Int. Cal. Appendix]. If $x<1$ it vanishes when $x=0$. If $x>1$ it vanishes when $x=\infty$. If then $x<1$ (7) can be written

$$z = A(x^2-1)^m + B(x^2-1)^m \int_0^x \frac{dx}{(x^2-1)^{m+1}} \tag{8}$$

and if $x>1$

$$z = A(x^2-1)^m + B(x^2-1)^m \int_x^\infty \frac{dx}{(x^2-1)^{m+1}} \tag{9}$$

and in these forms unnecessary arbitrary constants are avoided.

From (7) we can get the general solutions of (3), (4), (5), and (6).

$$z = A\frac{d^m(x^2-1)^m}{dx^m} + B\frac{d^m}{dx^m}\left[(x^2-1)^m \int \frac{dx}{(x^2-1)^{m+1}}\right] \tag{10}$$

is the general solution of (3).

$$z = A\frac{d^{2m}(x^2-1)^m}{dx^{2m}} + B\frac{d^{2m}}{dx^{2m}}\left[(x^2-1)^m \int \frac{dx}{(x^2-1)^{m+1}}\right] \tag{11}$$

is the general solution of (4).

$$z = A\frac{d^{m-n}(x^2-1)^m}{dx^{m-n}} + B\frac{d^{m-n}}{dx^{m-n}}\left[(x^2-1)^m \int \frac{dx}{(x^2-1)^{m+1}}\right] \tag{12}$$

is the general solution of (5).

$$z = A\frac{d^{m+n}(x^2-1)^m}{dx^{m+n}} + B\frac{d^{m+n}}{dx^{m+n}}\left[(x^2-1)^m \int \frac{dx}{(x^2-1)^{m+1}}\right] \tag{13}$$

is the general solution of (6).

In each of these forms A and B are arbitrary constants and the integral is to be taken from 0 to x if $x<1$ and from x to ∞ if $x>1$.

Of course (10) must be identical with the forms already obtained in Arts. 16 and 18 as general solutions of Legendre's Equation.

Equation (4) is so simple that it can be solved directly, and we get its solution in the form

$$z = A_1 + B_1 \int \frac{dx}{(x^2-1)^{m+1}} \tag{14}$$

which must be equivalent to (11).

Comparing (14) with (7), the solution of (2), we see that every solution of (4) can be obtained from a solution of (2) by dividing the latter by $(x^2-1)^m$, or in other words that if we write (2)

$$(1-x^2)\frac{d^2z}{dx^2}+2(m-1)x\frac{dz}{dx}+2mz=0, \tag{2}$$

and (4) as

$$(1-x^2)\frac{d^2z_1}{dx^2}-2(m+1)x\frac{dz_1}{dx}=0 \tag{4}$$

$z=z_1(x^2-1)^m$; and the substitution of this value in (2) will give (4), and the substitution of $z_1=\dfrac{z}{(x^2-1)^m}$ in (4) will give (2).

We have, then, two ways of obtaining (4) from (2); we may differentiate (2) $2m$ times with respect to x, or we may replace z in (2) by $z_1(x^2-1)^m$.

If we use the first method we have seen that Legendre's Equation (3) is midway between (2) and (4). That is if we differentiate (2) m times we get (3) and if we then differentiate (3) m times we get (4). Let us see if the half-way equation in our second process is Legendre's Equation.

If
$$z=y(x^2-1)^{\frac{m}{2}}$$
and
$$y=z_1(x^2-1)^{\frac{m}{2}}$$
$$z=z_1(x^2-1)^m.$$

So that if in (2) we replace z by $y(x^2-1)^{\frac{m}{2}}$ and then repeat the operation on the resulting equation we shall get (4). Making the first substitution we find,

$$(1-x^2)\frac{d^2y}{dx^2}-2x\frac{dy}{dx}+\left[m(m+1)-\frac{m^2}{1-x^2}\right]y=0, \tag{15}$$

not Legendre's Equation but a somewhat more general form. Of course its solution is

$$y=A(x^2-1)^{\frac{m}{2}}+B(x^2-1)^{\frac{m}{2}}\int\frac{dx}{(x^2-1)^{m+1}}. \tag{16}$$

(2) and (4) are special forms of (5) and (6). Let us try the experiment of substituting in (5) $z=y(1-x^2)^{\frac{n}{2}}$ and in (6) $z=\dfrac{y}{(1-x^2)^{\frac{n}{2}}}$. We find that both substitutions give the same equation

$$(1-x^2)\frac{d^2y}{dx^2}-2x\frac{dy}{dx}+\left[m(m+1)-\frac{n^2}{1-x^2}\right]y=0. \tag{17}$$

The solution of (17) can be obtained from either (12) or (13) and is

$$y=\frac{1}{(1-x^2)^{\frac{n}{2}}}\left\{A\frac{d^{m-n}(x^2-1)^m}{dx^{m-n}}+B\frac{d^{m-n}}{dx^{m-n}}\left[(x^2-1)^m\int\frac{dx}{(x^2-1)^{m+1}}\right]\right\} \tag{18}$$

or

$$y=(1-x^2)^{\frac{n}{2}}\left\{A_1\frac{d^{m+n}(x^2-1)^m}{dx^{m+n}}+B_1\frac{d^{m+n}}{dx^{m+n}}\left[(x^2-1)^m\int\frac{dx}{(x^2-1)^{m+1}}\right]\right\} \tag{19}$$

which of course must be equivalent.

86. In addition to the value of $P_m(x)$ given in (1) Art. 83 there is another important derivative form which we shall proceed to obtain. It is

$$P_m(\cos\theta) = \frac{(-1)^m}{m!} r^{m+1} D_x^m \left(\frac{1}{r}\right). \tag{1}$$

We have seen in Art. 75 that $\frac{1}{r}\frac{1}{\sqrt{1 - 2\frac{r_1}{r}\cos\theta + \frac{r_1^2}{r^2}}}$ can be developed into a convergent series if $r_1 < r$ and that the $(m+1)$st term of that series is $\frac{P_m(\cos\theta) r_1^m}{r^{m+1}}$. Let us obtain this term by Taylor's Theorem.

$$\frac{1}{r}\frac{1}{\sqrt{1 - 2\frac{r_1}{r}\cos\theta + \frac{r_1^2}{r^2}}} = \frac{1}{\sqrt{r^2 - 2r_1 r\cos\theta + r_1^2}} = \frac{1}{\sqrt{x^2 + y^2 + z^2 - 2xr_1 + r_1^2}}$$

$$= \frac{1}{\sqrt{(x - r_1)^2 + y^2 + z^2}}$$

Regarding this as a function of $(x - r_1)$ and developing according to powers of r_1 by Taylor's Theorem we get as the $(m+1)$st term

$$\frac{(-1)^m}{m!} r_1^m D_x^m \left[\frac{1}{\sqrt{x^2 + y^2 + z^2}}\right] \quad \text{or} \quad \frac{(-1)^m}{m!} r_1^m D_x^m \left(\frac{1}{r}\right).$$

Hence
$$\frac{P_m(\cos\theta)}{r^{m+1}} = \frac{(-1)^m}{m!} D_x^m \left(\frac{1}{r}\right).$$

87. We have now obtained four different forms for our *zonal harmonic*, a polynomial in x, an expression involving cosines of multiples of θ, a form involving an ordinary mth derivative with respect to x, and a form involving a partial mth derivative with respect to x. We shall now get a form due to Laplace, involving a definite integral.

$$\int_0^\pi \frac{d\phi}{a - b\cos\phi} = \frac{\pi}{(a^2 - b^2)^{\frac{1}{2}}} \tag{1}$$

if $a^2 > b^2$ [v. Int. Cal. page 68].

$\frac{1}{(1 - 2xz + z^2)^{\frac{1}{2}}}$ can be expressed in the form $\frac{1}{(a^2 - b^2)^{\frac{1}{2}}}$ by taking $a = 1 - zx$ and $b = z\sqrt{x^2 - 1}$ and no matter what value x may have z can be taken so small that a^2 will be greater than b^2. Then by (1)

$$\frac{1}{(1-2xz+z^2)^{\frac{1}{2}}} = \frac{1}{\pi}\int_0^\pi \frac{d\phi}{1 - zx - z\sqrt{x^2-1}.\cos\phi}$$
$$= \frac{1}{\pi}\int_0^\pi \frac{d\phi}{1 - z(x + \sqrt{x^2-1}.\cos\phi)}$$
$$= \frac{1}{\pi}\int_0^\pi [1 + (x+\sqrt{x^2-1}.\cos\phi)z + (x+\sqrt{x^2-1}.\cos\phi)^2 z^2$$
$$+ (x+\sqrt{x^2-1}.\cos\phi)^3 z^3 + \cdots]d\phi$$

if z is taken so small that the modulus of $z(x + \sqrt{x^2-1}.\cos\phi)$ is less than 1. But by Art. 77 (2) $P_m(x)$ is the coefficient of z^m in the development of $\dfrac{1}{(1-2xz+z^2)^{\frac{1}{2}}}$,

hence $$P_m(x) = \frac{1}{\pi}\int_0^\pi [x + \sqrt{x^2-1}.\cos\phi]^m d\phi. \tag{2}$$

By replacing ϕ by $\pi - \phi$ in (2) we get

$$P_m(x) = \frac{1}{\pi}\int_0^\pi [x - \sqrt{x^2-1}.\cos\phi]^m d\phi. \tag{3}$$

$\dfrac{1}{(1-2xz+z^2)^{\frac{1}{2}}} = \dfrac{1}{z}\dfrac{1}{\left(1 - 2x\dfrac{1}{z} + \dfrac{1}{z^2}\right)^{\frac{1}{2}}}$ and if mod $\dfrac{1}{z} < 1$ or in other words if mod $z > 1$ $\dfrac{1}{\left(1 - 2x\dfrac{1}{z} + \dfrac{1}{z^2}\right)^{\frac{1}{2}}}$ can be developed into a convergent series involving powers of $\dfrac{1}{z}$, and the coefficient of $\left(\dfrac{1}{z}\right)^m$ will be $P_m(x)$; but this will be the coefficient of z^{-m-1} in the development of $\dfrac{1}{(1-2xz+z^2)^{\frac{1}{2}}}$ according to descending powers of z, mod z being greater than 1.

If now we let $a = zx - 1$ and $b = z\sqrt{x^2-1}$, $a^2 - b^2 = 1 - 2xz + z^2$ and z may be taken so great that $a^2 - b^2 > 0$. Then by (1)

$$\frac{1}{(1-2xz+z^2)^{\frac{1}{2}}} = \frac{1}{\pi}\int_0^\pi \frac{d\phi}{zx-1-z\sqrt{x^2-1}.\cos\phi}$$

$$= \frac{1}{\pi}\int_0^\pi \frac{d\phi}{z(x-\sqrt{x^2-1}.\cos\phi)\left[1-\dfrac{1}{z(x-\sqrt{x^2-1}.\cos\phi)}\right]}$$

$$= \frac{1}{\pi}\int_0^\pi \frac{1}{(x-\sqrt{x^2-1}.\cos\phi)}\left[z^{-1}+\frac{1}{(x-\sqrt{x^2-1}.\cos\phi)}z^{-2}\right.$$

$$\left.+\frac{1}{(x-\sqrt{x^2-1}.\cos\phi)^2}z^{-3}+\cdots\right]d\phi$$

and the coefficient of z^{-m-1} is $\dfrac{1}{\pi}\displaystyle\int_0^\pi \frac{d\phi}{[x-\sqrt{x^2-1}.\cos\phi]^{m+1}}$.

Hence
$$P_m(x) = \frac{1}{\pi}\int_0^\pi \frac{d\phi}{[x-\sqrt{x^2-1}.\cos\phi]^{m+1}}. \tag{4}$$

Replace ϕ by $\pi-\phi$ and we get

$$P_m(x) = \frac{1}{\pi}\int_0^\pi \frac{d\phi}{[x+\sqrt{x^2-1}.\cos\phi]^{m+1}}. \tag{5}$$

88. In the problems in which we have already used *Zonal Harmonics* (v. Arts. 78–81) we have been able to start with the value of the Potential Function at any point on the axis of X, and it has been necessary to develop the expression for V on that axis in terms of ascending or descending powers of x. If, however, we start with the value of V in terms of θ for some given value of r, that is on the surface of some sphere, we must develop the function of θ in terms of *zonal harmonics* of $\cos\theta$ (v. Art. 10), and our problem becomes the following:—To develop a given function of $\cos\theta$ in terms of zonal harmonics of $\cos\theta$, or to develop a given function of x in terms of the functions $P_m(x)$, x lying between 1 and -1.

The problem resembles closely that of developing in a Fourier's series, which we have already considered at such length.

Let
$$f(x) = A_0P_0(x) + A_1P_1(x) + A_2P_2(x) + A_3P_3(x) + \cdots \tag{1}$$

for all values of x from -1 to 1 and let it be required to determine the coefficients.

If $f(x)$ is single-valued and has only finite discontinuities between $x=-1$ and $x=1$ we may proceed as in Art. 19.

Let us take $n+1$ terms of (1) and attempt to determine the coefficients. Take $n+1$ values of x at equal intervals Δx between $x=-1$ and $x=1$ so that

$(n+2)\Delta x = 2$; $f(-1+\Delta x)$, $f(-1+2\Delta x)$, $f(-1+3\Delta x)$, $\cdots$ $f[-1+(n+1)\Delta x]$ will be the corresponding values of $f(x)$. Substitute these values in (1) and we have

$$\left.\begin{aligned} f(-1+\Delta x) &= A_0P_0(-1+\Delta x) + A_1P_1(-1+\Delta x) \\ &\qquad + A_2P_2(-1+\Delta x) + \cdots + A_nP_n(-1+\Delta x) \\ f(-1+2\Delta x) &= A_0P_0(-1+2\Delta x) + A_1P_1(-1+2\Delta x) \\ &\qquad + A_2P_2(-1+2\Delta x) + \cdots + A_nP_n(-1+2\Delta x) \\ \vdots \quad & \quad \vdots \quad \vdots \quad \vdots \quad \vdots \quad \vdots \quad \vdots \\ f(1-\Delta x) &= A_0P_0(1-\Delta x) + A_1P_1(1-\Delta x) + A_2P_2(1-\Delta x) + \cdots \\ &\qquad + A_nP_n(1-\Delta x), \end{aligned}\right\} \tag{2}$$

that is, $n+1$ equations from which in theory the $n+1$ coefficients A_0, A_1, $\cdots$ A_n can be determined.

Following the analogy of Art. 24 let us multiply the first equation by $P_m(-1+\Delta x).\Delta x$, the second by $P_m(-1+2\Delta x).\Delta x$, the third by $P_m(-1+3\Delta x).\Delta x$, &c., and add the equations. The first member of the resulting equation is

$$\sum_{k=1}^{k=n+1} f(-1+k\Delta x)P_m(-1+k\Delta x).\Delta x, \tag{3}$$

and the coefficient of any A as A_l in the second member is

$$\sum_{k=1}^{k=n+1} P_m(-1+k\Delta x)P_l(-1+k\Delta x).\Delta x. \tag{4}$$

If now n is indefinitely increased (3) approaches as its limiting value

$$\int_{-1}^{1} f(x)P_m(x)dx \tag{5}$$

and (4) approaches

$$\int_{-1}^{1} P_m(x)P_l(x)dx. \tag{6}$$

We have now to find the value of the integral (6) or as we shall write it for the sake of greater convenience

$$\int_{-1}^{1} P_m(x)P_n(x)dx.$$

89. $\int_{-1}^{1} P_m(x)P_n(x)dx = \frac{1}{2^{m+n}m!n!}\int_{-1}^{1} \frac{d^m(x^2-1)^m}{dx^m}.\frac{d^n(x^2-1)^n}{dx^n}dx$ by (1) Art. 83.

$$\int_{-1}^{1} \frac{d^m(x^2-1)^m}{dx^m}.\frac{d^n(x^2-1)^n}{dx^n}dx = \left[\frac{d^m(x^2-1)^m}{dx^m}.\frac{d^{n-1}(x^2-1)^n}{dx^{n-1}}\right]_{-1}^{1} - \int_{-1}^{1} \frac{d^{m+1}(x^2-1)^m}{dx^{m+1}}.\frac{d^{n-1}(x^2-1)^n}{dx^{n-1}}dx \quad (1)$$

by *integration by parts.*

Now if $z = X(x^2-1)^n$

$$\frac{dz}{dx} = 2nxX(x^2-1)^{n-1} + (x^2-1)^n\frac{dX}{dx}$$
$$= (x^2-1)^{n-1}\left[2nxX + (x^2-1)\frac{dX}{dx}\right]. \quad (2)$$

Hence the pth derivative with respect to x of any function of x containing $(x^2-1)^n$ as a factor will contain $(x^2-1)^{n-p}$ as a factor if $p < n$.

$\frac{d^{n-1}(x^2-1)^n}{dx^{n-1}}$, then, contains (x^2-1) as a factor and is zero when $x = 1$ and when $x = -1$, so that (1) reduces to

$$\int_{-1}^{1} \frac{d^m(x^2-1)^m}{dx^m}.\frac{d^n(x^2-1)^n}{dx^n}dx = -\int_{-1}^{1} \frac{d^{m+1}(x^2-1)^m}{dx^{m+1}}.\frac{d^{n-1}(x^2-1)^n}{dx^{n-1}}dx.$$

It follows that

$$\int_{-1}^{1} \frac{d^m(x^2-1)^m}{dx^m}.\frac{d^n(x^2-1)^n}{dx^n}dx$$
$$= (-1)^p\int_{-1}^{1} \frac{d^{m+p}(x^2-1)^m}{dx^{m+p}}.\frac{d^{n-p}(x^2-1)^n}{dx^{n-p}}dx$$
$$= (-1)^p\int_{-1}^{1} \frac{d^{m-p}(x^2-1)^m}{dx^{m-p}}.\frac{d^{n+p}(x^2-1)^n}{dx^{n+p}}dx. \quad (3)$$

If $m < n$ we get from (3)

$$\int_{-1}^{1} \frac{d^m(x^2-1)^m}{dx^m}.\frac{d^n(x^2-1)^n}{dx^n}dx = (-1)^m\int_{-1}^{1} \frac{d^{2m}(x^2-1)^m}{dx^{2m}}.\frac{d^{n-m}(x^2-1)^n}{dx^{n-m}}dx$$
$$= (-1)^m(2m)!\left[\frac{d^{n-m-1}(x^2-1)^n}{dx^{n-m-1}}\right]_{-1}^{1} = 0,$$

since
$$\frac{d^{2m}(x^2-1)^m}{dx^{2m}} = (2m)!.$$

If $m > n$

$$\int_{-1}^{1} \frac{d^m(x^2-1)^m}{dx^m}.\frac{d^n(x^2-1)^n}{dx^n}dx = (-1)^n \int_{-1}^{1} \frac{d^{m-n}(x^2-1)^m}{dx^{m-n}}.\frac{d^{2n}(x^2-1)^n}{dx^{2n}}dx$$
$$= (-1)^n(2n)!\left[\frac{d^{m-n-1}(x^2-1)^m}{dx^{m-n-1}}\right]_{-1}^{1} = 0.$$

If, then, m is not equal to n

$$\int_{-1}^{1} P_m(x)P_n(x)dx = 0. \tag{4}$$

If $m = n$ we have to find $\int_{-1}^{1} [P_m(x)]^2 dx$.

$$\int_{-1}^{1} [P_m(x)]^2 dx = \frac{1}{2^{2m}(m!)^2} \int_{-1}^{1} \frac{d^m(x^2-1)^m}{dx^m}.\frac{d^m(x^2-1)^m}{dx^m}dx.$$

$$\int_{-1}^{1} \frac{d^m(x^2-1)^m}{dx^m}.\frac{d^m(x^2-1)^m}{dx^m}dx = (-1)^m \int_{-1}^{1} \frac{d^{2m}(x^2-1)^m}{dx^{2m}}.(x^2-1)^m dx$$

by (3),
$$= (-1)^m(2m)! \int_{-1}^{1} (x^2-1)^m dx.$$

$$\int_{-1}^{1} (x^2-1)^m dx = \int_{-1}^{1} (x-1)^m(x+1)^m dx = -\frac{m}{m+1}\int_{-1}^{1} (x-1)^{m-1}(x+1)^{m+1}dx$$
$$= (-1)^m \frac{m!}{(m+1)(m+2)\cdots 2m}\int_{-1}^{1} (x+1)^{2m}dx$$
$$= (-1)^m \frac{2^{2m+1}m!}{(m+1)(m+2)\cdots(2m+1)}.$$

Hence
$$\int_{-1}^{1} [P_m(x)]^2 dx = \frac{1}{2^{2m}(m!)^2}\frac{(-1)^m(2m)!(-1)^m m! 2^{2m+1}}{(m+1)(m+2)\cdots(2m+1)}$$

or
$$\int_{-1}^{1} [P_m(x)]^2 dx = \frac{2}{2m+1}. \tag{5}$$

90. The solution of the problem in Art. 88 is now readily obtained, and we have

$$f(x) = A_0P_0(x) + A_1P_1(x) + A_2P_2(x) + \cdots \tag{1}$$

where
$$A_m = \frac{2m+1}{2}\int_{-1}^{1} f(x)P_m(x)dx. \tag{2}$$

The function and the series are equal for all values of x from $x = -1$ to $x = 1$, and $f(x)$ is subject to no conditions save those which would enable us to develop it in a Fourier's Series. [v. Chapter III.]

Of course (1) can be written

$$f(\cos\theta) = A_0P_0(\cos\theta) + A_1P_1(\cos\theta) + A_2P_2(\cos\theta) + \cdots$$

where
$$A_m = \frac{2m+1}{2}\int_{-1}^{1} f(\cos\theta)P_m(\cos\theta)d(\cos\theta)$$

or if $f(\cos\theta) = F(\theta)$

$$F(\theta) = A_0P_0(\cos\theta) + A_1P_1(\cos\theta) + A_2P_2(\cos\theta) + \cdots \tag{3}$$

where
$$A_m = \frac{2m+1}{2}\int_{0}^{\pi} F(\theta)P_m(\cos\theta)\sin\theta.d\theta \tag{4}$$

and the development holds good from $\theta = 0$ to $\theta = \pi$.

If $f(x)$ is an even function, that is, if $f(-x) = f(x)$ (1) and (2) can be somewhat simplified. For in that case it can be easily shown (v. Art. 77) that

$$\int_{-1}^{1} f(x)P_{2k}(x)dx = 2\int_{0}^{1} f(x)P_{2k}(x)dx,$$

and that
$$\int_{-1}^{1} f(x)P_{2k+1}(x)dx = 0;$$

so that if $f(-x) = f(x)$

$$f(x) = A_0P_0(x) + A_2P_2(x) + A_4P_4(x) + A_6P_6(x) + \cdots \tag{5}$$

where
$$A_{2k} = (4k+1)\int_{0}^{1} f(x)P_{2k}(x)dx. \tag{6}$$

If $f(x)$ is an odd function, that is, if $f(-x) = -f(x)$ it can be shown in like manner that

$$f(x) = A_1P_1(x) + A_3P_3(x) + A_5P_5(x) + A_7P_7(x) + \cdots \tag{7}$$

where
$$A_{2k+1} = (4k+3)\int_{0}^{1} f(x)P_{2k+1}(x)dx. \tag{8}$$

If it is only necessary that the development should hold for $0 < x < 1$ any function may be expressed in form (5) or (7) at pleasure.

91. We can establish the fact that $\int_{-1}^{1} P_m(x)P_n(x)dx = 0$ by a more general method than that used in Art. 89.

Let X_m be any solution of Legendre's Equation

$$\frac{d}{dx}\left[(1-x^2)\frac{dz}{dx}\right] + m(m+1)z = 0 \qquad \text{[v. (1) Art. 16].}$$

which with its first derivative with respect to x is finite, continuous, and single-valued for values of x between -1 and 1, -1 and 1 being included.

Then
$$\frac{d}{dx}\left[(1-x^2)\frac{dX_m}{dx}\right] + m(m+1)X_m = 0 \tag{1}$$
and
$$\frac{d}{dx}\left[(1-x^2)\frac{dX_n}{dx}\right] + n(n+1)X_n = 0 \tag{2}$$

Multiply (1) by X_n and (2) by X_m and subtract and integrate and we get

$$[m(m+1)-n(n+1)]\int_{-1}^{1} X_mX_n dx = \int_{-1}^{1} X_m \frac{d}{dx}\left[(1-x^2)\frac{dX_n}{dx}\right]dx - \int_{-1}^{1} X_n \frac{d}{dx}\left[(1-x^2)\frac{dX_m}{dx}\right]dx.$$

Integrate by parts,

$$[m(m+1)-n(n+1)]\int_{-1}^{1} X_mX_n dx = \left[X_m(1-x^2)\frac{dX_n}{dx} - X_n(1-x^2)\frac{dX_m}{dx}\right]_{x=-1}^{x=1}$$
$$- \int_{-1}^{1}(1-x^2)\frac{dX_n}{dx}\frac{dX_m}{dx}dx + \int_{-1}^{1}(1-x^2)\frac{dX_m}{dx}\frac{dX_n}{dx}dx. \tag{3}$$

Whence
$$\int_{-1}^{1} X_mX_n dx = 0 \tag{4}$$

unless $m = n$.

(3) gives at once the important formula

$$\int_{x}^{1} X_mX_n dx = \frac{(1-x^2)\left[X_n\frac{dX_m}{dx} - X_m\frac{dX_n}{dx}\right]}{m(m+1)-n(n+1)} \tag{5}$$

from which come as special cases

$$\int_x^1 P_m(x)P_n(x)dx = \frac{(1-x^2)\left[P_n(x)\dfrac{dP_m(x)}{dx} - P_m(x)\dfrac{dP_n(x)}{dx}\right]}{m(m+1)-n(n+1)} \tag{6}$$

and since $P_0(x) = 1$

$$\int_x^1 P_m(x)dx = \frac{(1-x^2)\dfrac{dP_m(x)}{dx}}{m(m+1)}, \tag{7}$$

unless $m = 0$.

EXAMPLES.

1. Show that

$$\int_0^1 P_m(x)dx = 0 \quad \text{if } m \text{ is even and is not zero.}$$

$$= (-1)^{\frac{m-1}{2}} \frac{1}{m(m+1)} \frac{3.5.7.\cdots m}{2.4.6.\cdots(m-1)} \quad \text{if } m \text{ is odd.}$$

v. Art. 91 (7) and Art. 77 (10).

2. Show that

$$\int_0^1 P_m(x)P_n(x)dx = 0 \quad \text{if } m \text{ and } n \text{ are both even or both odd.}$$

$$= (-1)^{\frac{m+n+1}{2}} \frac{m!\,n!}{2^{m+n-1}(m-n)(m+n+1)\left(\frac{m}{2}!\right)^2\left(\frac{n-1}{2}!\right)^2}$$

if m is even and n odd. v. Art. 91 (6) and Art. 77 (8), (9), and (10). cf. J. W. Strutt (Lord Rayleigh) Lond. Phil. Trans. 1870, page 579.

3. Show that $\displaystyle\int_0^1 [P_m(x)]^2 dx = \frac{1}{2m+1}$ v. Art. 89 (5).

92. Formula (4) Art. 91 can be obtained directly from Laplace's Equation by the aid of *Green's Theorem* (v. Peirce's Newt. Pot. Func. § 48).

Take the special form of Green's Theorem, [(148) § 48 Peirce's Newt. Pot. Func.]

$$\iiint (U\nabla^2 V - V\nabla^2 U)dxdydz = \int (UD_nV - VD_nU)ds \tag{1}$$

where ∇^2 stands for $(D_x^2 + D_y^2 + D_2^2)$, D_n is the partial derivative along the external normal, and the left-hand member is the space-integral through the space bounded by any closed surface, and the right-hand member is the surface integral taken over the same surface. (v. Int. Cal. Chapter XIV.)

If U and V are solutions of Laplace's Equation $\nabla^2 V = \nabla^2 U = 0$ and (1) reduces to

$$\int (UD_nV - VD_nU)ds = 0. \tag{2}$$

Now $r^m X_m$ and $r^n X_n$ are solutions of Laplace's Equation if $x = \cos\theta$ (v. Art. 16).

If the unit sphere is taken as the bounding surface and $U = r^m X_m$ and $V = r^n X_n$ (1) and (2) will hold good.

$$D_nU = D_r(r^m X_m) = mr^{m-1}X_m,$$
$$D_nV = nr^{n-1}X_n,$$
$$ds = \sin\theta.d\theta d\phi,$$

and (2) becomes $\int_0^{2\pi} d\phi \int_0^{\pi} (nX_mX_n - mX_mX_n)\sin\theta.d\theta = 0$

or
$$2\pi(n-m)\int_0^{\pi} X_mX_n \sin\theta.d\theta = 0. \tag{3}$$

Since $x = \cos\theta$, $\sin\theta.d\theta = -dx$ and (3) reduces to

$$\int_{-1}^{1} X_mX_ndx = 0^{4} \tag{4}$$

unless $m = n$.

93. We can now solve completely the problem of Art. 10 which was in that article carried to the point where it was only necessary to develop a certain function of θ in the form

$$A_0P_0(\cos\theta) + A_1P_1(\cos\theta) + A_2P_2(\cos\theta) + \cdots$$

given that $\quad f(\theta) = 1 \quad \text{from} \quad \theta = 0 \quad \text{to} \quad \theta = \frac{\pi}{2}$

and $\quad f(\theta) = 0 \quad \text{from} \quad \theta = \frac{\pi}{2} \quad \text{to} \quad \theta = \pi.$

This amounts to the same thing as developing $F(x)$ into the series

$$F(x) = A_0P_0(x) + A_1P_1(x) + A_2P_2(x) + A_3P_3(x) + \cdots$$

[4]It should be noted that this proof is no more general than that of the last article, for, in order that Green's Theorem should apply to $r^m X_m$ this function and its first derivatives must be finite continuous and single-valued within and on the surface of the unit sphere. (v. Peirce, Newt. Pot. Func. § 48.)

where $\qquad F(x) = 0$ from $x = -1$ to $x = 0$

and $\qquad F(x) = 1$ from $x = 0$ to $x = 1.$

By Art. 90 (1) and (2)

$$A_0 = \frac{1}{2}\int_0^1 P_0(x)dx = \frac{1}{2}\int_0^1 dx = \frac{1}{2},$$

and any coefficient $\qquad A_m = \dfrac{(2m+1)}{2}\displaystyle\int_0^1 P_m(x)dx.$

By Art. 91, Ex. 1

$$\begin{aligned}\int_0^1 P_m(x)dx &= 0 \quad \text{if } m \text{ is even}\\ &= (-1)^{\frac{m-1}{2}}\frac{1}{m(m+1)}\frac{3.5.7.\cdots m}{2.4.6.\cdots(m-1)} \quad \text{if } m \text{ is odd.}\end{aligned}$$

Hence
$$\begin{aligned}A_m &= 0 \quad \text{if } m \text{ is even}\\ &= (-1)^{\frac{m-1}{2}}\frac{2m+1}{2m+2}.\frac{1.3.5.\cdots(m-2)}{2.4.6.\cdots(m-1)} \quad \text{if } m \text{ is odd.}\end{aligned}$$

Then
$$F(x) = \frac{1}{2} + \frac{3}{4}P_1(x) - \frac{7}{8}.\frac{1}{2}P_3(x) + \frac{11}{12}.\frac{1.3}{2.4}P_5(x) - \cdots \tag{1}$$

and
$$u = \frac{1}{2} + \frac{3}{4}rP_1(\cos\theta) - \frac{7}{8}.\frac{1}{2}r^3P_3(\cos\theta) + \frac{11}{12}.\frac{1.3}{2.4}r^5P_5(\cos\theta) + \cdots \tag{2}$$

for any point within the sphere.

94. If in a problem on the Potential Function the value of V is given at every point of a spherical surface and has circular symmetry[5] about a diameter of that surface the value of V at any point in space can be obtained.

We have to solve Laplace's Equation in the form

$$rD_r^2(rV) + \frac{1}{\sin\theta}D_\theta(\sin\theta D_\theta V) = 0 \tag{1}$$

subject to the conditions

$$\begin{aligned}V &= f(\theta) \quad \text{when} \quad r = a\\ V &= 0 \qquad \text{“} \qquad r = \infty.\end{aligned}$$

We have $\qquad f(\theta) = A_0P_0(\cos\theta) + A_1P_1(\cos\theta) + A_2P_2(\cos\theta) + \cdots$

where $\qquad A_m = \dfrac{(2m+1)}{2}\displaystyle\int_0^\pi f(\theta)P_m(\cos\theta)\sin\theta.d\theta.$ $\qquad$ v. Art. 90 (4).

[5]See note on page 12.

Hence

$$V = A_0 + A_1 \left(\frac{r}{a}\right) P_1(\cos\theta) + A_2 \left(\frac{r}{a}\right)^2 P_2(\cos\theta) + A_3 \left(\frac{r}{a}\right)^3 P_3(\cos\theta) + \cdots \quad (2)$$

is the required solution for a point within the sphere, and

$$V = A_0 \left(\frac{a}{r}\right) + A_1 \left(\frac{a}{r}\right)^2 P_1(\cos\theta) + A_2 \left(\frac{a}{r}\right)^3 P_2(\cos\theta) + A_3 \left(\frac{a}{r}\right)^4 P_3(\cos\theta) + \cdots \quad (3)$$

is the required solution for an external point.

EXAMPLES.

1. If on the surface of a sphere of radius c V is constant and equal to a show that $V = a$ for any point within the sphere and $V = \frac{ac}{r}$ for any external point.

2. Two equal thin hemispherical shells of radius c placed together to form a spherical surface are separated by a thin non-conducting layer. Charges of statical electricity are placed on the two hemispheres one of which is then found to be at potential a and the other at potential b. Find the value of the potential function at any point.

$$V = \frac{a+b}{2} + (b-a)\left[\frac{3}{4}\frac{r}{c}P_1(\cos\theta) - \frac{7}{8}\cdot\frac{1}{2}\frac{r^3}{c^3}P_3(\cos\theta) + \frac{11}{12}\cdot\frac{1.3}{2.4}\frac{r^5}{c^5}P_5(\cos\theta) - \cdots\right]$$

for an internal point

$$V = \frac{a+b}{2}\cdot\frac{c}{r} + (b-a)\left[\frac{3}{4}\frac{c^2}{r^2}P_1(\cos\theta) - \frac{7}{8}\cdot\frac{1}{2}\frac{c^4}{r^4}P_3(\cos\theta) + \frac{11}{12}\cdot\frac{1.3}{2.4}\frac{c^6}{r^6}P_5(\cos\theta) - \cdots\right]$$

for an external point.

3. If $V_1 = f(\cos\theta)$ when $r = a$ and $V_1 = 0$ when $r = b$ show that for $a < r < b$

$$V_1 = \sum_{m=0}^{m=\infty} A_m \left(\frac{b^{m+1}}{r^{m+1}} - \frac{r^m}{b^m}\right)\left(\frac{b^{m+1}}{a^{m+1}} - \frac{a^m}{b^m}\right)^{-1} P_m(\cos\theta)$$

where
$$A_m = \frac{2m+1}{2}\int_{-1}^{1} f(x)P_m(x)dx.$$

4. If $V_2 = F(\cos\theta)$ when $r = b$ and $V_2 = 0$ when $r = a$ show that for $a < r < b$

$$V_2 = \sum_{m=0}^{m=\infty} B_m \left(\frac{r^m}{a^m} - \frac{a^{m+1}}{r^{m+1}}\right)\left(\frac{b^m}{a^m} - \frac{a^{m+1}}{b^{m+1}}\right)^{-1} P_m(\cos\theta)$$

where
$$B_m = \frac{2m+1}{2}\int_{-1}^{1} F(x)P_m(x)dx.$$

5. If the value of the potential function is given arbitrarily on the surfaces of a spherical shell but has circular symmetry[6] about a diameter $V = V_1 + V_2$ (v. Exs. 3 and 4).

6. Two concentric hollow spherical conductors are insulated and charged. The inner one of radius a is at potential p, and the outer one of radius b is at potential q. Find V for any point in space.

$$V = p \quad \text{if} \quad r < a,$$
$$V = \frac{pa}{b-a}\left(\frac{b}{r} - 1\right) + \frac{qb}{b-a}\left(1 - \frac{a}{r}\right) \quad \text{if} \quad a < r < b,$$
$$V = \frac{qb}{r} \quad \text{if} \quad r > b.$$

7. If $V = 0$ on the base of a hemisphere and $V = f(\cos\theta)$ on the convex surface, show that for a point within the hemisphere

$$V = \sum_{k=0}^{k=\infty} A_{2k+1}\left(\frac{r}{a}\right)^{2k+1} P_{2k+1}(\cos\theta)$$

where
$$A_{2k+1} = (4k+3)\int_{0}^{1} f(x)P_{2k+1}(x)dx$$
[v. Art. 90 (8)].

8. If the convex surface of a solid hemisphere of radius a is kept at the constant temperature unity and the base at the constant temperature zero show that after the permanent state of temperatures is set up the temperature of any internal point is

$$u = \frac{3}{2}\frac{r}{a}P_1(\cos\theta) - \frac{7}{4}\cdot\frac{1}{2}\frac{r^3}{a^3}P_3(\cos\theta) + \frac{11}{6}\cdot\frac{1.3}{2.4}\frac{r^5}{a^5}P_5(\cos\theta) - \cdots$$

9. A sphere of radius a and with blackened surface is exposed to the direct rays of the sun in air at the temperature zero. Find the *stationary temperature* of any internal point.

---

[6]See note on page 12

Suggestion: $D_r u + hu - Mf(\theta) = 0$ when $r = a$.

Let $$u = \sum A_m \frac{r^m}{a^m} P_m(\cos\theta), \quad \text{and} \quad f(\theta) = \sum B_m P_m(\cos\theta).$$

Then we have

$$\sum m\frac{A_m}{a} P_m(\cos\theta) + h\sum A_m P_m(\cos\theta) - M\sum B_m P_m(\cos\theta) = 0,$$

whence $$A_m = \frac{MB_m}{h + \dfrac{m}{a}}.$$

Here $f(\theta) = \cos\theta$ if $0 < \theta < \dfrac{\pi}{2}$ and $f(\theta) = 0$ if $\dfrac{\pi}{2} < \theta < \pi$.

$$f(\theta) = \frac{1}{4} + \frac{1}{2}P_1(\cos\theta) + \frac{5}{16}P_2(\cos\theta) - \frac{3}{32}P_4(\cos\theta) + \cdots$$
$$+ (-1)^{k+1}\frac{(4k+1)(2k)!}{(4k+4)(2k-1)2^{2k}(k!)^2}P_{2k}(\cos\theta) + \cdots$$

v. Art. 91 Exs. (2) and (3). cf. J. W. Strutt (Lord Rayleigh), Lond. Phil. Trans. vol. 160, page 587.

95. The formulas of Art. 90 enable us to develop a given function of x in terms of *Zonal Surface Harmonics*, the development holding true for values of x between -1 and $+1$. If, however, we can show by outside considerations that a given function of x can be expressed in Zonal Surface Harmonics, the development holding true for all values of x, the formulas of Art. 90 will give us the development in question.

For example if n is a positive integer x^n can be expressed in terms of Zonal Surface Harmonics no matter what the value of x, and no Harmonic of higher order than n will enter. For the formulas giving the values of $P_1(x)$, $P_2(x), \cdots P_n(x)$ (v. Art. 77) may be regarded as n algebraic equations of the first degree in terms of x, x^2, $x^3, \cdots x^n$ and $P_1(x)$, $P_2(x), \cdots P_n(x)$.

From these equations the $n-1$ quantities x, x^2, $x^3, \ldots \cdot x^{n-1}$, can be eliminated, and there will result an equation of the first degree in x^n and $P_1(x)$, $P_2(x), \cdots P_n(x)$, which will enable us to express x^n in the form

$$A_0 + A_1P_1(x) + A_2P_2(x) + \cdots + A_nP_n(x),$$

no matter what the value of x, and we shall have the same formula when $-1 < x < 1$ as when $x > 1$ or $x < -1$.

Let us obtain this development. By Art. 90 (1) and (2)

$$x^n = A_0P_0(x) + A_1P_1(x) + A_2P_2(x) + \cdots \tag{1}$$

where $$A_m = \frac{2m+1}{2}\int_{-1}^{1} x^n P_m(x)dx. \tag{2}$$

Then $$A_m = \frac{2m+1}{2}\frac{1}{2^m m!}\int_{-1}^{1} x^n \frac{d^m(x^2-1)^m}{dx^m}dx \qquad \text{by (1) Art. 83.}$$

By *integration by parts* we get

$$\int_{-1}^{1} x^n \frac{d^m(x^2-1)^m}{dx^m} dx = n(n-1)(n-2)\cdots(n-m+1)\int_{-1}^{1} x^{n-m}(1-x^2)^m dx,$$
$$\text{if} \quad m < n+1, \tag{3}$$
$$= 0 \quad \text{if} \quad m > n.$$

By *integration by parts* we readily obtain the reduction formula

$$\int_{-1}^{1} x^p(1-x^2)^q dx = \frac{2q}{p+1}\int_{-1}^{1} x^{p+2}(1-x^2)^{q-1}dx \qquad \text{whence}$$

$$\int_{-1}^{1} x^{n-m}(1-x^2)^m dx = \frac{2^m m!}{(n-m+1)(n-m+3)\cdots(n+m-1)}\int_{-1}^{1} x^{n+m}dx.$$
$$\int_{-1}^{1} x^{n+m}dx = \frac{2}{(n+m+1)} \quad \text{if} \quad n+m \quad \text{is even,}$$
$$= 0 \quad \text{if} \quad n+m \quad \text{is odd.}$$

Hence
$$A_m = \frac{(2m+1)n(n-1)(n-2)\cdots(n-m+1)}{(n-m+1)(n-m+3)(n-m+5)\cdots(n+m+1)}$$
$$\text{if } m < n+1 \text{ and } m+n \text{ is even,}$$
$$= 0 \quad \text{if } m > n \text{ or if } m+n \text{ is odd.}$$

Therefore

$$x^n = \frac{n!}{1.3.5\cdots(2n+1)}\Big[(2n+1)P_n(x) + (2n-3)\frac{(2n+1)}{2}P_{n-2}(x)$$
$$+ (2n-7)\frac{(2n+1)(2n-1)}{2.4}P_{n-4}(x)$$
$$+ (2n-11)\frac{(2n+1)(2n-1)(2n-3)}{2.4.6}P_{n-6}(x) + \cdots\Big] \tag{4}$$

the second member ending with the term $\frac{1}{n+1}P_0(x)$ if n is even and with the term $\frac{3}{n+2}P_1(x)$ if n is odd.

For convenience of reference we write out a few powers of x.

$$\left.\begin{aligned}
x^0 &= 1 = P_0(x)\\
x &= P_1(x)\\
x^2 &= \frac{2}{3}P_2(x) + \frac{1}{3}P_0(x)\\
x^3 &= \frac{2}{5}P_3(x) + \frac{3}{5}P_1(x)\\
x^4 &= \frac{8}{35}P_4(x) + \frac{4}{7}P_2(x) + \frac{1}{5}P_0(x)\\
x^5 &= \frac{8}{63}P_5(x) + \frac{4}{9}P_3(x) + \frac{3}{7}P_1(x)\\
x^6 &= \frac{16}{231}P_6(x) + \frac{24}{77}P_4(x) + \frac{10}{21}P_2(x) + \frac{1}{7}P_0(x)\\
x^7 &= \frac{16}{429}P_7(x) + \frac{8}{39}P_5(x) + \frac{14}{33}P_3(x) + \frac{1}{3}P_1(x)\\
x^8 &= \frac{128}{6435}P_8(x) + \frac{64}{495}P_6(x) + \frac{48}{143}P_4(x) + \frac{40}{99}P_2(x) + \frac{1}{9}P_0(x).
\end{aligned}\right\} \qquad (5)$$

If a given function of x can be expressed as a *terminating power series* it can be developed into a Zonal Harmonic Series by the aid of (4). Given that

$$f(x) = a_0 + a_1 x + a_2 x^2 + a_3 x^3 + \cdots,$$

let
$$f(x) = B_0 + B_1 P_1(x) + B_2 P_2(x) + B_3 P_3(x) + \cdots;$$

then picking out carefully the coefficient of $P_m(x)$ we have

$$B_m = \frac{m!}{1.3.5.\cdots(2m-1)}\left[a_m + \frac{(m+1)(m+2)}{2.(2m+3)}a_{m+2} + \frac{(m+1)(m+2)(m+3)(m+4)}{2.4.(2m+3)(2m+5)}a_{m+4} + \cdots\right]. \qquad (6)$$

96. The development of $\frac{dP_n(x)}{dx}$ is useful and is easily obtained.

Let
$$\frac{dP_n(x)}{dx} = A_0P_0(x) + A_1P_1(x) + A_2P_2(x) + \cdots$$

Then
$$A_m = \frac{2m+1}{2}\int_{-1}^{1} P_m(x)\frac{dP_n(x)}{dx}dx \qquad (1)$$

by Art. 90 (2);

$$\int_{-1}^{1} P_m(x)\frac{dP_n(x)}{dx}dx = \Big[P_m(x)P_n(x)\Big]_{x=-1}^{x=1} - \int_{-1}^{1} P_n(x)\frac{dP_m(x)}{dx}dx. \qquad (2)$$

$$\Big[P_m(x)P_n(x)\Big]_{x=-1}^{x=1} = 0 \quad \text{if } m+n \text{ is even}$$
$$= 2 \quad \text{if } m+n \text{ is odd.}$$

Since $P_n(x)$ is an algebraic polynomial of the nth degree in x, $\dfrac{dP_n(x)}{dx}$ is an algebraic polynomial of the $n-1$st degree in x. Therefore in (1) m is less than n; consequently $\dfrac{dP_m(x)}{dx}$ is an algebraic polynomial in x of lower degree than n and

$$\int_{-1}^{1} P_n(x)\frac{dP_m(x)}{dx}dx = 0 \qquad \text{by Art. 95 (3).}$$

We get then $\quad A_m = 2m+1 \quad$ if $m+n$ is odd and $m<n$,

$= 0 \quad$ if $m+n$ is even or $m>n-1$; and

$$\frac{dP_n(x)}{dx} = (2n-1)P_{n-1}(x) + (2n-5)P_{n-3}(x) + (2n-9)P_{n-5}(x) + \cdots \tag{3}$$

the second member ending with the term $3P_1(x)$ if n is even and with the term $P_0(x)$ if n is odd.

From (3) a number of simple formulas are readily obtained. For example

$$\frac{dP_{n+1}(x)}{dx} - \frac{dP_{n-1}(x)}{dx} = (2n+1)P_n(x) \tag{4}$$

$$\int_{x}^{1} P_n(x)dx = \frac{1}{2n+1}[P_{n-1}(x) - P_{n+1}(x)]. \tag{5}$$

$$(2n+1)x\frac{dP_n(x)}{dx} = n\frac{dP_{n+1}(x)}{dx} + (n+1)\frac{dP_{n-1}(x)}{dx} \tag{6}$$

[v. (4) and Article 77 (12)].

$$(x^2-1)\frac{dP_n(x)}{dx} = nxP_n(x) - nP_{n-1}(x) \tag{7}$$

[v. (5) and Article 91 (7).]

97. By the aid of the formulas of Art. 96 a number of valuable developments can be obtained.

Let us get $\cos n\theta$ and $\sin n\theta$, n being any positive real.

$z = \cos n\theta$ and $z = \sin n\theta$ are solutions of the equation

$$\frac{d^2z}{d\theta^2} + n^2 z = 0$$

or if we let $x = \cos\theta$, of the equation

$$(1-x^2)\frac{d^2z}{dx^2} - x\frac{dz}{dx} + n^2 z = 0. \tag{1}$$

Let $$a_0P_0(x) + a_1P_1(x) + a_2P_2(x) + \cdots$$

be the required development of $\cos n\theta$ or of $\sin n\theta$.

Then $$\sum_{m=0}^{m=\infty} a_m \left[(1-x^2)\frac{d^2P_m(x)}{dx^2} - x\frac{dP_m(x)}{dx} + n^2P_m(x)\right] = 0 \qquad \text{by (1).}$$

$z = P_m(x)$ is a solution of Legendre's Equation (v. Art. 77). Hence

$$(1-x^2)\frac{d^2P_m(x)}{dx^2} - x\frac{dP_m(x)}{dx} = x\frac{dP_m(x)}{dx} - m(m+1)P_m(x),$$

and (1) becomes

$$\sum_{m=0}^{m=\infty} a_m \left[x\frac{dP_m(x)}{dx} + [n^2 - m(m+1)]P_m(x)\right] = 0. \tag{2}$$

Formulas (4) and (6) of Art. 96 enable us to throw (2) into the form

$$\sum_{m=0}^{m=\infty} a_m \left[\frac{n^2-m^2}{2m+1}\frac{dP_{m+1}(x)}{dx} - \frac{n^2-(m+1)^2}{2m+1}\frac{dP_{m-1}(x)}{dx}\right] = 0. \tag{3}$$

(3) must be identically true. Therefore the coefficient of $\dfrac{dP_{m+1}(x)}{dx}$ must equal zero, and we have

$$a_{m+2} = \frac{2m+5}{2m+1}.\frac{n^2-m^2}{n^2-(m+3)^2}a_m. \tag{4}$$

If we are developing $\cos n\theta$

$$a_0 = \frac{1}{2}\int_0^\pi \cos n\theta \sin\theta . d\theta \qquad \text{by Art. 90 (4),}$$

$$= \frac{1}{4}\int_0^\pi [\sin(n+1)\theta - \sin(n-1)\theta]d\theta,$$

$$a_0 = -\frac{1}{2}.\frac{1+\cos n\pi}{n^2-1}; \tag{5}$$

and $$a_1 = \frac{3}{2}\int_0^\pi \cos n\theta \cos\theta \sin\theta . d\theta \qquad \text{by Art. 90 (4),}$$

$$a_1 = -\frac{3}{2}.\frac{1-\cos n\pi}{n^2-4}. \tag{6}$$

(4), (5), and (6) give us

$$\cos n\theta = -\frac{1+\cos n\pi}{2(n^2-1)}\Big[P_0(\cos\theta) + 5\frac{n^2}{n^2-3^2}P_2(\cos\theta) + 9\frac{n^2(n^2-2^2)}{(n^2-3^2)(n^2-5^2)}P_4(\cos\theta)+\cdots\Big] - \frac{1-\cos n\pi}{2(n^2-2^2)}\Big[3P_1(\cos\theta) + 7\frac{n^2-1^2}{n^2-4^2}P_3(\cos\theta) + 11\frac{(n^2-1^2)(n^2-3^2)}{(n^2-4^2)(n^2-6^2)}P_5(\cos\theta)+\cdots\Big]. \quad (7)$$

If n is a whole number $1+\cos n\pi$ or $1-\cos n\pi$ will vanish and the series will end with the term involving $P_n(\cos\theta)$. For this case (7) may be rewritten

$$\cos n\theta = \frac{1}{2}.\frac{2.4.6.\cdots 2n}{3.5.7.\cdots(2n+1)}\Big[(2n+1)P_n(\cos\theta) + (2n-3)\frac{n^2-(n+1)^2}{n^2-(n-2)^2}P_{n-2}(\cos\theta) + (2n-7)\frac{[n^2-(n+1)^2][n^2-(n-1)^2]}{[n^2-(n-2)^2][n^2-(n-4)^2]}P_{n-4}(\cos\theta)+\cdots\Big]. \quad (8)$$

If we are developing $\sin n\theta$

$$a_0 = \frac{1}{2}\int_0^\pi \sin n\theta\sin\theta.d\theta = -\frac{1}{2}.\frac{\sin n\pi}{n^2-1},$$

$$a_1 = \frac{3}{2}\int_0^\pi \sin n\theta\cos\theta\sin\theta.d\theta = \frac{3}{2}.\frac{\sin n\pi}{n^2-2^2}$$ and

$$\sin n\theta = -\frac{1}{2}.\frac{\sin n\pi}{n^2-1}\Big[P_0(\cos\theta) + 5\frac{n^2}{n^2-3^2}P_2(\cos\theta) + 9\frac{n^2(n^2-2^2)}{(n^2-3^2)(n^2-5^2)}P_4(\cos\theta)+\cdots\Big] + \frac{1}{2}.\frac{\sin n\pi}{n^2-2^2}\Big[3P_1(\cos\theta) + 7\frac{n^2-1^2}{n^2-4^2}P_3(\cos\theta) + 11\frac{(n^2-1^2)(n^2-3^2)}{(n^2-4^2)(n^2-6^2)}P_5(\cos\theta)+\cdots\Big]. \quad (9)$$

If n is a whole number $\sin n\pi = 0$, and all the terms of (9) vanish except those involving $P_{n-1}(\cos\theta)$, $P_{n+1}(\cos\theta)$, $P_{n+3}(\cos\theta)$ &c., which become indeterminate. For this case it is necessary to compute a_{n-1} independently.

We have

$$a_{n-1} = \frac{2n-1}{2}\int_0^\pi \sin n\theta P_{n-1}(\cos\theta)\sin\theta . d\theta$$

$$= \frac{2n-1}{4}\int_0^\pi [\cos(n-1)\theta - \cos(n+1)\theta]P_{n-1}(\cos\theta)d\theta.$$

Hence $$a_{n-1} = \frac{2n-1}{4}.\frac{1.3.5.\cdots(2n-3)}{2.4.6.\cdots(2n-2)}\pi \qquad \text{[v. Art. 82 (1)]},$$

and

$$\sin n\theta = \frac{\pi}{4}.\frac{1.3.\cdots(2n-3)}{2.4.\cdots(2n-2)}\Big[(2n-1)P_{n-1}(\cos\theta)$$
$$+ (2n+3)\frac{n^2-(n-1)^2}{n^2-(n+2)^2}P_{n+1}(\cos\theta)$$
$$+ (2n+7)\frac{[n^2-(n-1)^2][n^2-(n+1)^2]}{[n^2-(n+2)^2][n^2-(n+4)^2]}P_{n+3}(\cos\theta) + \cdots\Big]. \quad (10)$$

EXAMPLES.

1. Show that

$$\csc\theta = \frac{\pi}{2}\Big[1 + 5\Big(\frac{1}{2}\Big)^2 P_2(\cos\theta) + 9\Big(\frac{1.3}{2.4}\Big)^2 P_4(\cos\theta) + 13\Big(\frac{1.3.5}{2.4.6}\Big)^2 P_6(\cos\theta) + \cdots\Big]$$

whence

$$\frac{1}{\sqrt{1-x^2}} = \frac{\pi}{2}\Big[1 + 5\Big(\frac{1}{2}\Big)^2 P_2(x) + 9\Big(\frac{1.3}{2.4}\Big)^2 P_4(x) + 13\Big(\frac{1.3.5}{2.4.6}\Big)^2 P_6(x) + \cdots\Big]$$

[v. Art. 90 (4) and Art. 82].

2. Show that

$$\operatorname{ctn}\theta = \frac{\pi}{2}\Big[3\Big(\frac{1}{2}\Big)P_1(\cos\theta) + 7\Big(\frac{3}{4}\Big)\Big(\frac{1}{2}\Big)^2 P_3(\cos\theta) + 11\Big(\frac{5}{6}\Big)\Big(\frac{1.3}{2.4}\Big)^2 P_5(\cos\theta) + \cdots\Big]$$

whence

$$\frac{1}{\sqrt{1-x^2}} = \frac{\pi}{2}\Big[3\Big(\frac{1}{2}\Big)P_1(x) + 7\Big(\frac{3}{4}\Big)\Big(\frac{1}{2}\Big)^2 P_3(x) + 11\Big(\frac{5}{6}\Big)\Big(\frac{1.3}{2.4}\Big)^2 P_5(x) + \cdots\Big]$$

[v. Art. 90 (4) and Art. 82].

3. By integrating the result of Ex. 1 and simplifying by the aid of Art. 96 (5), obtain the development

$$\sin^{-1} x = \frac{\pi}{2}\Big[3\Big(\frac{1}{2}\Big)^2 P_1(x) + 7\Big(\frac{1}{2.4}\Big)^2 P_3(x)$$
$$+ 11\Big(\frac{1.3}{2.4.6}\Big)^2 P_5(x) + 15\Big(\frac{1.3.5}{2.4.6.8}\Big)^2 P_7(x) + \cdots\Big]$$

whence
$$\theta = \frac{\pi}{2}\Big[P_0(\cos\theta) - 3\Big(\frac{1}{2}\Big)^2 P_1(\cos\theta) - 7\Big(\frac{1}{2.4}\Big)^2 P_3(\cos\theta)$$
$$-11\Big(\frac{1.3}{2.4.6}\Big)^2 P_5(\cos\theta) - \cdots\Big].$$

4. By integrating the result of Ex. 2 and simplifying by the aid of Art. 96 (5) obtain

$$\sqrt{1-x^2} = \frac{\pi}{2}\Big[\frac{1}{2} - 5\Big(\frac{1}{4}\Big)\Big(\frac{1}{2}\Big)^2 P_2(x) - 9\Big(\frac{3}{6}\Big)\Big(\frac{1}{2.4}\Big)^2 P_4(x)$$
$$-13\Big(\frac{5}{8}\Big)\Big(\frac{1.3}{2.4.6}\Big)^2 P_6(x) + \cdots\Big]$$

whence

$$\sin\theta = \frac{\pi}{2}\Big[\frac{1}{2}P_0(\cos\theta) - 5\Big(\frac{1}{4}\Big)\Big(\frac{1}{2}\Big)^2 P_2(\cos\theta) - 9\Big(\frac{3}{6}\Big)\Big(\frac{1}{2.4}\Big)^2 P_4(\cos\theta) - \cdots\Big].$$

To make clearer the analogy of development in Zonal Harmonic Series with development in Fourier's Series we give on page 186 a cut representing the first seven Surface Zonal Harmonics $P_1(\cos\theta)$, $P_2(\cos\theta)$, $\cdots$ $P_7(\cos\theta)$, which are of course somewhat complicated Trigonometric curves resembling roughly $\cos\theta$, $\cos 2\theta$, $\cdots$ $\cos 7\theta$; and on page 187, the first four successive approximations to the Zonal Harmonic Series

$$\frac{1}{2} + \frac{3}{4}P_1(\cos\theta) - \frac{7}{8}.\frac{1}{2}P_3(\cos\theta) + \frac{11}{12}.\frac{1.3}{2.4}P_5(\cos\theta) - \cdots \qquad \text{[I]}$$

[v. (1) Art. 93], and

$$\frac{\pi}{2}\Big[P_0(\cos\theta) - 3\Big(\frac{1}{2}\Big)^2 P_1(\cos\theta) - 7\Big(\frac{1}{2.4}\Big)^2 P_3(\cos\theta)$$
$$- 11\Big(\frac{1.3}{2.4.6}\Big)^2 P_5(\cos\theta) - \cdots\Big] \qquad \text{[II]}$$

(v. Ex. 3 Art. 97).

[I] is equal to 1 from $\theta = 0$ to $\theta = \dfrac{\pi}{2}$, and to 0 from $\theta = \dfrac{\pi}{2}$ to $\theta = \pi$; and [II] is equal to θ from $\theta = 0$ to $\theta = \pi$.

The figures on page 187 are constructed on precisely the same principle as those on pages 64 and 65, with which they should be carefully compared.

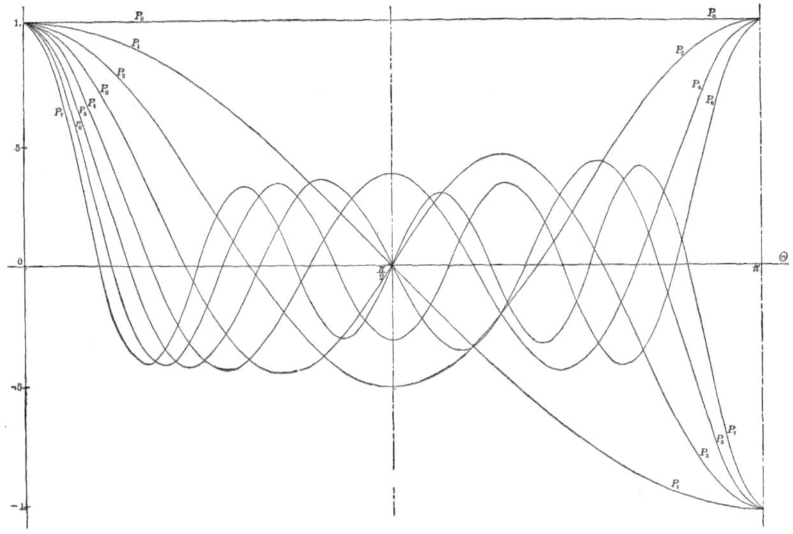

The curves $y = P_0(\cos\theta)$, $y = P_1(\cos\theta)$, $\ldots y = P_7(\cos\theta)$. (v. page 185.)

98. By applying *Gauss's Theorem* (B. O. Peirce, Newt. Pot. Func. § 31) or the special Form of *Green's Theorem*,

$$\iiint \nabla^2 V dxdydz = \int D_n V ds = -4\pi \iiint \rho dxdydz,$$

[Peirce, N. P. F. § 49 (149)] to a box cut from an infinitely thin shell of attracting matter by a tube of force whose end is an element of the surface of the shell we readily obtain the important result

$$4\pi\rho\kappa = D_n V_1 - D_n V_2. \tag{1}$$

where ρ is the density and κ the thickness of the shell, V_1 the value of the potential function due to the shell at an internal point and V_2 its value at an external point, and where D_n is the partial derivative along the external normal to the outer surface of the shell.

If we have to deal with a surface distribution of matter we have only to replace $\rho\kappa$ in (1) by σ where σ is the surface density, whence

$$4\pi\sigma = D_n V_1 - D_n V_2 \tag{2}$$

(v. Peirce, N. P. F. §§ 45, 46, and 47).

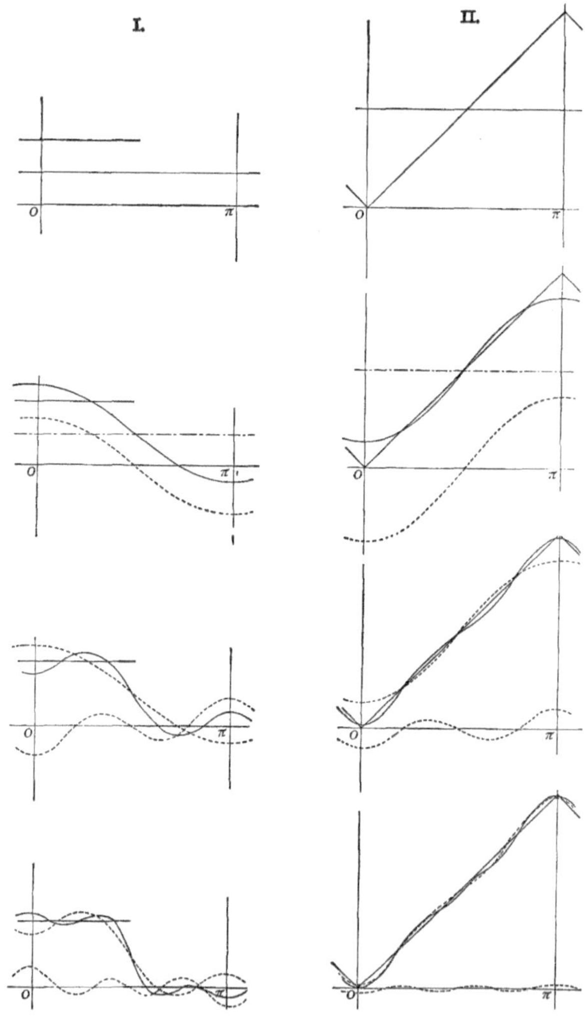

Formulas (1) and (2) enable us to solve problems in attraction when we know the density of the attracting mass, and problems in Statical Electricity when we know the distribution of the charge, by methods analogous to that of Art. 94.

For example let us find the value of the potential function due to a thin material spherical shell of density ρ and radius a.

Since V must be a solution of Laplace's Equation and must be finite both when $r = 0$ and $r = \infty$ we have

$$V_1 = \sum A_m r^m P_m(\cos\theta)$$

$$V_2 = \sum B_m \frac{1}{r^{m+1}} P_m(\cos\theta).$$

V_1 and V_2 must approach the same limiting values as r approaches a. Hence

$$\frac{B_m}{a^{m+1}} = A_m a^m$$

or

$$B_m = A_m a^{2m+1}.$$

$$D_n V_1 = D_r V_1 = \sum m r^{m-1} A_m P_m(\cos\theta),$$

$$D_n V_2 = D_r V_2 = -\sum (m+1) \frac{A_m a^{2m+1}}{r^{m+2}} P_m(\cos\theta).$$

Therefore by (1)

$$4\pi\rho\kappa = \sum (2m+1) A_m a^{m-1} P_m(\cos\theta)$$

if κ is the thickness of the shell.

Let

$$\rho = f(\cos\theta) = \sum C_m P_m(\cos\theta)$$

where

$$C_m = \frac{2m+1}{2} \int_{-1}^{1} f(x) P_m(x) dx \qquad \text{by Art. 90 (2).}$$

Then

$$4\pi\kappa C_m = (2m+1) A_m a^{m-1}, \qquad \text{and}$$

$$A_m = \frac{4\pi\kappa C_m}{(2m+1)a^{m-1}}, \quad \text{and} \quad B_m = \frac{4\pi\kappa}{2m+1} C_m a^{m+2},$$

and

$$V_1 = 4\pi a\kappa \sum \frac{C_m}{2m+1} \frac{r_m}{a_m} P_m(\cos\theta), \tag{3}$$

and

$$V_2 = 4\pi a\kappa \sum \frac{C_m}{2m+1} \frac{a^{m+1}}{r^{m+1}} P_m(\cos\theta). \tag{4}$$

99. We can now get the value of the potential function due to a spherical shell of finite thickness, provided that its density can be expressed as a sum of terms of the form $Cr^k P_m(\cos\theta)$.

Let a be the radius of the outer surface and b be the radius of the inner surface of the shell.

1st.—Let $\rho = Cr^k P_m(\cos\theta)$. Then for the shell of radius s and thickness ds

$$V_1 = 4\pi s ds \frac{Cs^k}{2m+1} \frac{r^m}{s^m} P_m(\cos\theta) \qquad \text{by (3) Art. 98,}$$

and

$$V_2 = 4\pi s ds \frac{Cs^k}{2m+1} \frac{s^{m+1}}{r^{m+1}} P_m(\cos\theta) \qquad \text{by (4) Art. 98.}$$

Then if $r < b$

$$V = \int_b^a V_1 = \frac{4\pi C}{(2m+1)} \frac{(a^{k-m+2} - b^{k-m+2})}{(k-m+2)} r^m P_m(\cos\theta), \tag{1}$$

if $r > a$

$$V = \int_b^a V_2 = \frac{4\pi C}{(2m+1)} \frac{(a^{k+m+3} - b^{k+m+3})}{(k+m+3)} \frac{P_m(\cos\theta)}{r^{m+1}}, \tag{2}$$

and if $b < r < a$

$$V = \int_b^r V_2 + \int_r^a V_1 = \frac{4\pi C}{2m+1} \left[\frac{r^{k+m+3} - b^{k+m+3}}{(k+m+3) r^{m+1}} + \frac{a^{k-m+2} - r^{k-m+2}}{(k-m+2)} r^m \right] P_m(\cos\theta). \tag{3}$$

2d.—If $\rho = \sum C_m r^k P_m(\cos\theta)$ the solutions will consist of sums of terms of the forms given in (1), (2), and (3).

EXAMPLES.

1. If the shell is homogeneous

$$V = 2\pi\rho(a^2 - b^2) \quad \text{if} \quad r < b,$$
$$V = \frac{4}{3}\pi\rho(a^3 - b^3)\frac{1}{r} = \frac{M}{r} \quad \text{if} \quad r > a,$$
$$V = 2\pi\rho\left[a^2 - \frac{2b^3}{3r} - \frac{r^2}{3}\right] \quad \text{if} \quad b < r < a.$$

2. If the density is any given function of the distance from the centre $V = \dfrac{M}{r}$ if $r > a$, and $V =$ a constant if $r < b$.

3. If the density at any point of a solid sphere is proportional to the square of the distance from a diametral plane

$$V = \frac{M}{a}\left[\frac{a}{r} + \frac{2}{7}\frac{a^3}{r^3} P_2(\cos\theta)\right] \quad \text{if} \quad r > a.$$

4. If the density at any point of a solid sphere is proportional to its distance from a diametral plane

$$V = \frac{M}{a}\left[\frac{a}{r} + \frac{1}{6}\frac{a^3}{r^3}P_2(\cos\theta) - \frac{1.1}{6.8}\frac{a^5}{r^5}P_4(\cos\theta) + \frac{1.1.3}{6.8.10}\frac{a^7}{r^7}P_6(\cos\theta) - \cdots\right]$$

if $r > a$. Compare Ex. 2 Art. 80.

100. We have seen in Art. 18 (c) (3) that

$$Q_m(x) = CP_m(x)\int \frac{dx}{(1-x^2)[P_m(x)]^2}, \tag{1}$$

no constant term being understood with $\int \frac{dx}{(1-x^2)[P_m(x)]^2}$.

$\frac{1}{(1-x^2)[P_m(x)]^2}$ is a rational fraction and becomes infinite only for $x = 1$, $x = -1$, and for the roots of $P_m(x) = 0$, all of which are real and lie between -1 and 1, as can be proved by the aid of the relation $P_m(x) = \frac{1}{2^m m!}\frac{d^m(x^2-1)^m}{dx^m}$.

If $x^2 > 1$ $\int\limits_x^\infty \frac{dx}{(1-x^2)[P_m(x)]^2}$ is finite and determinate and contains no constant term. Hence if $x^2 > 1$

$$Q_m(x) = -P_m(x)\int\limits_x^\infty \frac{dx}{(1-x^2)[P_m(x)]^2} = P_m(x)\int\limits_x^\infty \frac{dx}{(x^2-1)[P_m(x)]^2} \tag{2}$$

for the constant factor of $Q_m(x)$ has been chosen so that $C = -1$.

If $x^2 < 1$ the second member of (2) is not finite and determinate, and we are thrown back to the form (1), and C proves to be unity.

(1) gives us readily

$$Q_0(x) = \frac{1}{2}\log\frac{1+x}{1-x} \tag{3}$$

$$Q_1(x) = -1 + \frac{x}{2}\log\frac{1+x}{1-x} \tag{4}$$

if $x^2 < 1$.

(2) gives us

$$Q_0(x) = \frac{1}{2}\log\frac{x+1}{x-1} \tag{5}$$

$$Q_1(x) = -1 + \frac{x}{2}\log\frac{x+1}{x-1} \tag{6}$$

if $x^2 > 1$.

From Art. 85 (10) it follows that

$$Q_m(x) = C\frac{d^m}{dx^m}\left[(x^2-1)^m\int\limits_0^x \frac{dx}{(x^2-1)^{m+1}}\right] \quad \text{if} \quad x^2 < 1,$$

$$= C\frac{d^m}{dx^m}\left[(x^2-1)^m\int\limits_x^\infty \frac{dx}{(x^2-1)^{m+1}}\right] \quad \text{if} \quad x^2 > 1.$$

C can be determined and is equal to $\frac{(-1)^{m+1}2^m m!}{(2m)!}$ if $x^2 < 1$, and is equal to $\frac{(-1)^m 2^m m!}{(2m)!}$ if $x^2 > 1$.

Hence
$$Q_m(x) = \frac{(-1)^{m+1}2^m m!}{(2m)!}\frac{d^m}{dx^m}\left[(x^2-1)^m\int\limits_0^x \frac{dx}{(x^2-1)^{m+1}}\right] \tag{7}$$

if $x^2 < 1$,

and
$$Q_m(x) = \frac{(-1)^m 2^m m!}{(2m)!}\frac{d^m}{dx^m}\left[(x^2-1)^m \int_x^\infty \frac{dx}{(x^2-1)^{m+1}}\right] \tag{8}$$

if $x^2 > 1$.

(7) and (8) give us for $Q_0(x)$ and $Q_1(x)$ the values already written in (3), (4), (5), and (6).

By the repeated application of the formula

$$(m+1)Q_{m+1}(x) - (2m+1)xQ_m(x) + mQ_{m-1}(x) = 0, \tag{9}$$

which may be obtained for the case where $x^2 < 1$ from Art. 16 (13) and (14), and for the case where $x^2 > 1$ from Art. 16 (9), any Surface Zonal Harmonic of the Second Kind can be obtained from $Q_0(x)$ and $Q_1(x)$ as given in (3), (4), (5), and (6).

Analogous formulas for $p_m(x)$ and $q_m(x)$ can be obtained without difficulty from Art. 16 (4) and (5). They are

$$(m+1)^2 q_{m+1}(x) - (2m+1)xp_m(x) - m^2 q_{m-1}(x) = 0 \tag{10}$$

and
$$p_{m+1}(x) + (2m+1)xq_m(x) - p_{m-1}(x) = 0 \tag{11}$$

and they hold good for any value of m.

EXAMPLES.

1. Confirm the values of $Q_0(x)$ and $Q_1(x)$ given in Art. 100 (3), (4), (5), and (6) by expanding them and comparing them with Art. 16 (13), (14), and (9).

2. If the value of V on the surface of a cone of revolution can be expressed in terms of whole powers positive or negative of r, V can be found for any point in space, cf. Art. 81.

If $V = \sum\left(A_m r^m + \frac{B_m}{r^{m+1}}\right)$ when $\theta = \alpha$ then

$$V = \sum\left(A_m r^m + \frac{B_m}{r^{m+1}}\right)\frac{P_m(\cos\theta)}{P_m(\cos\alpha)}.$$

3. If $V = \sum\left(A_m r^m + \frac{B_m}{r^{m+1}}\right)$ when $\theta = \alpha$, and $V = 0$ when $\theta = \beta$,

$$V = \sum\left(A_m r^m + \frac{B_m}{r^{m+1}}\right)\left[\frac{Q_m(\cos\beta)P_m(\cos\theta) - P_m(\cos\beta)Q_m(\cos\theta)}{P_m(\cos\alpha)Q_m(\cos\beta) - P_m(\cos\beta)Q_m(\cos\alpha)}\right].$$

4. Find V for points corresponding to values of θ between α and β when V can be given in terms of whole powers of r for $\theta = \alpha$ and for $\theta = \beta$.

5. Find by the method of Art. 16 solutions of Legendre's Equation of the form

$$z = {}_1P_m(x) = 1 + \frac{m(m+1)}{2}(x-1) + \frac{(m-1)m(m+1)(m+2)}{2^2(2!)^2}(x-1)^2 + \frac{(m-2)(m-1)m(m+1)(m+2)(m+3)}{2^3(3!)^2}(x-1)^3 + \cdots,$$

$$z = {}_{-1}P_m(x) = 1 - \frac{m(m+1)}{2}(x+1) + \frac{(m-1)m(m+1)(m+2)}{2^2(2!)^2}(x+1)^2 + \frac{(m-2)(m-1)m(m+1)(m+2)(m+3)}{2^3(3!)^2}(x+1)^3 + \cdots.$$

If m is a whole number, ${}_1P_m(x) = P_m(x)$ and ${}_{-1}P_m(x) = (-1)^m P_m(x)$. No matter what the value of m, ${}_1P_m(x)$ is absolutely convergent for $-1 < x < 3$, and ${}_{-1}P_m(x)$ is absolutely convergent for $-3 < x < 1$.

6. By the aid of (7) Art. 16 show that

$$V = \frac{1}{\sqrt{r}}\sin(n\log r)k_n(\cos\theta), \qquad V = \frac{1}{\sqrt{r}}\sin(n\log r)l_n(\cos\theta),$$

$$V = \frac{1}{\sqrt{r}}\cos(n\log r)k_n(\cos\theta), \qquad V = \frac{1}{\sqrt{r}}\cos(n\log r)l_n(\cos\theta),$$

are solutions of Laplace's Equation

$$rD_r^2(rV) + \frac{1}{\sin\theta}D_\theta(\sin\theta D_\theta V) = 0,$$

if

$$k_n(x) = p_{-\frac{1}{2}+ni}(x) = 1 + \frac{n^2 + \left(\frac{1}{2}\right)^2}{2!}x^2 + \frac{\left[n^2 + \left(\frac{1}{2}\right)^2\right]\left[n^2 + \left(\frac{5}{2}\right)^2\right]}{4!}x^4 + \frac{\left[n^2 + \left(\frac{1}{2}\right)^2\right]\left[n^2 + \left(\frac{5}{2}\right)^2\right]\left[n^2 + \left(\frac{9}{2}\right)^2\right]}{6!}x^6 + \cdots$$

and

$$l_n(x) = -q_{-\frac{1}{2}+ni}(x) = x + \frac{n^2 + \left(\frac{3}{2}\right)^2}{3!}x^3 + \frac{\left[n^2 + \left(\frac{3}{2}\right)^2\right]\left[n^2 + \left(\frac{7}{2}\right)^2\right]}{5!}x^5 + \frac{\left[n^2 + \left(\frac{3}{2}\right)^2\right]\left[n^2 + \left(\frac{7}{2}\right)^2\right]\left[n^2 + \left(\frac{11}{2}\right)^2\right]}{7!}x^7 + \cdots$$

$k_n(x)$ and $l_n(x)$ are convergent if $x^2 < 1$, but are divergent if $x^2 = 1$.

7. Show by the aid of Example 5 that

$$V = \frac{1}{\sqrt{r}}\sin(n\log r)K_n(\cos\theta), \qquad V = \frac{1}{\sqrt{r}}\sin(n\log r)K_n(-\cos\theta),$$

$$V = \frac{1}{\sqrt{r}}\cos(n\log r)K_n(\cos\theta), \qquad V = \frac{1}{\sqrt{r}}\cos(n\log r)K_n(-\cos\theta),$$

are solutions of $$rD_r^2(rV) + \frac{1}{\sin\theta}D_\theta(\sin\theta D_\theta V) = 0$$

if $$K_n(x) = {}_1P_{-\frac{1}{2}+ni}(x) = 1 - \frac{n^2 + \left(\frac{1}{2}\right)^2}{2}(x-1) + \frac{\left[n^2 + \left(\frac{1}{2}\right)^2\right]\left[n^2 + \left(\frac{3}{2}\right)^2\right]}{2^2(2!)^2}(x-1)^2 - \frac{\left[n^2 + \left(\frac{1}{2}\right)^2\right]\left[n^2 + \left(\frac{3}{2}\right)^2\right]\left[n^2 + \left(\frac{5}{2}\right)^2\right]}{2^2(3!)^2}(x-1)^3 + \cdots$$

and

$$K_n(-x) = {}_{-1}P_{-\frac{1}{2}+ni}(x) = 1 + \frac{n^2 + \left(\frac{1}{2}\right)^2}{2}(x+1) + \frac{\left[n^2 + \left(\frac{1}{2}\right)^2\right]\left[n^2 + \left(\frac{3}{2}\right)^2\right]}{2^2(2!)^2}(x+1)^2 + \frac{\left[n^2 + \left(\frac{1}{2}\right)^2\right]\left[n^2 + \left(\frac{3}{2}\right)^2\right]\left[n^2 + \left(\frac{5}{2}\right)^2\right]}{2^3(3!)^2}(x+1)^3 + \cdots.$$

$K_n(\cos\theta)$ is convergent except for $\theta = \pi$, and $K_n(-\cos\theta)$ is convergent except for $\theta = 0$.

$k_n(x)$, $l_n(x)$, $K_n(x)$, and $K_n(-x)$ are sometimes called *Conal Harmonics.* They are particular values of z which satisfy Legendre's Equation written in the form

$$(1 - x^2)\frac{d^2z}{dx^2} - 2x\frac{dz}{dx} - \left(n^2 + \frac{1}{4}\right)z = 0.$$

For an elaborate treatment of them see E. W. Hobson on "A Class of Spherical Harmonics of Complex Degree." Trans. Camb. Phil. Soc., Vol. XIV.

8. If $V = f(r)$ when $\theta = \beta$,

$$V = \frac{1}{\pi\sqrt{r}}\int_{-\infty}^{\infty} d\lambda \int_0^{\infty} e^{\frac{\lambda}{2}} f(e^\lambda)\frac{K_\alpha(\cos\theta)}{K_\alpha(\cos\beta)}\cos[\alpha(\lambda - \log r)]d\alpha; \quad \text{if} \quad \theta < \beta.$$

9. If $V = f(r)$ when $\theta = \beta$ and $r < a$, and $V = 0$ when $r = a$,

$$V = \frac{2}{\pi}\sqrt{\frac{a}{r}}\int_{-\infty}^{0} d\lambda \int_0^{\infty} e^{\frac{\lambda}{2}} f(ae^\lambda)\frac{K_\alpha(\cos\theta)}{K_\alpha(\cos\beta)}\sin\alpha\lambda \sin\left(\alpha\log\frac{r}{a}\right) d\alpha; \quad \text{if} \quad \theta < \beta.$$

10. If $V = f(r)$ when $\theta = \beta$ and $a < r < b$, and $V = 0$ when $r = a$ and when $r = b$,

$$V = \sum_{m=1}^{m=\infty} A_m \frac{K_{m'}(\cos\theta)}{K_{m'}(\cos\beta)} \sin\left[\frac{m\pi(\log r - \log a)}{\log b - \log a}\right]$$

where
$$m' = \frac{m\pi}{\log b - \log a}$$
and

$$A_m = \frac{2}{\log b - \log a}\sqrt{\frac{a}{r}} \int\limits_0^{\log\frac{b}{a}} e^{\frac{x}{2}} f(ae^x) \sin\frac{m\pi x}{\log b - \log a} dx; \quad \text{if} \quad \theta < \beta.$$

11. If $\theta > \beta$ $\cos\theta$ must be replaced by $(-\cos\theta)$ in examples 8, 9, and 10.

12. If $V = f(r)$ when $\theta = \beta$, and $V = 0$ when $\theta = \gamma$,

$$V = \frac{1}{\pi\sqrt{r}} \int\limits_{-\infty}^{\infty} d\lambda \int\limits_0^{\infty} e^{\frac{\lambda}{2}} f(e^\lambda)$$
$$\frac{k_\alpha(\cos\theta)l_\alpha(\cos\gamma) - k_\alpha(\cos\gamma)l_\alpha(\cos\theta)}{k_\alpha(\cos\beta)l_\alpha(\cos\gamma) - k_\alpha(\cos\gamma)l_\alpha(\cos\beta)} \cos[\alpha(\lambda - \log r)]d\alpha;$$

if $\beta < \theta < \gamma$.

13. If $V = f(r)$ when $\theta = \beta$ and $a < r < b$, $V = 0$ when $\theta = \gamma$ and $a < r < b$, and $V = 0$ when $r = a$ and when $r = b$,

$$V = \sum_{m=1}^{m=\infty} A_m \frac{k_{m'}(\cos\theta)l_{m'}(\cos\gamma) - k_{m'}(\cos\gamma)l_{m'}(\cos\theta)}{k_{m'}(\cos\beta)l_{m'}(\cos\gamma) - k_{m'}(\cos\gamma)l_{m'}(\cos\beta)} \sin\frac{m\pi(\log r - \log a)}{\log b - \log a},$$

where
$$m' = \frac{m\pi}{\log b - \log a}$$
and

$$A_m = \frac{2}{\log b - \log a}\sqrt{\frac{a}{r}} \int\limits_0^{\log\frac{b}{a}} e^{\frac{x}{2}} f(ae^x) \sin\frac{m\pi x}{\log b - \log a} dx;$$

if $\beta < \theta < \gamma$ and $a < r < b$.

14. If $V = f(r)$ when $\theta = \beta$ and $a < r < b$, and $V = 0$ when $r = a$ and $D_r V + hV = 0$ when $r = b$,

$$V = \sum_{m=1}^{m=\infty} A_m \frac{K_{\alpha_m}(\cos\theta)}{K_{\alpha_m}(\cos\beta)} \sin\left(\alpha_m \log\frac{r}{a}\right),$$
where

$$A_m = \frac{2(\alpha_m^2 + h^2b^2)}{\alpha_m^2(\log b - \log a) + hb[hb(\log b - \log a) + 1]} \int\limits_0^{\log\frac{b}{a}} e^{\frac{x}{2}} f(ae^x) \sin\alpha_m x . dx$$

and α_m is a root of the equation

$$\alpha \cos\left(\alpha \log \frac{b}{a}\right) + hb \sin\left(\alpha \log \frac{b}{a}\right) = 0 \qquad \text{v. Art. 68 Ex. 5.}$$

CHAPTER VI.

SPHERICAL HARMONICS.

101. When we are dealing with problems in finding the *potential function* due to forces which have not circular symmetry[1] about an axis and are using Spherical Coördinates, we have to solve Laplace's Equation in the form

$$rD_r^2(rV)+\frac{1}{\sin\theta}D_\theta(\sin\theta D_\theta V)+\frac{1}{\sin^2\theta}D_\phi^2 V=0 \tag{1}$$

[v. (XIII) Art. 1].

To get a particular solution of (1) we shall assume as usual that V is a product of functions each of which involves but a single variable.

Let $V=R.\Theta.\Phi$; where R involves r only, Θ involves θ only, and Φ ϕ only. Substitute in (1) and we get

$$\frac{r}{R}\frac{d^2(rR)}{dr^2}+\frac{1}{\Theta\sin\theta}\frac{d\left(\sin\theta\frac{d\Theta}{d\theta}\right)}{d\theta}+\frac{1}{\Phi\sin^2\theta}\frac{d^2\Phi}{d\phi^2}=0 \tag{2}$$

or

$$\frac{r\sin^2\theta}{R}\frac{d^2(rR)}{dr^2}+\frac{\sin\theta}{\Theta}\frac{d\left(\sin\theta\frac{d\Theta}{d\theta}\right)}{d\theta}=-\frac{1}{\Phi}\frac{d^2\Phi}{d\phi^2}.$$

As the first member does not contain ϕ the second member cannot contain ϕ, and as it contains no other variable it must be constant; call it n^2. Equation (2) is then equivalent to the two equations

$$\frac{d^2\Phi}{d\phi^2}+n^2\Phi=0 \tag{3}$$

and

$$\frac{r}{R}\frac{d^2(rR)}{dr^2}+\frac{1}{\Theta\sin\theta}\frac{d\left[\sin\theta\frac{d\Theta}{d\theta}\right]}{d\theta}-\frac{n^2}{\sin^2\theta}=0 \tag{4}$$

(3) has been solved before and gives us

$$\Phi=A\cos n\phi+B\sin n\phi \tag{5}$$

[v. Art. 13(a)].

The first term of (4) does not involve θ and the second and third terms do not involve r.

$\frac{r}{R}\frac{d^2(rR)}{dr^2}$ must, then, be a constant; we shall call it $m(m+1)$ as in Art. 13(c).

Then (4) breaks up into

$$r\frac{d^2(rR)}{dr^2}=m(m+1)R \tag{6}$$

[1] See note, page 12.

and $$\frac{1}{\sin\theta}\frac{d\left[\sin\theta\frac{d\Theta}{d\theta}\right]}{d\theta}+\left[m(m+1)-\frac{n^2}{\sin^2\theta}\right]\Theta=0. \tag{7}$$

(6) was solved in Art. 13(c) and gives

$$R=A_1r^m+B_1r^{-m-1}. \tag{8}$$

If in (7) we replace $\cos\theta$ by μ we get

$$\frac{d}{d\mu}\left[(1-\mu^2)\frac{d\Theta}{d\mu}\right]+\left[m(m+1)-\frac{n^2}{1-\mu^2}\right]\Theta=0, \tag{9}$$

the equivalent of

$$(1-x^2)\frac{d^2z}{dx^2}-2x\frac{dz}{dx}+\left[m(m+1)-\frac{n^2}{1-x^2}\right]z=0, \tag{10}$$

[v. (17) Art. 85], which was solved in Art. 85 for the case where m and n are positive integers and $n<m+1$. v. (18) and (19) Art. 85.

From (19) Art. 85 we get as a particular solution of (9)

$$\Theta=(1-\mu^2)^{\frac{n}{2}}\frac{d^nP_m(\mu)}{d\mu^n}=\sin^n\theta\frac{d^nP_m(\mu)}{d\mu^n}, \tag{11}$$

if we restrict ourselves to whole positive values of m and n, as we shall do hereafter unless the contrary is explicitly stated, and suppose m not less than n.

A second but less useful particular solution of (9) is

$$\Theta=(1-\mu^2)^{\frac{n}{2}}\frac{d^nQ_m(\mu)}{d\mu^n}.$$

Combining our results we have as important particular solutions of (1)

$$V=r^m(A\cos n\phi+B\sin n\phi)\sin^n\theta\frac{d^nP_m(\mu)}{d\mu^n}, \tag{12}$$

and $$V=\frac{1}{r^{m+1}}(A\cos n\phi+B\sin n\phi)\sin^n\theta\frac{d^nP_m(\mu)}{d\mu^n}, \tag{13}$$

where m and n are positive integers and $n<m+1$.

102. $\sin^n\theta\frac{d^nP_m(\mu)}{d\mu^n}$ or $(1-\mu^2)^{\frac{n}{2}}\frac{d^nP_m(\mu)}{d\mu^n}$ is a new function of μ, that is of $\cos\theta$, and we shall represent it by $P_m^n(\mu)$[2] and shall call it an *associated function* of the nth order and mth degree. It is a value of Θ satisfying equation (9) Art. 101.

[2] Most of the English writers represent this function by $T_m^n(\mu)$.

By differentiating the value of $P_m(x)$ given in (9) Art. 74 we get the formula

$$P_m^n(\mu) = \frac{(2m)! \sin^n \theta}{2^m m!(m-n)!} \left[\mu^{m-n} - \frac{(m-n)(m-n-1)}{2.(2m-1)} \mu^{m-n-2} \right. \\ \left. + \frac{(m-n)(m-n-1)(m-n-2)(m-n-3)}{2.4.(2m-1)(2m-3)} \mu^{m-n-4} - \dots \right] \quad (1)$$

the expression in the parenthesis ending with the term involving μ^0 if $m-n$ is even and with the term involving μ if $m-n$ is odd.

For convenience of reference we give on the next page a table from which $P_m^n(\mu)$ can be readily obtained for values of m and n from 1 to 8.

$\cos n\phi P_m^n(\mu)$ and $\sin n\phi P_m^n(\mu)$, that is,

$$\cos n\phi \sin^n \theta \frac{d^n P_m(\mu)}{d\mu^n} \quad \text{and} \quad \sin n\phi \sin^n \theta \frac{d^n P_m(\mu)}{d\mu^n}$$

are called *Tesseral Harmonics* of the mth degree and nth order, and are values of V which satisfy the equation

$$m(m+1)V + \frac{1}{\sin\theta} D_\theta(\sin\theta D_\theta V) + \frac{1}{\sin^2\theta} D_\phi^2 V = 0 \quad (2)$$

or its equivalent

$$m(m+1)V + D_\mu[(1-\mu^2)D_\mu V] + \frac{1}{1-\mu^2} D_\phi^2 V = 0. \quad (3)$$

There are obviously $2m+1$ Tesseral Harmonics of the mth degree, namely

$$\begin{array}{lll} P_m(\mu), & \cos\phi \sin\theta \frac{dP_m(\mu)}{d\mu}, & \sin\phi \sin\theta \frac{dP_m(\mu)}{d\mu} \\ & \cos 2\phi \sin^2\theta \frac{d^2 P_m(\mu)}{d\mu^2}, & \sin 2\phi \sin^2\theta \frac{d^2 P_m(\mu)}{d\mu^2} \\ & \cos 3\phi \sin^3\theta \frac{d^3 P_m(\mu)}{d\mu^3}, & \sin 3\phi \sin^3\theta \frac{d^3 P_m(\mu)}{d\mu^3} \\ & \dots\dots & \dots\dots \\ & \cos m\phi \sin^m\theta \frac{d^m P_m(\mu)}{d\mu^m}, & \sin m\phi \sin^m\theta \frac{d^m P_m(\mu)}{d\mu^m} \end{array}$$

If each of these is multiplied by a constant and their sum taken, this sum is called a *Surface Spherical Harmonic* of the mth degree, and is a solution of equations (2) and (3). We shall represent it by $Y_m(\mu, \phi)$ or by $Y_m(\theta, \phi)$.

Table for $\csc^n \theta P_m^n(\mu) = \dfrac{d^n P_m(\mu)}{d\mu^n}$.

m	$n = 1.$	$n = 2.$
1	1	
2	3μ	3
3	$\frac{3}{2}(5\mu^2 - 1)$	15μ
4	$\frac{5}{2}(7\mu^3 - 3\mu)$	$\frac{15}{2}(7\mu^2 - 1)$
5	$\frac{15}{8}(21\mu^4 - 14\mu^2 + 1)$	$\frac{105}{2}(3\mu^3 - \mu)$
6	$\frac{21}{8}(33\mu^5 - 30\mu^3 + 5\mu)$	$\frac{105}{8}(33\mu^4 - 18\mu^2 + 1$
7	$\frac{7}{16}(429\mu^6 - 495\mu^4 + 135\mu^2 - 5)$	$\frac{63}{8}(143\mu^5 - 110\mu^3 + 15\mu)$
8	$\frac{9}{16}(715\mu^7 - 1001\mu^5 + 385\mu^3 - 35\mu)$	$\frac{315}{16}(143\mu^6 - 143\mu^4 + 33\mu^2 - 1)$

m	$n = 3.$	$n = 4.$
1		
2		
3	15	
4	105μ	105
5	$\frac{105}{2}(9\mu^2 - 1)$	945μ
6	$\frac{315}{2}(11\mu^3 - 3\mu)$	$\frac{945}{2}(11\mu^2 - 1)$
7	$\frac{315}{8}(143\mu^4 - 66\mu^2 + 3)$	$\frac{3465}{2}(13\mu^3 - 3\mu)$
8	$\frac{3465}{8}(39\mu^5 - 26\mu^3 + 3\mu)$	$\frac{10395}{8}(65\mu^4 - 26\mu^2 + 1)$

m	$n = 5.$	$n = 6.$	$n = 7.$	$n = 8.$
1				
2				
3				
4				
5	945			
6	10395μ	10395		
7	$\frac{10395}{2}(13\mu^2 - 1)$	135135μ	135135	
8	$\frac{135135}{2}(5\mu^3 - \mu)$	$\frac{135135}{2}(15\mu^2 - 1)$	2027025μ	2027025

$r^m Y_m(\mu, \phi)$ and $\frac{1}{r^{m+1}} Y_m(\mu, \phi)$ are called *Solid Spherical Harmonics* of the mth degree, and are solutions of Laplace's Equation (1) Art. 101.

To formulate:—

$$Y_m(\mu, \phi) = \sum_{n=0}^{n=m} \left[A_n \cos n\phi \sin^n \theta \frac{d^n P_m(\mu)}{d\mu^n} + B_n \sin n\phi \sin^n \theta \frac{d^n P_m(\mu)}{d\mu^n} \right] \tag{4}$$

or $$Y_m(\mu, \phi) = A_0 P_m(\mu) + \sum_{n=1}^{n=m} [A_n \cos n\phi P_m^n(\mu) + B_n \sin n\phi P_m^n(\mu)] \tag{5}$$

is a Surface Spherical Harmonic of the mth degree.

A Tesseral Harmonic is a special case of a Surface Spherical Harmonic, and a Zonal Harmonic a special case of a Tesseral Harmonic; $P_m(\mu)$ being the Tesseral Harmonic of the zeroth order and the mth degree; it might be written $P_m^0(\mu)$.

EXAMPLES.

1. Show that

$$(1 - x^2)\frac{d^2 z}{dx^2} - 2x\frac{dz}{dx} + \left[m(m+1) - \frac{n^2}{1 - x^2}\right] z = 0$$

reduces to

$$(1 - x^2)\frac{d^2 y}{dx^2} - 2(n+1)x\frac{dy}{dx} + [m(m+1) - n(n+1)]y = 0$$

if we substitute $(1 - x^2)^{\frac{n}{2}} y$ for z, even when m and n are unrestricted.

2. Show that if in the second equation of Ex. 1 we let $y = \sum a_k x^k$ we get

$$a_{k+2} = -\frac{(m-n-k)(m+n+1+k)}{(k+1)(k+2)} a_k \qquad \text{(v. Art. 16)}$$

whence $z = p_m^n(x)$ and $z = q_m^n(x)$ are solutions of the first equation of Ex. 1, no matter what the values of m and n, if

$$p_m^n(x) = (1-x^2)^{\frac{n}{2}} \left[1 - \frac{(m-n)(m+n+1)}{2!} x^2 + \frac{(m-n)(m-n-2)(m+n+1)(m+n+2)}{4!} x^4 - \cdots \right]$$

and

$$q_m^n(x) = (1-x^2)^{\frac{n}{2}} \left[x - \frac{(m-n-1)(m+n+2)}{3!} x^3 + \frac{(m-n-1)(m-n-3)(m+n+2)(m+n+4)}{5!} x^5 - \cdots \right].$$

If $m - n$ is a positive integer, $p_m^n(x)$ or $q_m^n(x)$ will terminate with the term involving x^{m-n}, and in that case

$$z = (1-x^2)^{\frac{n}{2}} \left[x^{m-n} - \frac{(m-n)(m-n-1)}{2.(2m-1)} x^{m-n-2} + \frac{(m-n)(m-n-1)(m-n-2)(m-n-3)}{2.4.(2m-1)(2m-3)} x^{m-n-4} - \cdots \right].$$

the parenthesis ending with a term involving x^0 if $m-n$ is even and x if $m-n$ is odd, is a solution of the first equation of Ex. 1. If m and n are integers this value of z is $\dfrac{2^m m!(m-n)!}{(2m)!} P_m^n(x)$.

103. We have seen in the last chapter that in many problems it is important to be able to express a given function of $\cos\theta$, that is of μ, in terms of Zonal Harmonics of μ. So it is often desirable to express a given function of μ and ϕ in terms of Tesseral Harmonics of μ and ϕ.

If, for example, we are trying to find the *Potential Function* due to certain forces and have the value of the function given for some given value of r, that is, on the surface of some given sphere whose centre is at the origin of coördinates, of course the given value will be a function of θ and ϕ and if we can express it in terms of Spherical Harmonics of θ and ϕ we have only to multiply each term by the proper power of r to get the required solution of the problem. For we shall then have a value of V satisfying Laplace's Equation and reducing to the given function of θ and ϕ on the surface of the given sphere.

104. Suppose that we have a function of μ and ϕ given for all points on the unit sphere, that is, for all values of μ from -1 to 1 and for all values of ϕ from 0 to 2π, μ and ϕ being independent variables, and that we wish to express it in terms of Surface Spherical Harmonics.

Assume that

$$f(\mu,\phi)=\sum_{m=0}^{m=\infty}\left[A_{0,m}P_m(\mu)+\sum_{n=1}^{n=m}\left(A_{n,m}\cos n\phi P_m^n(\mu)+B_{n,m}\sin n\phi P_m^n(\mu)\right)\right]. \quad (1)$$

Let us consider first a finite case, and attempt to determine the coefficients so that

$$f(\mu,\phi)=\sum_{m=0}^{m=p}\left[A_{0,m}P_m(\mu)+\sum_{n=1}^{n=m}\left(A_{n,m}\cos n\phi P_m^n(\mu)+B_{n,m}\sin n\phi P_m^n(\mu)\right)\right] \quad (2)$$

shall hold good at as many points of the sphere as possible. The expression in brackets in the second member of (2) is a Surface Spherical Harmonic of the mth degree and contains $2m+1$ constant coefficients. The whole number of coefficients to be determined is then the sum of an Arithmetical Progression of $p+1$ terms the first term of which is 1 and the last is $2p+1$, and is therefore equal to $(p+1)^2$.

Let the interval from $\mu=-1$ to $\mu=1$ be divided into $p+2$ parts each of which is $\Delta\mu$ so that $(p+2)\Delta\mu=2$, and let the interval from $\phi=0$ to $\phi=2\pi$ be divided into $p+2$ parts each of which is $\Delta\phi$ so that $(p+2)\Delta\phi=2\pi$.

Then if we substitute in equation (2) in turn the values $(-1+\Delta\mu,\ \Delta\phi)$, $(-1+2\Delta\mu,\ \Delta\phi)$, $\cdots[-1+(p+1)\Delta\mu,\ \Delta\phi]$; $(-1+\Delta\mu,\ 2\Delta\phi)$, $(-1+2\Delta\mu,\ 2\Delta\phi)$, $\cdots[-1+(p+1)\Delta\mu,\ 2\Delta\phi]$; $\cdots[-1+\Delta\mu,\ (p+1)\Delta\phi]$, $[-1+2\Delta\mu,\ (p+1)\Delta\phi]$, $\cdots[-1+(p+1)\Delta\mu, (p+1)\Delta\phi]$; since the first member in each case will be known we shall have $(p+1)^2$ equations of the first degree containing no unknown except the $(p+1)^2$ coefficients, and from them the coefficients can be determined. When they are substituted in equation (2) it will hold good at the $(p+1)^2$ points of the unit sphere where $p+1$ circles of latitude whose planes are equidistant intersect $p+1$ meridians which divide the equator into equal arcs. If now p is indefinitely increased the limiting values of the coefficients will be the coefficients in equation (1), and (1) will hold good all over the surface of the unit sphere.

To determine any particular constant we multiply each of our $(p+1)^2$ equations by $\Delta\mu\Delta\phi$ times the coefficient of the constant in question in that equation and add the equations and then investigate the limiting form approached by the resulting equation as p is indefinitely increased.

As p is indefinitely increased the summation in question will approach an integration; and since $d\mu d\phi=-\sin\theta.d\theta d\phi$ is the element of surface of the unit sphere, and as the limits -1 and 1 of μ correspond to π and 0 of θ the integration is a *surface integration* over the surface of the unit sphere.

In determining any coefficient as $A_{n,m}$ in (1) the first member of the limiting form of our resulting equation will be

$$\int_0^{2\pi} d\phi \int_{-1}^{1} f(\mu,\phi)\cos n\phi P_m^n(\mu)d\mu.$$

In the second member we shall come across terms of the forms

$$\int_0^{2\pi} d\phi \int_{-1}^{1} \sin l\phi \cos n\phi P_m^l(\mu) P_m^n(\mu) d\mu, \quad \int_0^{2\pi} d\phi \int_{-1}^{1} \cos l\phi \cos n\phi P_m^l(\mu) P_m^n(\mu) d\mu,$$

$$\int_0^{2\pi} d\phi \int_{-1}^{1} \sin n\phi \cos n\phi [P_m^n(\mu)]^2 d\mu, \quad \int_0^{2\pi} d\phi \int_{-1}^{1} \cos^2 n\phi [P_m^n(\mu)]^2 d\mu,$$

and other terms all of which come under the form

$$\int_0^{2\pi} d\phi \int_{-1}^{1} Y_l(\mu,\phi) Y_m(\mu,\phi) d\mu,$$

where $Y_m(\mu,\phi)$ and $Y_l(\mu,\phi)$ are Surface Spherical Harmonics of different degrees.

If we are determining a coefficient $B_{n,m}$ the only difference is that $\sin n\phi$ and $\cos n\phi$ will be interchanged in the forms just specified.

105. *The integral over the surface of the unit sphere of the product of two Surface Spherical Harmonics of different degrees is zero.*

That is

$$\int_0^{2\pi} d\phi \int_{-1}^{1} Y_l(\mu,\phi) Y_m(\mu,\phi) d\mu = 0. \tag{1}$$

For as we have seen $U = r^l Y_l(\mu,\phi)$ and $V = r^m Y_m(\mu,\phi)$ are solutions of Laplace's Equation. Hence by *Green's Theorem*

$$\int (UD_nV - VD_nU) ds = 0 \qquad \text{v. Art. 92.}$$

$$D_nV = D_rV = mr^{m-1} Y_m(\mu,\phi),$$

$$D_nU = D_rU = lr^{l-1} Y_l(\mu,\phi);$$

$$UD_nV - VD_nU = (m-l) r^{l+m-1} Y_l(\mu,\phi) Y_m(\mu,\phi),$$

$$= (m-l) Y_l(\mu,\phi) Y_m(\mu,\phi)$$

on the surface of the unit sphere; and

$$(m-l)\int Y_l(\mu,\phi) Y_m(\mu,\phi) ds = (m-l)\int_0^{2\pi} d\phi \int_1^{1} Y_l(\mu,\phi) Y_m(\mu,\phi) d\mu = 0.$$

Hence unless $l = m$

$$\int_0^{2\pi} d\phi \int_{-1}^{1} Y_l(\mu,\phi) Y_m(\mu,\phi) d\mu = 0.$$

EXAMPLES.

1. Obtain (1) Art. 105 directly from the equation

$$m(m+1)Y_m(\mu,\phi) + D_\mu[(1-\mu^2)D_\mu Y_m(\mu,\phi)] + \frac{1}{1-\mu^2}D_\phi^2 Y_m(\mu,\phi) = 0$$

v. (3) Art. 102, and Art. 91.

2. Show that the integral over the surface of the unit sphere of the product of two Tesseral Harmonics of the same degree but of different orders is zero.

Suggestion:

$$\int_0^{2\pi} \sin k\phi \cos l\phi . d\phi = \int_0^{2\pi} \sin k\phi \sin l\phi . d\phi = \int_0^{2\pi} \cos k\phi \cos l\phi . d\phi = 0.$$

106.

$$\int_{-1}^{1} P_l^n(\mu) P_m^n(\mu) d\mu = 0 \quad \text{unless} \quad l = m$$

$$= \frac{2}{2m+1}\frac{(m+n)!}{(m-n)!} \quad \text{if} \quad l = m.$$

For

$$\int_{-1}^{1} P_l^n(\mu)P_m^n(\mu)d\mu = \int_{-1}^{1}(1-\mu^2)^n \frac{d^n P_l(\mu)}{d\mu^n}.\frac{d^n P_m(\mu)}{d\mu^n}d\mu$$

$$= (1-\mu^2)^n \frac{d^n P_m(\mu)}{d\mu^n}.\frac{d^{n-1}P_l(\mu)}{d\mu^{n-1}}\Bigg]_{-1}^{1}$$

$$- \int_{-1}^{1} \frac{d^{n-1}P_l(\mu)}{d\mu^{n-1}}.\frac{d}{d\mu}\left[(1-\mu^2)^n \frac{d^n P_m(\mu)}{d\mu^n}\right] d\mu,$$

$$= -\int_{-1}^{1} \frac{d^{n-1}P_l(\mu)}{d\mu^{n-1}}.\frac{d}{d\mu}\left[(1-\mu^2)^n \frac{d^n P_m(\mu)}{d\mu^n}\right] d\mu;$$

by *integration by parts.*

Replacing n by $n-1$ in equation (2) Art. 84 and remembering that $\frac{d^{n-1}P_m(x)}{dx^{n-1}}$ is a possible value of $z^{(n-1)}$ we get

$$(1-\mu^2)\frac{d^{n+1}P_m(\mu)}{d\mu^{n+1}} - 2n\mu\frac{d^n P_m(\mu)}{d\mu^n} + [m(m+1) - n(n-1)]\frac{d^{n-1}P_m(\mu)}{d\mu^{n-1}} = 0,$$

or if we multiply by $(1-\mu^2)^{n-1}$

$$(1-\mu^2)^n \frac{d^{n+1}P_m(\mu)}{d\mu^{n+1}} - 2n\mu(1-\mu^2)^{n-1}\frac{d^n P_m(\mu)}{d\mu^n}$$

$$+ (m+n)(m-n+1)(1-\mu^2)^{n-1}\frac{d^{n-1}P_m(\mu)}{d\mu^{n-1}} = 0,$$

or

$$\frac{d}{d\mu}\left[(1-\mu^2)^n\frac{d^nP_m(\mu)}{d\mu^n}\right] = -(m+n)(m-n+1)(1-\mu^2)^{n-1}\frac{d^{n-1}P_m(\mu)}{d\mu^{n-1}}.$$

Hence follows the *reduction formula*

$$\int_{-1}^{1}(1-\mu^2)^n\frac{d^nP_l(\mu)}{d\mu^n}.\frac{d^nP_m(\mu)}{d\mu^n}d\mu$$

$$= (m+n)(m-n+1)\int_{-1}^{1}(1-\mu^2)^{n-1}\frac{d^{n-1}P_l(\mu)}{d\mu^{n-1}}.\frac{d^{n-1}P_m(\mu)}{d\mu^{n-1}}d\mu.$$

Using this formula n times we get

$$\int_{-1}^{1}P_l^n(\mu)P_m^n(\mu)d\mu = \frac{(m+n)!}{(m-n)!}\int_{-1}^{1}P_l(\mu)P_m(\mu)d\mu$$

$$= 0 \quad \text{unless} \quad l = m$$

$$= \frac{2}{2m+1}\frac{(m+n)!}{(m-n)!} \quad \text{if} \quad l = m$$

v. Art. 89 (4) and (5).

107. We are now able to complete the solution of the problem in Art. 104 and since $\int_0^{2\pi}\cos^2 n\phi.d\phi = \int_0^{2\pi}\sin^2 n\phi.d\phi = \pi$ and $\int_0^{2\pi}d\phi = 2\pi$ we get as the coefficients in (1) Art. 104

$$A_{0,m} = \frac{2m+1}{4\pi}\int_0^{2\pi}d\phi\int_{-1}^{1}f(\mu,\phi)P_m(\mu)d\mu, \tag{1}$$

$$A_{n,m} = \frac{2m+1}{2\pi}.\frac{(m-n)!}{(m+n)!}\int_0^{2\pi}d\phi\int_{-1}^{1}f(\mu,\phi)\cos n\phi P_m^n(\mu)d\mu, \tag{2}$$

$$B_{n,m} = \frac{2m+1}{2\pi}.\frac{(m-n)!}{(m+n)!}\int_0^{2\pi}d\phi\int_{-1}^{1}f(\mu,\phi)\sin n\phi P_m^n(\mu)d\mu, \tag{3}$$

whence

$$f(\mu,\phi) = \sum_{m=0}^{m=\infty}\left[A_{0,m}P_m(\mu) + \sum_{n=1}^{n=m}(A_{n,m}\cos n\phi + B_{n,m}\sin n\phi)P_m^n(\mu)\right] \tag{4}$$

and the development holds good for all values of μ and ϕ corresponding to points on the unit sphere, provided only that the given function satisfies the conditions that would have to be satisfied if it were to be developed into a *Fourier's Series.*

If we use μ_1 and ϕ_1 in place of μ and ϕ in (1), (2), and (3), we can write (4) in the form

$$f(\mu,\phi) = \frac{1}{2\pi}\sum_{m=0}^{m=\infty}(2m+1)\left[\frac{1}{2}\int_0^{2\pi} d\phi_1 \int_{-1}^{1} f(\mu_1,\phi_1)P_m(\mu)P_m(\mu_1)d\mu_1 \right.$$
$$\left. + \sum_{n=1}^{n=m}\frac{(m-n)!}{(m+n)!}\int_0^{2\pi} d\phi_1 \int_{-1}^{1} f(\mu_1,\phi_1)P_m^n(\mu)P_m^n(\mu_1)\cos n(\phi-\phi_1)d\mu_1\right]. \quad (5)$$

Formulas (1), (2), (3), and (4) are convenient for actual work; (5) is rather more compactly written.

108. As an example let us express $\sin^2\theta\cos^2\theta\sin\phi\cos\phi$ in terms of Surface Spherical Harmonics.

Here $$f(\mu,\phi) = \frac{1}{2}\mu^2(1-\mu^2)\sin 2\phi.$$

$$A_{0,m} = \frac{2m+1}{8\pi}\int_{-1}^{1}\mu^2(1-\mu^2)P_m(\mu)d\mu\int_0^{2\pi}\sin 2\phi.d\phi = 0,$$

$$A_{n,m} = \frac{2m+1}{4\pi}.\frac{(m-n)!}{(m+n)!}\int_{-1}^{1}\mu^2(1-\mu^2)P_m^n(\mu)d\mu\int_0^{2\pi}\sin 2\phi\cos n\phi.d\phi = 0,$$

$$B_{n,m} = \frac{2m+1}{4\pi}.\frac{(m-n)!}{(m+n)!}\int_0^{1}\mu^2(1-\mu^2)P_m^n(\mu)d\mu\int_0^{2\pi}\sin 2\phi\sin n\phi.d\phi,$$
$$= 0 \quad \text{unless} \quad n = 2.$$

If $n = 2$ $$\int_0^{2\pi}\sin 2\phi\sin n\phi.d\phi = \int_0^{2\pi}\sin^2 2\phi.d\phi = \pi,$$ and

$$B_{2,m} = \frac{2m+1}{4}.\frac{(m-2)!}{(m+2)!}\int_{-1}^{1}\mu^2(1-\mu^2)^2\frac{d^2P_m(\mu)}{d\mu^2}d\mu$$
$$= \frac{1}{2^m m!}\frac{2m+1}{4}\frac{(m-2)!}{(m+2)!}\int_{-1}^{1}\mu^2(1-\mu^2)^2\frac{d^{m+2}(\mu^2-1)^m}{d\mu^{m+2}}d\mu.$$

$$\int_{-1}^{1}\mu^2(1-\mu^2)^2\frac{d^{m+2}(\mu^2-1)^m}{d\mu^{m+2}}d\mu = 720\int_{-1}^{1}\frac{d^{m-4}(\mu^2-1)^m}{d\mu^{m-4}}d\mu$$

by repeated integration by parts,

$$= 0 \quad \text{if} \quad m > 4,$$

$$= 720 \int_{-1}^{1} (\mu^2 - 1)^4 d\mu = \frac{4096}{7} \quad \text{if} \quad m = 4,$$

and $$B_{2,4} = \frac{1}{2^4 4!} \cdot \frac{9}{4} \cdot \frac{2!}{6!} \cdot \frac{4096}{7} = \frac{1}{105}.$$

By a like process we find

$$B_{2,3} = 0 \quad \text{and} \quad B_{2,2} = \frac{1}{42}.$$

Hence

$$\sin^2\theta \cos^2\theta \sin\phi \cos\phi = \frac{1}{42} P_2^2(\mu) \sin 2\phi + \frac{1}{105} P_4^2(\mu) \sin 2\phi, \tag{1}$$

$$= \frac{1}{42} \sin 2\phi \sin^2\theta \frac{d^2 P_2(\mu)}{d\mu^2} + \frac{1}{105} \sin 2\phi \sin^2\theta \frac{d^2 P_4(\mu)}{d\mu^2}, \tag{2}$$

$$= \frac{1}{14} \sin^2\theta \sin 2\phi + \frac{1}{14} \sin^2\theta (7\mu^2 - 1) \sin 2\phi. \tag{3}$$

The required expression might have been obtained without using the formulas of Art. 107, by a very simple device, as follows:

$$\sin^2\theta \cos^2\theta \sin\phi \cos\phi = \frac{1}{2} \mu^2 \sin^2\theta \sin 2\phi. \tag{4}$$

If now we can express μ^2 in the form $\sum \frac{d^2 P_m(\mu)}{d\mu^2}$ the work will be done.

$$\mu^2 = \frac{1}{4.3} \frac{d^2(\mu^4)}{d\mu^2},$$

$$\mu^4 = \frac{8}{35} P_4(\mu) + \frac{4}{7} P_2(\mu) + \frac{1}{5} P_0(\mu), \qquad \text{(5) Art. 95.}$$

$$\frac{d^2(\mu^4)}{d\mu^2} = \frac{8}{35} \frac{d^2 P_4(\mu)}{d\mu^2} + \frac{4}{7} \frac{d^2 P_2(\mu)}{d\mu^2};$$

whence $$\mu^2 = \frac{2}{105} \frac{d^2 P_4(\mu)}{d\mu^2} + \frac{1}{21} \frac{d^2 P_2(\mu)}{d\mu^2},$$

and substituting this value in (4) we get (2).

EXAMPLES.

1. Show that

$$\cos^3\theta \sin^3\theta \sin\phi \cos^2\phi = \left[\frac{1}{6930} P_6^3(\mu) + \frac{1}{1540} P_4^3(\mu)\right] \sin 3\phi$$
$$- \left[\frac{2}{693} P_6^1(\mu) - \frac{1}{770} P_4^1(\mu) - \frac{1}{63} P_2^1(\mu)\right] \sin\phi.$$

2. Show that

$$\cos 2\phi = 2\cos 2\phi\left[\frac{5}{4!}P_2^2(\mu) + \frac{9.2!}{6!}P_4^2(\mu) + \frac{13.4!}{8!}P_6^2(\mu) + \cdots\right].$$

3. If in a problem on the Potential Function $V = f(\mu,\phi)$ when $r = a$, we shall obviously have

$$V = \sum_{m=0}^{m=\infty}\frac{r^m}{a^m}\left[A_{0,m}P_m(\mu) + \sum_{n=1}^{n=m}(A_{n,m}\cos n\phi + B_{n,m}\sin n\phi)P_m^n(\mu)\right]$$

at an internal point and

$$V = \sum_{m=0}^{m=\infty}\frac{a^{m+1}}{r^{m+1}}\left[A_{0,m}P_m(\mu) + \sum_{n=1}^{n=m}(A_{n,m}\cos n\phi + B_{n,m}\sin n\phi)P_m^n(\mu)\right]$$

at an external point, where $A_{0,m}$, $A_{n,m}$, and $B_{n,m}$ have the values given in (1), (2), and (3) Art. 107.

4. Solve problems (3), (4), and (5) of Art. 94 for the case where V is not symmetrical with respect to an axis.

109. Any Solid Spherical Harmonic $r^mY_m(\mu,\phi)$ being a value of V that satisfies Laplace's Equation in Spherical Coördinates will transform into a function of x, y, and z satisfying $\nabla^2V = 0$ if we change to a set of rectangular axes having the same origin and the same axis of X as the polar system. Moreover the new function will be a homogeneous rational integral Algebraic function of x, y, z, of the mth degree.

For each term of $r^m\cos n\phi P_m^n(\mu)$ is of the form

$$Cr^m\cos^{n-2k}\phi\sin^{2k}\phi\sin^n\theta\cos^{m-2l-n}\theta$$

where $\qquad 2k < n+1 \quad\text{and}\quad 2l < m-n+1.$

This may be written

$$Cr^{2l}.r^{m-2l-n}\cos^{m-2l-n}\theta.r^{n-2k}\sin^{n-2k}\theta\cos^{n-2k}\phi.r^{2k}\sin^{2k}\theta\sin^{2k}\phi$$

which becomes $\qquad C(x^2+y^2+z^2)^lx^{m-2l-n}y^{n-2k}z^{2k},$

and is a homogeneous rational integral Algebraic function of x, y, and z of the mth degree. The same thing may be shown of each term of $r^m\sin n\phi P_m^n(\mu)$. Consequently $r^mY_m(\mu,\phi)$ is a homogeneous rational integral Algebraic function of the mth degree in x, y, and z.

110. Any homogeneous rational integral Algebraic function $S_m(x, y, z)$ of the mth degree in x, y, and z, which is a value of V satisfying $\nabla^2 V = 0$ contains $2m + 1$ arbitrary constant coefficients.

For $S_m(x, y, z)$ will in general consist of $\frac{(m+1)(m+2)}{2}$ terms and will therefore contain $\frac{(m+1)(m+2)}{2}$ coefficients.

$\nabla^2 S_m(x, y, z)$ will be homogeneous of the $(m-2)$d degree and will contain $\frac{m(m-1)}{2}$ coefficients, which, of course, will be functions of the coefficients in $S_m(x, y, z)$. Since $\nabla^2 S_m(x, y, z) = 0$ independently of the numerical values of x, y, and z the $\frac{m(m-1)}{2}$ coefficients in $\nabla^2 S_m(x, y, z)$ must be separately zero, and that fact will give us $\frac{m(m-1)}{2}$ equations of condition between the $\frac{(m+1)(m+2)}{2}$ original coefficients and will leave $\frac{(m+1)(m+2)}{2} - \frac{m(m-1)}{2}$ or $2m + 1$ of them undetermined. $S_m(x, y, z)$ contains, then, the same number of arbitrary coefficients as $r^m Y_m(\mu, \phi)$.

We can then choose the coefficients in $r^m Y_m(\mu, \phi)$ so that it will transform into any given $S_m(x, y, z)$.

Consequently a Solid Spherical Harmonic of the mth degree might be defined as *a homogeneous rational integral Algebraic function of x, y, and z, $S_m(x, y, z)$, of the mth degree satisfying the equation $\nabla^2 S_m(x, y, z) = 0$*; and a Surface Spherical Harmonic of the mth degree as such a function divided by $(x^2 + y^2 + z^2)^{\frac{m}{2}}$, that is by r^m.

EXAMPLES.

1. Show that if $S_m(x, y, z)$ is a Solid Spherical Harmonic of the mth degree

$$\nabla^2[r^n S_m(x, y, z)] = n(2m + n + 1)r^{n-2} S_m(x, y, z).$$

Suggestion:

$$\nabla^2 S_m = 0. \quad \nabla^2 r = \frac{2}{r}. \quad D_r S_m = \frac{m S_m}{r}. \quad (D_x r)^2 + (D_y r)^2 + (D_z r)^2 = 1.$$

2. Show that if $f_n(x, y, z)$ is a rational integral homogeneous function of x, y, and z of the nth degree it can be expressed in the form

$$f_n(x, y, z) = S_n(x, y, z) + r^2 S_{n-2}(x, y, z) + r^4 S_{n-4}(x, y, z) + \cdots, \tag{1}$$

terminating with $r^{n-1} S_1(x, y, z)$ if n is odd, and with $r^n S_0(x, y, z)$ if n is even.

Suggestion: If a term rS_{n-1} were present in the second member of (1), and we were to operate with ∇^2 on both members we should by Ex. 1 have a term $\frac{2n}{r} S_{n-1}$ which would be irrational when all the other terms of the resulting equation were rational. No such term, then, could occur. In the same way it may be shown by operating twice on (1) with ∇^2 that there can be no term $r^3 S_{n-3}$ in (1); and thus step by step we can reach the result formulated in (1).

3. Express x^2yz in the form $S_4 + r^2S_2 + r^4S_0$.

Suggestion: let
$$x^2yz = S_4 + r^2S_2 + r^4S_0$$
and take ∇^2 of both members we get
$$2yz = 14S_2 + 20r^2S_0.$$
Operate again with ∇^2.
$$0 = 120S_0.$$
Whence
$$S_0 = 0, \quad S_2 = \frac{1}{7}yz, \quad \text{and} \quad S_4 = \frac{1}{7}(6x^2 - y^2 - z^2)yz.$$

4. Express $\sin^2\theta\cos^2\theta\sin\phi\cos\phi$ in terms of Surface Spherical Harmonics.

Suggestion:
$$\sin^2\theta\cos^2\theta\sin\phi\cos\phi = \frac{x^2yz}{r^4}.$$

For result v. Art. 108 (3).

111. A transformation of coördinates to a new set of axes having the same origin as the old set will change a given Surface Spherical Harmonic into another of the same degree. For such a transformation does not change the form of Laplace's Equation $\nabla^2V = 0$ if both sets of axes are rectangular, and it is effected by replacing x, y, and z in the Solid Harmonic corresponding to the given Surface Harmonic by $x\cos\alpha_1 + y\cos\alpha_2 + z\cos\alpha_3$, $x\cos\beta_1 + y\cos\beta_2 + z\cos\beta_3$ and $x\cos\gamma_1 + y\cos\gamma_2 + z\cos\gamma_3$ respectively where the cosines are the *direction cosines* of the new axes, and it will leave the function a homogeneous function of the mth degree in the new variables, and on dividing this by the mth power of the unchanged radius vector we shall have a Surface Spherical Harmonic of the mth degree.

112. We have seen in Art. 75 that if (x_1, y_1, z_1) are the coördinates of a given point
$$V = \frac{1}{\sqrt{(x-x_1)^2 + (y-y_1)^2 + (z-z_1)^2}} \tag{1}$$
is a solution of Laplace's Equation $\nabla^2V = 0$, and transforming to spherical coördinates that
$$V = \frac{1}{\sqrt{r^2 - 2rr_1[\cos\theta\cos\theta_1 + \sin\theta\sin\theta_1\cos(\phi-\phi_1)] + r_1^2}} \tag{2}$$
is a solution of
$$rD_r^2(rV) + \frac{1}{\sin\theta}D_\theta(\sin\theta D_\theta V) + \frac{1}{\sin^2\theta}D_\phi^2V = 0. \tag{3}$$

If γ is the angle between the radii vectores r and r_1 of the points (x, y, z) and (x_1, y_1, z_1) (1) can be written

$$V = \frac{1}{\sqrt{r^2 - 2rr_1 \cos\gamma + r_1^2}} \tag{4}$$

which must be equivalent to (2), and hence

$$\cos\gamma = \cos\theta\cos\theta_1 + \sin\theta\sin\theta_1\cos(\phi - \phi_1).$$

(4) which is a solution of (3) is of the same form as (5) Art. 75 and by developing it as we developed (5) Art. 75 we find that

$$V = P_m(\cos\gamma)$$

is a solution of the equation

$$m(m+1)V + \frac{1}{\sin\theta}D_\theta(\sin\theta D_\theta V) + \frac{1}{\sin^2\theta}D_\phi^2 V = 0 \tag{5}$$

and that $$V = r^m P_m(\cos\gamma) \quad \text{and} \quad V = \frac{1}{r^{m+1}}P_m(\cos\gamma)$$

are solutions of (3).

If we transform our coördinates keeping the origin unchanged and taking as our new polar axis the radius vector of (x_1, y_1, z_1) γ becomes our new θ and $P_m(\cos\gamma)$ reduces to $P_m(\cos\theta)$, a Surface Zonal Harmonic, or a *Legendrian*,[3] of the mth degree. It is then a Legendrian having for its axis not the original polar axis but the radius vector of (x_1, y_1, z_1). Since a Legendrian is a Surface Spherical Harmonic,

$$P_m(\cos\gamma) = P_m[\cos\theta\cos\theta_1 + \sin\theta\sin\theta_1\cos(\phi - \phi_1)]$$

is a Surface Spherical Harmonic of the mth degree.

It is, however, of very special form, since being a determinate function of μ, ϕ, μ_1, and ϕ_1 it contains but two arbitrary constants if we regard it as a function of μ and ϕ, instead of containing $2m+1$.

It is known as a *Laplace's Coefficient*, or briefly as a Laplacian, of the mth degree.

We shall soon express it in the regulation form of a Surface Spherical Harmonic.

The radius vector of (x_1, y_1, z_1) is called the axis of the Laplacian and the point where the axis cuts the surface of the unit sphere is the *pole* of the Laplacian.

We shall represent the Laplacian $P_m(\cos\gamma)$ by $L_m(\mu, \phi, \mu_1, \phi_1)$. Of course $L_m(\mu, \phi, 1, \phi_1) = P_m(\mu) = P_m(\cos\theta)$, and is really independent of ϕ.

[3] v. Art. 74.

113. *If the product of a Surface Spherical Harmonic of the mth degree by a Laplacian of the same degree is integrated over the surface of the unit sphere, the result is equal to* $\frac{4\pi}{2m+1}$ *multiplied by the value of the Spherical Harmonic at the pole of the Laplacian.*

That is,

$$\int_0^{2\pi} d\phi \int_{-1}^{1} Y_m(\mu,\phi)L_m(\mu,\phi,\mu_1,\phi_1)d\mu = \frac{4\pi}{2m+1}Y_m(\mu_1,\phi_1). \qquad (1)$$

Transform to the axis of the Laplacian as a new polar axis, and let $Z_m(\mu,\phi)$ be the transformed Spherical Harmonic. $L_m(\mu,\phi,\mu_1,\phi_1)$ will become $P_m(\mu)$, and (1) will be proved if we can show that

$$\int_0^{2\pi} d\phi \int_{-1}^{1} Z_m(\mu,\phi)P_m(\mu)d\mu = \frac{4\pi}{2m+1}Z_m(1,0). \qquad (2)$$

$$Z_m(\mu,\phi)P_m(\mu) = A_0[P_m(\mu)]^2 + \sum_{n=1}^{n=m}(A_n\cos n\phi + B_n\sin n\phi)P_m^n(\mu)P_m(\mu)$$

(v. (5) Art. 102).

$$\int_0^{2\pi} Z_m(\mu,\phi)P_m(\mu)d\phi = 2\pi A_0[P_m(\mu)]^2 \qquad \text{and}$$

$$\int_{-1}^{1} d\mu \int_0^{2\pi} Z_m(\mu,\phi)P_m(\mu)d\phi = \frac{4\pi}{2m+1}A_0 \qquad \text{(v. (5) Art. 89).}$$

But $Z_m(1,0) = A_0$, since $P_m(1) = 1$ and $P_m^n(1)$ contains $(1-1)^{\frac{n}{2}}$ as a factor and is equal to zero.

Hence (2) is proved.

114. We can now express a Laplacian in the regulation form as a Spherical Harmonic, by the formulas of Art. 107.

$$\begin{aligned} L_m(\mu,\phi,\mu_1,\phi_1) &= P_m(\cos\gamma) = P_m[\cos\theta\cos\theta_1 + \sin\theta\sin\theta_1\cos(\phi-\phi_1)] \\ &= \sum_{k=0}^{k=\infty}\left[A_{0,k}P_k(\mu) + \sum_{n=1}^{n=k}(A_{n,k}\cos n\phi + B_{n,k}\sin n\phi)P_k^n(\mu)\right] \end{aligned}$$

where

$$\begin{aligned} A_{0,m} &= \frac{2m+1}{4\pi}\int_0^{2\pi} d\phi \int_{-1}^{1} L_m(\mu,\phi,\mu_1,\phi_1)P_m(\mu)d\mu. \\ &= \frac{2m+1}{4\pi}\frac{4\pi}{2m+1}P_m(\mu_1) = P_m(\mu_1) \qquad \text{by (1) Art. 113,} \end{aligned}$$

$$A_{n,m} = \frac{2m+1}{2\pi}\frac{(m-n)!}{(m+n)!}\int_0^{2\pi} d\phi \int_{-1}^{1} L_m(\mu,\phi,\mu_1,\phi_1)\cos n\phi P_m^n(\mu)d\mu$$

$$= \frac{2(m-n)!}{(m+n)!}\cos n\phi_1 P_m^n(\mu_1) \qquad \text{by (1) Art. 113, and}$$

$$B_{n,m} = \frac{2m+1}{2\pi}\frac{(m-n)!}{(m+n)!}\int_0^{2\pi} d\phi \int_{-1}^{1} L_m(\mu,\phi,\mu_1,\phi_1)\sin n\phi P_m^n(\mu)d\mu$$

$$= \frac{2(m-n)!}{(m+n)!}\sin n\phi_1 P_m^n(\mu_1) \qquad \text{by (1) Art. 113,}$$

and $A_{0,k} = A_{n,k} = B_{n,k} = 0$ by Art. 105 unless $k = m$. Hence

$$L_m(\mu,\phi,\mu_1,\phi_1) = P_m(\mu)P_m(\mu_1) + 2\sum_{n=1}^{n=m}\left[\frac{(m-n)!}{(m+n)!}P_m^n(\mu)P_m^n(\mu_1)\cos n(\phi-\phi_1)\right]. \qquad (1)$$

Each term of a Laplacian involves a numerical coefficient, a factor which is a function of μ, a second factor which is the same function of μ_1, and a third factor which is of the form $\cos k(\phi-\phi_1)$. We give below a table of the first few Laplacians, taken from Minchin's Statics, omitting in each term for the sake of brevity the function of μ_1.

By the aid of (1) we can write (5) Art. 107 more compactly. It becomes

$$f(\mu,\phi) = \frac{1}{4\pi}\sum_{m=0}^{m=\infty}(2m+1)\int_0^{2\pi} d\phi_1 \int_{-1}^{1} f(\mu_1,\phi_1)L_m(\mu,\phi,\mu_1,\phi_1)d\mu_1 \qquad (2)$$

or

$$F(\theta,\phi) = \frac{1}{4\pi}\sum_{m=0}^{m=\infty}(2m+1)\int_0^{2\pi} d\phi_1 \int_0^{\pi} F(\theta_1,\phi_1)P_m(\cos\gamma)\sin\theta_1 d\theta_1. \qquad (3)$$

LAPLACIANS.

	coef. of $\cos 0(\phi-\phi_1)$	coef. of $\cos(\phi-\phi_1)$	coef. of $\cos 2(\phi-\phi_1)$
L_0	1		
L_1	μ	$(1-\mu^2)^{\frac{1}{2}}$	
L_2	$\frac{1}{4}(3\mu^2-1)$	$3\mu(1-\mu^2)^{\frac{1}{2}}$	$\frac{3}{4}(1-\mu^2)$
L_3	$\frac{1}{4}(5\mu^3-3\mu)$	$\frac{3}{8}(1-\mu^2)^{\frac{1}{2}}(5\mu^2-1)$	$\frac{15}{4}\mu(1-\mu^2)$
L_4	$\frac{1}{64}(35\mu^4-30\mu^2+3)$	$\frac{5}{8}(1-\mu^2)^{\frac{1}{2}}(7\mu^3-3\mu)$	$\frac{5}{16}(1-\mu^2)(7\mu^2-1)$

	coef. of $\cos 3(\phi - \phi_1)$	coef. of $\cos 4(\phi - \phi_1)$
L_0		
L_1		
L_2		
L_3	$\frac{5}{8}(1-\mu^2)^{\frac{3}{2}}$	
L_4	$\frac{35}{8}\mu(1-\mu^2)^{\frac{3}{2}}$	$\frac{35}{64}(1-\mu^2)^2$

EXAMPLE.

Work the problems of Art. 108 and Art. 108 Exs. 1 and 2 by the aid of (3) Art. 114.

115. Such problems as we have handled in Arts. 98 and 99, and also problems differing from them in not having circular symmetry about an axis, can now be solved by direct integration.

For instance let it be required to find the value at an external point of the potential function due to the attraction of a solid sphere whose density at any point is proportional to the product of any power of the radius vector by a Surface Spherical Harmonic.

Let $$\rho = C r_1^k Y_m(\mu_1, \phi_1).$$

Then using our ordinary notation we have

$$V = \int_0^a dr_1 \int_0^{2\pi} d\phi_1 \int_{-1}^{1} \frac{C r_1^k Y_m(\mu_1,\phi_1) r_1^2 d\mu_1}{\sqrt{r^2 - 2rr_1\cos\gamma + r_1^2}}.$$

But by (3) Art. 77

$$\frac{1}{\sqrt{r^2 - 2rr_1\cos\gamma + r_1^2}} = \frac{1}{r}\Big[P_0(\cos\gamma) + \frac{r_1}{r}P_1(\cos\gamma) + \frac{r_1^2}{r^2}P_2(\cos\gamma) + \cdots + \frac{r_1^m}{r^m}P_m(\cos\gamma) + \cdots\Big]$$

if $r > r_1$.

Consequently since

$$\int_0^{2\pi} d\phi_1 \int_{-1}^{1} Y_m(\mu_1,\phi_1) Y_n(\mu_1,\phi_1) d\mu_1 = 0,$$

V reduces to the single term

$$V = \frac{C}{r^{m+1}} \int_0^a r_1^{m+k+2} dr_1 \int_0^{2\pi} d\phi_1 \int_{-1}^{1} Y_m(\mu_1, \phi_1) P_m(\cos\gamma) d\mu_1$$

$$= \frac{C}{r^{m+1}} \int_0^a r_1^{m+k+2} \left(\frac{4\pi}{2m+1} Y_m(\mu, \phi) \right) dr_1 \qquad \text{by Art. 113.}$$

$$\therefore V = \frac{4\pi C}{2m+1} \cdot \frac{a^{m+k+3}}{m+k+3} \cdot \frac{Y_m(\mu, \phi)}{r^{m+1}}.$$

EXAMPLES.

1. Solve by direct integration the problems worked in Arts. 98 and 99 and Examples 1, 2, 3, and 4 of Art. 99.

2. The density of a solid sphere is proportional to the product of the squares of the distances from two mutually perpendicular diametral planes; find the value of the potential function at an external point.

Ans. $\rho = kr_1^4 \cos^2\theta_1 \sin^2\theta_1 \cos^2\phi_1$

$$= kr_1^4 \left[\frac{1}{15} P_0(\mu_1) + \frac{1}{21} P_2(\mu_1) + \frac{1}{42} \cos 2\phi_1 P_2^2(\mu_1) - \frac{4}{35} P_4(\mu_1) + \frac{1}{105} \cos 2\phi_1 P_4^2(\mu_1) \right].$$

$$V = \frac{M}{a} \left[\frac{a}{r} + \frac{a^3}{r^3} \left(\frac{1}{9} P_2(\mu) + \frac{1}{18} \cos 2\phi P_2^2(\mu) \right) - \frac{a^5}{r^5} \left(\frac{4}{33} P_4(\mu) - \frac{1}{99} \cos 2\phi P_4^2(\mu) \right) \right].$$

3. Solve Example 2 by an extension of the method of Arts. 98 and 99.

4. A conducting sphere of radius a connected with the ground by a wire is placed in the field of force due to an electrified point at which m units of electricity are concentrated. Find the value of the potential function due to the induced charge.

Suggestion: Let V_1 be the potential function due to the point, and V_2 that due to the induced charge, and let b be the distance of the point from the centre of the sphere. Then

$$V_1 = \frac{m}{\sqrt{b^2 - 2br\cos\theta + r^2}}$$

$$= \frac{m}{b}\left[P_0(\cos\theta) + \frac{r}{b}P_1(\cos\theta) + \frac{r^2}{b^2}P_2(\cos\theta) + \cdots\right] \quad \text{if} \quad r < b.$$

$$= \frac{m}{r}\left[P_0(\cos\theta) + \frac{b}{r}P_1(\cos\theta) + \frac{b^2}{r^2}P_2(\cos\theta) + \cdots\right] \quad \text{if} \quad r > b.$$

$$V_2 = A_0P_0(\cos\theta) + A_1\frac{r}{a}P_1(\cos\theta) + A_2\frac{r^2}{a^2}P_2(\cos\theta) + \cdots \quad \text{if} \quad r < a.$$

$$= A_0\frac{a}{r}P_0(\cos\theta) + A_1\frac{a^2}{r^2}P_1(\cos\theta) + A_2\frac{a^3}{r^3}P_2(\cos\theta) + \cdots \quad \text{if} \quad r > a.$$

When $r = a$ $V_1 + V_2 = 0$. Hence

$$A_0 = -\frac{m}{b}, \quad A_1 = -\frac{ma}{b^2}, \quad A_2 = -\frac{ma^2}{b^3}, \cdots$$

and

$$V_2 = -\frac{m}{b}\left[P_0(\cos\theta) + \frac{r}{b}P_1(\cos\theta) + \frac{r^2}{b^2}P_2(\cos\theta) + \cdots\right] \quad \text{if} \quad r < a$$

$$= -\frac{ma}{br}\left[P_0(\cos\theta) + \frac{a^2}{br}P_1(\cos\theta) + \frac{a^4}{b^2r^2}P_2(\cos\theta) + \cdots\right] \quad \text{if} \quad r > a.$$

Hence the effect of the induced charge is precisely the same at an external point as if the sphere were replaced by $\frac{ma}{b}$ units of negative electricity concentrated at the point $r = \frac{a^2}{b}$, $\theta = 0$. v. Peirce, Newt. Pot. Func., § 66.

116. If the two points P and P' are taken on the line OH whose direction cosines are λ, μ, and ν, and if u and u' are the values at P and P' of any continuous function of the space coördinates, then $\lim\limits_{PP' \doteq 0}\left[\frac{u' - u}{PP'}\right]$ is called the *partial derivative* of u along the line OH and will be represented by $D_h u$.

Let x, y, z be the coördinates of P and $x+\Delta x$, $y+\Delta y$, $z+\Delta z$ the coördinates of P'; then

$$u' - u = D_x u.\Delta x + D_y u.\Delta y + D_z u.\Delta z + \epsilon$$

where ϵ is an infinitesimal of higher order than the first if Δx, Δy, and Δz are infinitesimal (v. Dif. Cal. Art. 198).

Hence $$\frac{u' - u}{PP'} = D_x u.\frac{\Delta x}{PP'} + D_y u.\frac{\Delta y}{PP'} + D_z u.\frac{\Delta z}{PP'} + \frac{\epsilon}{PP'}.$$

But $$\frac{\Delta x}{PP'} = \lambda, \quad \frac{\Delta y}{PP'} = \mu, \quad \text{and} \quad \frac{\Delta z}{PP'} = \nu.$$

Therefore $$D_h u = \lambda D_x u + \mu D_y u + \nu D_z u. \qquad (1)$$

If $\nabla^2 u = 0$, $D_x^p D_y^q D_z^r u$ is a solution of Laplace's Equation.

For
$$\nabla^2(D_x^p D_y^q D_z^r u) = D_x^p D_y^q D_z^r(\nabla^2 u) = 0.$$

Hence if $\nabla^2 u = 0$ $D_h u$ is a solution of Laplace's Equation, and if OH_1, OH_2, OH_3, $\cdots$ are a set of lines through the origin $D_{h_1} D_{h_2} D_{h_3} \cdots u$ is a solution of Laplace's Equation.

117. If H_k is a rational integral homogeneous Algebraic function of x, y, and z of the kth degree

$$\begin{aligned} D_x\left(\frac{H_k}{r^l}\right) &= D_r\left(\frac{H_k}{r^l}\right) D_x r + \frac{1}{r^l} D_x(H_k) \\ &= -\frac{lxH_k}{r^{l+2}} + \frac{H_{k-1}}{r^l} = -\frac{lxH_k}{r^{l+2}} + \frac{r^2 H_{k-1}}{r^{l+2}}, \end{aligned}$$

and is of the form $\dfrac{H_{k+1}}{r^{l+2}}$.

The same thing can be proved of $D_y\left(\dfrac{H_k}{r^l}\right)$ and $D_z\left(\dfrac{H_k}{r^l}\right)$ and therefore holds good of $D_h\left(\dfrac{H_k}{r^l}\right)$.

If u is a homogeneous function of x, y, and z of the degree $-m-1$ and $\nabla^2 u = 0$ then $\nabla^2(r^{2m+1}u) = 0$.

$$\begin{aligned} \nabla^2(r^{2m+1}u) &= (2m+1)(2m+2)r^{2m-1}u \\ &\quad + 2(2m+1)r^{2m-1}(xD_x u + yD_y u + zD_z u) + r^{2m+1}\nabla^2 u \\ &= 0, \end{aligned}$$

since
$$xD_x u + yD_y u + zD_z u = -(m+1)u$$

by Euler's Theorem (v. Dif. Cal. Art. 220).

118. $\dfrac{M}{r} = \dfrac{M}{\sqrt{x^2+y^2+z^2}}$ is a solution of Laplace's Equation (v. Art. 75) and is of the form $\dfrac{H_0}{r}$.

$D_{h_1} D_{h_2} D_{h_3} \cdots D_{h_m}\left(\dfrac{M}{r}\right)$ is then a solution of Laplace's Equation by Art. 116; it is of the form $\dfrac{H_m}{r^{2m+1}}$ by Art. 117 and is a homogeneous function of the degree $-m-1$.

Therefore $r^{2m+1} D_{h_1} D_{h_2} D_{h_3} \cdots D_{h_m}\left(\dfrac{M}{r}\right)$ is a solution of Laplace's Equation, and is a rational integral homogeneous Algebraic function of x, y, and z of the mth degree, and is consequently a Solid Spherical Harmonic of the mth degree (v. Art. 110); and $r^{m+1} D_{h_1} D_{h_2} D_{h_3} \cdots D_{h_m}\left(\dfrac{M}{r}\right)$ is a Surface Spherical Harmonic of the mth degree.

Moreover since the direction of each of the lines OH_1, OH_2, $\cdots$ OH_m depends upon two angles which may be taken at pleasure, these angles and M are $2m+1$ arbitrary constants and may be so chosen that $r^{m+1}D_{h_1}D_{h_2}D_{h_3}\cdots D_{h_m}\left(\frac{M}{r}\right)$ may be any given Surface Spherical Harmonic.

Consequently any given Surface Spherical Harmonic may be regarded as formed by differentiating $\frac{M}{r}$ successively along m determinate lines OH_1, OH_2 $\cdots$ OH_m, and is given except for the undetermined factor M when these lines are given.

The lines OH_1, OH_2, OH_3, $\cdots OH_m$ are called the *axes* of the Harmonic, and the points where they meet the surface of the unit sphere the *poles* of the Harmonic. The m axes of a Zonal Harmonic coincide with the axis of coördinates (v. Art. 86) and consequently the m axes of a Laplacian coincide with what we have called the axis of the Laplacian (v. Art. 112).

119. Any *Surface Zonal Harmonic* $P_m(\mu)$ is equal to zero for m real and distinct values of μ which lie between -1 and 1; and any *Associated Function* $P_m^n(\mu)$ is equal to zero for $m-n$ real and distinct values of μ, which lie between -1 and 1.

$$P_m(\mu)=\frac{1}{2^m m!}\cdot\frac{d^m(\mu^2-1)^m}{d\mu^m}. \qquad \text{v. Art. 83 (1).}$$

$\frac{d^k(\mu^2-1)^m}{d\mu^k}$ contains $(\mu^2-1)^{m-k}$ as a factor. v. Art. 89.

From Rolle's Theorem, "If $f(x)$ is continuous and single-valued and is equal to zero for the real values a and b of x, $\frac{df(x)}{dx}$ is equal to zero for at least one real value of x between a and b," (v. Dif. Cal. Art. 126) it follows that since $(\mu^2-1)^m=0$ when $\mu=-1$ and when $\mu=1$ $\frac{d(\mu^2-1)^m}{d\mu}=0$ for at least one value of μ between -1 and 1. $\frac{d(\mu^2-1)^m}{d\mu}$ cannot be equal to zero for more than one value of μ between -1 and 1, for it contains $(\mu^2-1)^{m-1}$ as a factor and is a rational Algebraic polynomial of the $2m-1$st degree.

In like manner we can show that $\frac{d^2(\mu^2-1)^m}{d\mu^2}=0$ has $m-2$ roots equal to -1, $m-2$ roots equal to 1 and two real roots between -1 and 1 which separate the three distinct roots of $\frac{d(\mu^2-1)^m}{d\mu}=0$; and in general if $k<m+1$ that $\frac{d^k(\mu^2-1)^m}{d\mu^k}=0$ has $m-k$ roots equal to -1, $m-k$ roots equal to 1, and k real roots separating the $k+1$ distinct roots of $\frac{d^{k-1}(\mu^2-1)^m}{d\mu^{k-1}}=0$.

Hence $P_m(\mu)=0$ or $\frac{1}{2^m m!}\cdot\frac{d^m(\mu^2-1)^m}{d\mu^m}=0$ has m real and distinct roots between -1 and 1, and it has no more since it is of the mth degree.

The same reasoning shows that $\dfrac{d^{m+n}(\mu^2-1)^m}{d\mu^{m+n}} = 0$ has $m-n$ distinct real roots between -1 and 1, and therefore that $P_m^n(\mu)$ is equal to zero for $m-n$ distinct real values of μ between -1 and 1. Since $P_m^n(\mu)$ contains $\sin^n\theta$ as a factor it is also equal to zero when $\mu=-1$ and when $\mu=1$.

$\cos n\phi$ is equal to zero for $2n$ equidistant values of ϕ, and $\sin n\phi$ is equal to zero for $2n$ values of ϕ. Hence any *Tesseral Harmonic* $\sin n\phi P_m^n(\mu)$ or $\cos n\phi P_m^n(\mu)$ is equal to zero for $2n$ equidistant values of ϕ, for $\mu = 1$, for $\mu=-1$, and for $m-n$ real and different values of μ between -1 and 1.

It follows that the value of any Surface Zonal Harmonic $P_m(\mu)$ at a point on the surface of the unit sphere will have the same sign so long as the point remains on one of the *zones* into which the surface of the sphere is divided by the m circles of latitude corresponding to the m roots of $P_m(\mu)=0$, and will change sign whenever the point passes from one of these zones into an adjoining one; and that the value of any Tesseral Harmonic $\sin n\phi P_m^n(\mu)$ at a point on the surface of the unit sphere will have the same sign so long as the point remains on any one of the *tesserae* into which the surface of the sphere is divided by the $m-n$ circles of latitude corresponding to the roots of $P_m^n(\mu)=0$ and the $2n$ meridians corresponding to the roots of $\sin n\phi = 0$, and will change sign whenever the point passes from one of these tesserae into an adjoining one.

CHAPTER VII.[1]

CYLINDRICAL HARMONICS (BESSEL'S FUNCTIONS).

120. In Arts. 11 and 17 we obtained

$$z = AJ_0(x) + BK_0(x) \tag{1}$$

as the general solution of *Fourier's Equation*

$$\frac{d^2z}{dx^2} + \frac{1}{x}\frac{dz}{dx} + z = 0, \tag{2}$$

where
$$J_0(x) = 1 - \frac{x^2}{2^2} + \frac{x^4}{2^2.4^2} - \frac{x^6}{2^2.4^2.6^2} + \cdots \tag{3}$$

and is called a *Cylindrical Harmonic* or *Bessel's Function* of the zeroth order; and where

$$K_0(x) = J_0(x)\log x + \frac{x^2}{2^2} - \frac{x^4}{2^2.4^2}\left(\frac{1}{1} + \frac{1}{2}\right) + \frac{x^6}{2^2.4^2.6^2}\left(\frac{1}{1} + \frac{1}{2} + \frac{1}{3}\right) - \cdots \tag{4}$$

and is called a *Cylindrical Harmonic* or *Bessel's Function* of the Second Kind, and of the zeroth order.

In Art. 17 we found that $z = J_n(x)$
is a particular solution of *Bessel's Equation*

$$\frac{d^2z}{dx^2} + \frac{1}{x}\frac{dz}{dx} + \left(1 - \frac{n^2}{x^2}\right)z = 0, \tag{5}$$

where if n is unrestricted in value

$$J_n(x) = \frac{x^n}{2^n\Gamma(n+1)}\left[1 - \frac{x^2}{2^2(n+1)} + \frac{x^4}{2^4.2!(n+1)(n+2)} - \frac{x^6}{2^6.3!(n+1)(n+2)(n+3)} + \cdots\right] \tag{6}$$

and is called a *Cylindrical Harmonic* or *Bessel's Function* of the nth order; and that unless n is an integer

$$z = AJ_n(x) + BJ_{-n}(x)$$

is the general solution of Bessel's Equation.

If n is an integer it can be shown that

$$J_n(x) = (-1)^n J_{-n}(x),$$

[1] The student should re-read carefully Arts. 11, 17, and 18(d) before beginning this chapter.

(v. Forsyth's Diff. Eq. Art. 102), and then

$$z = AJ_n(x) + B\{K_n(x)\}$$

is the general solution of Bessel's Equation and

$$\{K_n(x)\} = J_n(x)\log x - \frac{1}{2}\left(\frac{x}{2}\right)^{-n}\sum_{k=0}^{k=n-1}\frac{(n-k-1)!}{k!}\left(\frac{x}{2}\right)^{2k}$$
$$-\frac{1}{2}\left(\frac{x}{2}\right)^{n}\sum_{k=0}^{k=\infty}\frac{(-1)^k}{(n+k)!k!}\left[1+\frac{1}{2}+\frac{1}{3}+\cdots+\frac{1}{k}\right.$$
$$\left.+1+\frac{1}{2}+\frac{1}{3}+\cdots+\frac{1}{n+k}\right]\left(\frac{x}{2}\right)^{2k} \qquad (7)$$

v. M. Bôcher, Ann. Math. Vol. VI, No. 4.

121. A useful expression for $J_n(x)$ as a definite integral can be obtained without difficulty from Bessel's Equation [(5) Art. 120] by a slight modification of the method given by Forsyth (Diff. Eq. Art. 136).

It was shown in Art. 17 that $z = x^n v$ is a solution of Bessel's Equation if v satisfies the equation

$$\frac{d^2v}{dx^2} + \frac{2n+1}{x}\frac{dv}{dx} + v = 0. \qquad (1)$$

Assume
$$v = \int_a^b T\cos(xt)dt \qquad (2)$$

where x and t are independent, T is an unknown function of t, and a and b are at present undetermined.

Then
$$\frac{dv}{dx} = -\int_a^b tT\sin(xt)dt$$

and
$$\frac{d^2v}{dx^2} = -\int_a^b t^2T\cos(xt)dt.$$

Substituting in (1) after multiplying through by x, we have

$$\int_a^b (1-t^2)Tx\cos(xt)dt - \int_a^b (2n+1)tT\sin(xt)dt = 0. \qquad (3)$$

By *integration by parts* we find that

$$\int_a^b (1-t^2)Tx\cos(xt)dt = \Big[(1-t^2)T\sin(xt)\Big]_a^b$$
$$-\int_a^b\left[(1-t^2)\frac{dT}{dt} - 2tT\right]\sin(xt)dt,$$

and (3) reduces to

$$\Big[(1-t^2)T\sin(xt)\Big]_a^b - \int_a^b \Big[(1-t^2)\frac{dT}{dt} + (2n-1)tT\Big]\sin(xt)dt = 0. \tag{4}$$

If we determine T so that

$$(1-t^2)\frac{dT}{dt} + (2n-1)tT = 0, \tag{5}$$

and a and b so that
$$\Big[(1-t^2)T\sin(xt)\Big]_a^b = 0 \tag{6}$$

(4) will be satisfied and our problem will be solved. (5) gives

$$T = C(1-t^2)^{n-\frac{1}{2}}, \tag{7}$$

and (6) will obviously be satisfied if $a = -1$ and $b = 1$.

Hence
$$v = C\int_{-1}^{1} \frac{(1-t^2)^n\cos(xt)dt}{\sqrt{1-t^2}}$$
is a solution of (1),

and
$$z = Cx^n \int_{-1}^{1} \frac{(1-t^2)^n\cos(xt)dt}{\sqrt{1-t^2}} \tag{8}$$

is a solution of Bessel's Equation.

If we let $t = \cos\phi$ in (8) we get

$$z = Cx^n \int_0^{\pi} \sin^{2n}\phi\cos(x\cos\phi)d\phi.$$

Expand $\cos(x\cos\phi)$ into a series involving powers of $x\cos\phi$, integrate term by term by the aid of the formulas

$$\int_0^{\frac{\pi}{2}} \sin^n x.dx = \frac{\sqrt{\pi}}{2}.\frac{\Gamma\left(\frac{n+1}{2}\right)}{\Gamma\left(\frac{n}{2}+1\right)}$$ [Int. Cal. (1) Art. 99],

$$\int_0^{\frac{\pi}{2}} \sin^n x\cos^m x.dx = \frac{\Gamma\left(\frac{m+1}{2}\right)\Gamma\left(\frac{n+1}{2}\right)}{2\Gamma\left(\frac{m+n}{2}+1\right)}$$

(Int. Cal. Art. 99 Ex. 2), and compare with (6) Art. 120, and we get

$$J_n(x) = \frac{x^n}{2^n\sqrt{\pi}\,\Gamma\left(n+\frac{1}{2}\right)}\int_0^{\pi} \sin^{2n}\phi\cos(x\cos\phi)d\phi. \tag{9}$$

If n is a positive integer (9) reduces to

$$J_n(x) = \frac{1}{\pi} \cdot \frac{x^n}{1.3.5.\cdots(2n-1)} \int_0^\pi \sin^{2n}\phi \cos(x\cos\phi)d\phi. \tag{10}$$

Let $n = 0$ in (9) or (10) and we get

$$J_0(x) = \frac{1}{\pi} \int_0^\pi \cos(x\cos\phi)d\phi. \tag{11}$$

EXAMPLES.

1. Obtain Formula (11) directly from Fourier's Equation, (2) Art. 120.

2. Prove by *integration by parts* that if $n > -\frac{1}{2}$

$$\int_0^\pi \sin^{2n}\phi \cos\phi \sin(x\cos\phi)d\phi = \frac{x}{2n+1} \int_0^\pi \sin^{2n+2}\phi \cos(x\cos\phi)d\phi.$$

3. Prove by *integration by parts* that if $n > \frac{1}{2}$

$$\int_0^\pi \sin^{2n}\phi \cos\phi \sin(x\cos\phi)d\phi$$

$$= \frac{1}{x} \int_0^\pi [2n\sin^{2n}\phi - (2n-1)\sin^{2n-2}\phi] \cos(x\cos\phi)d\phi.$$

122. We can now readily obtain a number of useful formulas. Differentiate (11) Art. 121 with respect to x and we get

$$\frac{dJ_0(x)}{dx} = -\frac{1}{\pi} \int_0^\pi \cos\phi \sin(x\cos\phi)d\phi$$

$$= -\frac{x}{\pi} \int_0^\pi \sin^2\phi \cos(x\cos\phi)d\phi \qquad \text{by Ex. 2 Art. 121.}$$

Hence by (10) Art. 121 $$\frac{dJ_0(x)}{dx} = -J_1(x). \tag{1}$$

In like manner by the aid of Exs. 3 and 2, Art. 121, we can obtain the relations

$$\frac{d[x^n J_n(x)]}{dx} = x^n J_{n-1}(x) \tag{2}$$

if $n > \frac{1}{2}$,

$$\frac{d[x^{-n}J_n(x)]}{dx} = -x^{-n}J_{n+1}(x) \tag{3}$$

if $n > -\frac{1}{2}$.

(2) can be written

$$\int_0^x x^n J_{n-1}(x)dx = x^n J_n(x) \tag{4}$$

if $n > \frac{1}{2}$.

(2) and (3) can be written

$$x^n \frac{dJ_n(x)}{dx} + nx^{n-1}J_n(x) = x^n J_{n-1}(x)$$

and

$$x^{-n}\frac{dJ_n(x)}{dx} - nx^{-n-1}J_n(x) = -x^{-n}J_{n+1}(x),$$

or

$$\frac{dJ_n(x)}{dx} = J_{n-1}(x) - \frac{n}{x}J_n(x) \tag{5}$$

and

$$\frac{dJ_n(x)}{dx} = -J_{n+1}(x) + \frac{n}{x}J_n(x); \tag{6}$$

whence

$$2\frac{dJ_n(x)}{dx} = J_{n-1}(x) - J_{n+1}(x) \tag{7}$$

and

$$\frac{2n}{x}J_n(x) = J_{n-1}(x) + J_{n+1}(x). \tag{8}$$

The repeated use of formula (8) will enable us to get from $J_0(x)$ and $J_1(x)$ any of Bessel's Functions whose order is a positive integer. For example, we have

$$J_2(x) = \frac{2}{x}J_1(x) - J_0(x)$$

$$J_3(x) = \left(\frac{8}{x^2} - 1\right)J_1(x) - \frac{4}{x}J_0(x).$$

From a table giving the values of $J_0(x)$ and $J_1(x)$, then, tables for the functions of higher order are readily constructed. Such a table taken from Rayleigh's Sound (Vol. I., page 265) will be found in the Appendix (Table VI.).

By the aid of (5) and (6) any derivative of $J_n(x)$ can be expressed in terms of $J_n(x)$ and $J_{n+1}(x)$. For example

$$\frac{d^2J_n(x)}{dx^2} = \left[\frac{n(n-1)}{x^2} - 1\right]J_n(x) + \frac{1}{x}J_{n+1}(x).$$

If we write $J_0(x)$ for z in Fourier's Equation [(2) Art. 120], then multiply through by xdx and integrate from zero to x, simplifying the resulting equation by *integration by parts*, we get

$$x\frac{dJ_0(x)}{dx}+\int_0^x xJ_0(x)dx=0;$$

whence by (1)
$$\int_0^x xJ_0(x)dx=xJ_1(x). \tag{9}$$

If we write $J_0(x)$ for z in Fourier's Equation, then multiply through by $x^2\dfrac{dJ_0(x)}{dx}dx$ and integrate from zero to x, simplifying by *integration by parts* we get

$$\frac{x^2}{2}\left[\left(\frac{dJ_0(x)}{dx}\right)^2+(J_0(x))^2\right]-\int_0^x x(J_0(x))^2dx=0;$$

whence by (1)
$$\int_0^x x(J_0(x))^2dx=\frac{x^2}{2}[(J_0(x))^2+(J_1(x))^2]. \tag{10}$$

In like manner we can get from Bessel's Equation [(5) Art. 120] the formula

$$\int_0^x x(J_n(x))^2dx=\frac{1}{2}\left[x^2\left(\frac{dJ_n(x)}{dx}\right)^2+(x^2-n^2)(J_n(x))^2\right] \tag{11}$$

which (6) enables us to reduce to the form

$$\int_0^x x(J_n(x))^2dx=\frac{x^2}{2}[(J_n(x))^2+(J_{n+1}(x))^2]-nxJ_n(x)J_{n+1}(x). \tag{12}$$

Formulas (9), (10), (11), and (12) will prove useful when we attempt to develop in terms of *Cylindrical Harmonics*.

Values of $J_n(x)$ for larger values of x than those given in Table VI., Appendix, may be computed very easily from the formula

$$\begin{aligned}J_n(x)=\sqrt{\frac{2}{\pi x}}&\left[1-\frac{(1^2-4n^2)(3^2-4n^2)}{2!(8x)^2}\right.\\&\left.+\frac{(1^2-4n^2)(3^2-4n^2)(5^2-4n^2)(7^2-4n^2)}{4!(8x)^4}-\cdots\right]\cos\left(x-\frac{\pi}{4}-n\frac{\pi}{2}\right)\\+\sqrt{\frac{2}{\pi x}}&\left[\frac{1^2-4n^2}{1!8x}\right.\\&\left.-\frac{(1^2-4n^2)(3^2-4n^2)(5^2-4n^2)}{3!(8x)^3}+\cdots\right]\sin\left(x-\frac{\pi}{4}-n\frac{\pi}{2}\right).\end{aligned} \tag{13}$$

v. Lommel, Studien über die Bessel'schen Functionen, page 59.

The series terminates if $2n$ is an odd integer, but otherwise it is divergent. It can be proved, however, that in any case the sum of m terms differs from $J_n(x)$ by less than the last term included, and consequently the formula can safely be used for numerical computation.

EXAMPLES.

1. Confirm (1), (2), and (3), Art. 122, by obtaining them from (3) and (6), Art. 120.

2. Confirm (1), Art. 122, by showing that Fourier's Equation will differentiate into the special form assumed by Bessel's Equation when $n = 1$.

3. Show that (9), Art. 122, is a special case of (4), Art. 122.

4. Show that the limit approached by $J_n(x)$ as n increases indefinitely is zero, and by the aid of this fact and of (8), Art. 122, prove that

$$J_{n-1}(x) = \frac{2}{x}[nJ_n(x) - (n+2)J_{n+2}(x) + (n+4)J_{n+4}(x) + \cdots].$$

5. Prove that

$$\frac{dJ_n(x)}{dx} = \frac{2}{x}[\tfrac{1}{2}nJ_n(x) - (n+2)J_{n+2}(x) + (n+4)J_{n+4}(x) - \cdots].$$

6. Show that the substitution of $\left(1 - \frac{y^2}{n^2}\right)^{\frac{1}{2}}$ for x in Legendre's Equation will reduce it to the form

$$\left(1 - \frac{y^2}{n^2}\right)\frac{d^2z}{dy^2} + \left(\frac{1}{y} - \frac{2y}{n^2}\right)\frac{dz}{dy} + \left(1 + \frac{1}{n}\right)z = 0,$$

and that the limiting form approached by this equation as n is indefinitely increased is Fourier's Equation, and hence that $J_0(x)$ can be regarded as some constant factor multiplied by the limiting value approached by $P_n\left(1 - \frac{x^2}{n^2}\right)^{\frac{1}{2}}$ as n is indefinitely increased.

123. To complete the solution of the drumhead problem taken up in Art. 11, we found that it would be necessary to develop a given function of r in the form

$$f(r) = A_1J_0(\mu_1 r) + A_2J_0(\mu_2 r) + A_3J_0(\mu_3 r) + \cdots$$

where μ_1, μ_2, μ_3, &c., are the roots of the transcendental equation $J_0(\mu a) = 0$; and in Art. 11, Ex. the development of unity in a series of precisely the same form was needed.

(*a*) Let us consider another problem.

The convex surface and one base of a cylinder of radius a and length b are kept at the constant temperature zero, the temperature at each point of the other base is a given function of the distance of the point from the centre of the base; required the temperature of any point of the cylinder after the permanent temperatures have been established.

Here we have to solve Laplace's Equation in Cylindrical Coördinates ([XIV] Art. 1).

$$D_r^2 u + \frac{1}{r} D_r u + \frac{1}{r^2} D_\phi^2 u + D_z^2 u = 0 \tag{1}$$

subject to the conditions

$$\begin{aligned} u &= 0 \quad \text{when} \quad z = 0 \\ u &= 0 \quad \text{"} \quad r = a \\ u &= f(r) \quad \text{"} \quad z = b, \end{aligned}$$

and from the symmetry of the problem we know that $D_\phi^2 u = 0$.

Assuming as usual $u = R.Z$ we break (1) up into the equations

$$\frac{d^2 Z}{dz^2} - \mu^2 Z = 0$$

$$\frac{d^2 R}{dr^2} + \frac{1}{r}\frac{dR}{dr} + \mu^2 R = 0,$$

whence

$$u = \sinh(\mu z) J_0(\mu r) \tag{2}$$

and

$$u = \cosh(\mu z) J_0(\mu r) \tag{3}$$

are particular solutions of (1).

If μ_k is a root of

$$J_0(\mu a) = 0 \tag{4}$$

$$u = \sinh(\mu_k z) J_0(\mu_k r)$$

satisfies (1) and two of the three equations of condition.

If then

$$f(r) = A_1 J_0(\mu_1 r) + A_2 J_0(\mu_2 r) + A_3 J_0(\mu_3 r) + \cdots \tag{5}$$

μ_1, μ_2, μ_3, &c., being roots of (4),

$$u = A_1 \frac{\sinh(\mu_1 z)}{\sinh(\mu_1 b)} J_0(\mu_1 r) + A_2 \frac{\sinh(\mu_2 z)}{\sinh(\mu_2 b)} J_0(\mu_2 r) + A_3 \frac{\sinh(\mu_3 z)}{\sinh(\mu_3 b)} J_0(\mu_3 r) + \cdots \tag{6}$$

satisfies (1) and all of the equations of condition, and is the required solution.

(*b*) If instead of keeping the convex surface of the cylinder at the temperature zero we surround it by a jacket impervious to heat, the equation of condition $u = 0$ when $r = a$ will be replaced by $D_r u = 0$ when $r = a$, or if

$$u = \sinh(\mu z) J_0(\mu r),$$

by
$$\frac{dJ_0(\mu r)}{dr} = 0 \qquad \text{when } r = a,$$

that is by
$$\mu J_0'(\mu a) = 0^2 \qquad \text{or (v. (1) Art. 122)}$$

by
$$J_1(\mu a) = 0. \tag{7}$$

If now in (5) and (6) μ_1, μ_2, μ_3, &c., are roots of (7), (6) will be the solution of our new problem.

(*c*) If instead of keeping the convex surface of the cylinder at the temperature zero we allow it to cool in air at the temperature zero, the condition $u = 0$ when $r = a$ will be replaced by $D_r u + hu = 0$ when $r = a$, or if

$$u = \sinh(\mu z) J_0(\mu r)$$

by
$$\mu J_0'(\mu r) + h J_0(\mu r) = 0 \qquad \text{when} \quad r = a$$

that is by
$$\mu a J_0'(\mu a) + ah J_0(\mu a) = 0 \qquad \text{or (v. (1) Art. 122)}$$

by
$$\mu a J_1(\mu a) - ah J_0(\mu a) = 0. \tag{8}$$

If now in (5) and (6) μ_1, μ_2, μ_3, &c., are roots of (8), (6) will be the solution of our present problem.

124. It can be shown that
$$J_0(x) = 0 \tag{1}$$
$$J_1(x) = 0 \tag{2}$$
and
$$x J_0'(x) + \lambda J_0(x) = 0 \tag{3}$$

have each an infinite number of real positive roots (v. Riemann, Par. Dif. Gl., § 97). The earlier roots of these equations can be computed without serious difficulty from the table for the values of $J_0(x)$ (Table VI., Appendix).

The first twelve roots of $J_0(x) = 0$ and $J_1(x) = 0$ are given in Table IV., Appendix, a table due to Stokes. Large roots of $J_0(x) = 0$ and of $J_1(x) = 0$ may be very easily computed from the formulas

$$\frac{x_0^{(s)}}{\pi} = s - .25 + \frac{.050661}{4s-1} - \frac{.053041}{(4s-1)^3} + \frac{.262051}{(4s-1)^5} - \cdots \tag{4}$$

$$\frac{x_1^{(s)}}{\pi} = s + .25 - \frac{.151982}{4s+1} + \frac{.015399}{(4s+1)^3} - \frac{.245270}{(4s+1)^5} + \cdots \tag{5}$$

---

[2]We shall find it convenient to use the familiar notation of $f'(x) = \frac{df(x)}{dx}$ (v. Dif. Cal., p. 119).

given by Stokes in Camb. Phil. Trans., Vol. IX., $x_0^{(s)}$ representing the sth root of $J_0(x) = 0$, and $x_1^{(s)}$ the sth root of $J_1(x) = 0$.

125. We have seen in Art. 123 that $U = \sinh(\mu_k z)J_0(\mu_k r)$ and $V = \sinh(\mu_l z)J_0(\mu_l r)$ are solutions of $\nabla^2 U = 0$ and $\nabla^2 V = 0$ if we express Laplace's Equation in terms of Cylindrical Coördinates (v. (1) Art. 123).

Hence, if $\int dS$ represents the surface integral over any closed surface, we have

$$\int (UD_nV - VD_nU)dS = 0$$

by Green's Theorem (v. Art. 92).

If we take the cylinder of Art. 123 as our surface, and perform the integrations and simplify the resulting equation, we find

$$\int_0^a rJ_0(\mu_k r)J_0(\mu_l r)dr = \frac{-1}{\mu_k{}^2 - \mu_l{}^2}[\mu_k a J_0(\mu_l a)J_0'(\mu_k a) - \mu_l a J_0(\mu_k a)J_0'(\mu_l a)]$$

$$= \frac{-1}{\mu_l{}^2 - \mu_k{}^2}[\mu_k a J_0(\mu_l a)J_1(\mu_k a) - \mu_l a J_0(\mu_k a)J_1(\mu_l a)]. \quad (1)$$

Hence if μ_k and μ_l are different roots of

$$J_0(\mu a) = 0,$$

or of

$$J_1(\mu a) = 0,$$

or of

$$\mu a J_1(\mu a) - \lambda J_0(\mu a) = 0,$$

then

$$\int_0^a rJ_0(\mu_k r)J_0(\mu_l r)dr = 0. \quad (2)$$

EXAMPLE.

Obtain (1) Art. 125 directly from Fourier's Equation

$$\frac{d^2 J_0(\mu r)}{dr^2} + \frac{1}{r}\frac{dJ_0(\mu r)}{dr} + \mu^2 J_0(\mu r) = 0.$$

126. We are now able to obtain the developments called for in Art. 123.

Let $$f(r) = A_1J_0(\mu_1 r) + A_2J_0(\mu_2 r) + A_3J_0(\mu_3 r) + \cdots \quad (1)$$

μ_1, μ_2, μ_3, &c., being roots of $J_0(\mu a) = 0$, or of $J_1(\mu a) = 0$, or of

$$\mu a J_1(\mu a) - \lambda J_0(\mu a) = 0.$$

To determine any coefficient A_k multiply (1) by $rJ_0(\mu_k r)dr$ and integrate from zero to a. The first member will become

$$\int_0^a rf(r)J_0(\mu_k r)dr.$$

Every term of the second member will vanish by (2) Art. 125 except the term

$$A_k \int_0^a r(J_0(\mu_k r))^2 dr.$$

$$\int_0^a r(J_0(\mu_k r))^2 dr = \frac{1}{\mu_k^2} \int_0^{\mu_k a} x(J_0(x))^2 dx = \frac{a^2}{2}[(J_0(\mu_k a))^2 + (J_1(\mu_k a))^2]$$

by (10) Art. 122.

Hence $$A_k = \frac{2}{a^2[(J_0(\mu_k a))^2 + (J_1(\mu_k a))^2]} \int_0^a r f(r) J_0(\mu_k r) dr. \quad (2)$$

The development (1) holds good from $r = 0$ to $r = a$ (v. Arts. 24, 25, and 88).

If μ_1, μ_2, μ_3, &c., are roots of $J_0(\mu a) = 0$, (2) reduces to

$$A_k = \frac{2}{a^2(J_1(\mu_k a))^2} \int_0^a r f(r) J_0(\mu_k r) dr. \quad (3)$$

If μ_1, μ_2, μ_3, &c., are roots of $J_1(\mu a) = 0$, (2) reduces to

$$A_k = \frac{2}{a^2(J_0(\mu_k a))^2} \int_0^a r f(r) J_0(\mu_k r) dr. \quad (4)$$

If μ_1, μ_2, μ_3, &c., are roots of $\mu a J_1(\mu a) - \lambda J_0(\mu a) = 0$, (2) reduces to

$$A_k = \frac{2\mu_k^2}{(\lambda^2 + \mu_k^2 a^2)(J_0(\mu_k a))^2} \int_0^a r f(r) J_0(\mu_k r) dr. \quad (5)$$

For the important case where $f(r) = 1$

$$\int_0^a r f(r) J_0(\mu_k r) dr = \int_0^a r J_0(\mu_k r) dr = \frac{1}{\mu_k^2} \int_0^{\mu_k a} x J_0(x) dx = \frac{a}{\mu_k} J_1(\mu_k a) \quad (6)$$

by (9) Art. 122, and (3) reduces to

$$A_k = \frac{2}{\mu_k a J_1(\mu_k a)}, \quad (7)$$

(4) reduces to $A_k = 0$ except for $k = 1$ when $\mu_k = 0$ and we have $A_1 = 1$,

(5) reduces to $$A_k = \frac{2\lambda}{(\lambda^2 + \mu_k^2 a^2) J_0(\mu_k a)}. \quad (8)$$

EXAMPLES.

1. Show that in (12) Art. 11 any coefficient A_k has the value given in (3) Art. 126; and in the answer to Art. 11, Ex. the value given in (7) Art. 126.

2. Show that if a drumhead be initially distorted so that it has circular symmetry, it will not in general give a musical note; that it may be initially distorted so as to give a musical note; that in this case the vibration will be a *steady* vibration; that the frequencies of the various musical notes that can be given when the distortion has circular symmetry are proportional to the roots of $J_0(x) = 0$; that the possible nodes for such vibrations are concentric circles whose radii are proportional to the roots of $J_0(x) = 0$.

3. A cylinder of radius one meter and altitude one meter has its upper surface kept at the temperature 100°, and its base and convex surface at the temperature 15°, until the *stationary temperature* is set up. Find the temperature at points on the axis 25 cm., 50 cm., and 75 cm. from the base, and also at a point 25 cm. from the base and 50 cm. from the axis.

Ans., 29°.6; 47°.6; 71°.2; 25°.8.

4. An iron cylinder one meter long and twenty centimeters in diameter has its convex surface covered with a so-called non-conducting cement one centimeter thick. One end and the convex surface of the cylinder thus coated are kept at the temperature zero, the other end at the temperature of 100°. Find to the nearest tenth of a degree the temperature of the middle point of the axis, and of the points of the axis twenty centimeters from each end after the temperatures have ceased to change. Given that the conductivity of iron is 0.185 and of cement 0.000162 in C. G. S. units. Find also the temperature of a point on the surface midway between the ends, and of points on the surface twenty centimeters from each end. Find the temperatures of the three points of the axis, supposing the coating a perfect non-conductor, and again, supposing the coating absent. Neglect the curvature of the coating.

Ans., 15°.4; 40°.85; 72°.8; 15°.3; 40°.7; 72°.5; 0°.0; 0°.0; 1°.3.

127. If instead of considering the cooling of a cylinder as in Art. 123 we have to deal with a cylindrical shell whose curved surfaces are co-axial cylinders, we are obliged to use the Bessel's Functions of the second kind. Let our equations of condition be

$$u = 0 \quad \text{when} \quad z = 0, \qquad u = 0 \quad \text{when} \quad r = a,$$
$$u = f(r) \quad \text{"} \quad z = b, \qquad u = 0 \quad \text{"} \quad r = c.$$

Then (v. Art. 123)

$$u = \sinh(\mu_k z)\left[J_0(\mu_k r) - \frac{J_0(\mu_k c)}{K_0(\mu_k c)} K_0(\mu_k r)\right]$$

where μ_k is a root of the equation

$$J_0(\mu a) - \frac{J_0(\mu c)}{K_0(\mu c)} K_0(\mu a) = 0 \tag{1}$$

will satisfy Laplace's Equation [(1) Art. 123] and all of the equations of condition except the second.

Hence $$u = \sum_{k=1}^{k=\infty} A_k \frac{\sinh(\mu_k z)}{\sinh(\mu_k b)} \left[J_0(\mu_k r) - \frac{J_0(\mu_k c)}{K_0(\mu_k c)} K_0(\mu_k r)\right] \tag{2}$$

is the required solution if

$$f(r) = \sum_{k=1}^{k=\infty} A_k \left[J_0(\mu_k r) - \frac{J_0(\mu_k c)}{K_0(\mu_k c)} K_0(\mu_k r)\right]. \tag{3}$$

The development (3) is easily obtained.

Call the parenthesis for the sake of brevity $B_0(\mu_k r)$. Then by the method of Art. 125 we get if we integrate over our cylindrical shell

$$\int_a^c r B_0(\mu_k r) B_0(\mu_l r) dr = 0 \tag{4}$$

if μ_k and μ_l are roots of (1); and by an easy extension of (10) Art. 122

$$\int_a^c r[B_0(\mu_k r)]^2 dr = \tfrac{1}{2}\{c^2[B_0'(\mu_k c)]^2 - a^2[B_0'(\mu_k a)]^2\}. \tag{5}$$

Determining the coefficients in (3) as in Art. 124 and simplifying by the aid of (4) we have

$$A_k = \frac{2\int_a^c r f(r)\left[J_0(\mu_k r) - \frac{J_0(\mu_k c)}{K_0(\mu_k c)} K_0(\mu_k r)\right] dr}{c^2\left[J_0'(\mu_k c) - \frac{J_0(\mu_k c)}{K_0(\mu_k c)} K_0'(\mu_k c)\right]^2 - a^2\left[J_0'(\mu_k a) - \frac{J_0(\mu_k c)}{K_0(\mu_k c)} K_0'(\mu_k a)\right]^2}. \tag{6}$$

EXAMPLE.

If a membrane bounded by concentric circles of radius a and radius b, and fastened at the edges, is initially distorted into a form symmetrical with respect to the centre, and then allowed to vibrate

$$y = \sum_{k=1}^{k=\infty} A_k \cos(\mu_k ct) \left[J_0(\mu_k r) - \frac{J_0(\mu_k b)}{K_0(\mu_k b)} K_0(\mu_k r)\right]$$

where A_k is obtained from (6) Art. 127 by replacing c by b.

128. If in the cooling of a cylinder $u = 0$ when $z = 0$, $u = 0$ when $z = b$, and $u = f(z)$ when $r = a$, the problem is easily solved.

If in (2) and (3) Art. 123 μ is replaced by μi we can readily obtain

$$z = \sin(\mu z) J_0(\mu r i)$$

and

$$z = \cos(\mu z) J_0(\mu r i)$$

as particular solutions of Laplace's Equation [(1) Art. 123]; and

$$J_0(xi) = 1 + \frac{x^2}{2^2} + \frac{x^4}{2^2.4^2} + \frac{x^6}{2^2.4^2.6^2} + \cdots \tag{1}$$

and is real.

$$f(z) = \sum_{k=1}^{k=\infty} A_k \sin \frac{k\pi z}{b}$$

where

$$A_k = \frac{2}{b} \int_a^b f(z) \sin \frac{k\pi z}{b} dz \tag{2}$$

by Art. 31 (7) and (8).

Hence

$$u = \sum_{k=1}^{k=\infty} A_k \sin \frac{k\pi z}{b} \frac{J_0\left(\frac{k\pi r i}{b}\right)}{J_0\left(\frac{k\pi a i}{b}\right)} \tag{3}$$

is our required solution.

EXAMPLES.

1. If the cylinder is hollow and we have $u = 0$ when $z = 0$, $u = 0$ when $z = b$, $u = 0$ when $r = c$, and $u = f(z)$ when $r = a$; then

$$u = \sum_{k=1}^{k=\infty} A_k \sin \frac{k\pi z}{b} \left[\frac{J_0\left(\frac{k\pi r i}{b}\right)}{J_0\left(\frac{k\pi c i}{b}\right)} - \frac{\overline{K_0}\left(\frac{k\pi r i}{b}\right)}{\overline{K_0}\left(\frac{k\pi c i}{b}\right)} \right] \div \left[\frac{J_0\left(\frac{k\pi a i}{b}\right)}{J_0\left(\frac{k\pi c i}{b}\right)} - \frac{\overline{K_0}\left(\frac{k\pi a i}{b}\right)}{\overline{K_0}\left(\frac{k\pi c i}{b}\right)} \right]$$

where A_k has the value given in (2) Art. 128, and

$$\overline{K_0}(xi) = K_0(xi) - J_0(xi)\log i$$
$$= J_0(xi)\log x - \frac{x^2}{2^2} - \frac{x^4}{2^2.4^2}\left(\tfrac{1}{1}+\tfrac{1}{2}\right) - \frac{x^6}{2^2.4^2.6^2}\left(\tfrac{1}{1}+\tfrac{1}{2}+\tfrac{1}{3}\right) - \cdots$$

[v. (4) Art. 120], and is real.

2. A hollow cylinder 6 feet long whose inner surface has the radius 3 inches, and whose outer surface has the radius 1 foot, has its bases and outer surface kept at the temperature 0°, and its inner surface at the temperature 100°, until the permanent state of temperatures is established; find the temperatures of two points in a plane parallel to the bases and half-way between them, one of which is 6 inches and the other 9 inches from the axis. *Ans.*, 49°.6; 20°.2.

129. If in the problem of Art. 123 the temperatures of the points of the upper base of the cylinder are unsymmetrical so that $u = f(r, \theta)$ when $z = b$, we have to get particular solutions of Laplace's Equation [(1) Art. 123] for the case where $D_\phi^2 u$ is not equal to zero. We readily find that

$$u = \sinh\,(\mu z)[A\cos n\phi + B\sin n\phi]J_n(\mu r)$$

and

$$u = \cosh\,(\mu z)[A\cos n\phi + B\sin n\phi]J_n(\mu r)$$

are such solutions, and that

$$u = \sum_{n=0}^{n=\infty}\sum_{k=1}^{k=\infty}\frac{\sinh\mu_k z}{\sinh\mu_k b}[A_{n,k}\cos n\phi + B_{n,k}\sin n\phi]J_n(\mu_k r) \qquad (1)$$

is the solution of the given problem if

$$f(r,\phi) = \sum_{n=0}^{n=\infty}\sum_{k=1}^{k=\infty}(A_{n,k}\cos n\phi + B_{n,k}\sin n\phi)J_n(\mu_k r) \qquad (2)$$

where μ_k is a root of the equation

$$\frac{J_n(\mu a)}{\mu^n a^n} = 0. \qquad (3)$$

EXAMPLES.

1. Show that

$$\int_0^a rJ_n(\mu_k r)J_n(\mu_l r)dr$$
$$= \frac{a}{{\mu_k}^2 - {\mu_l}^2}[\mu_l J_n(\mu_k a)J_n'(\mu_l a) - \mu_k J_n(\mu_l a)J_n'(\mu_k a)]$$
$$= \frac{a}{{\mu_k}^2 - {\mu_l}^2}[\mu_k J_n(\mu_l a)J_{n+1}(\mu_k a) - \mu_l J_n(\mu_k a)J_{n+1}(\mu_l a)].$$

2. Show that

$$\int_0^a r[J_n(\mu_k r)]^2 dr$$
$$= \frac{1}{2}\Big[a^2(J_n'(\mu_k a))^2 + \Big(a^2 - \frac{n^2}{{\mu_k}^2}\Big)(J_n(\mu_k a))^2\Big]$$
$$= \frac{a^2}{2}[(J_n(\mu_k a))^2 + (J_{n+1}(\mu_k a))^2] - \frac{na}{\mu_k}J_n(\mu_k a)J_{n+1}(\mu_k a).$$

3. Show that in Art. 129

$$A_{0,k} = \frac{1}{\pi}\frac{\int_0^{2\pi} d\phi \int_0^a rf(r,\phi)J_0(\mu_k r)dr}{a^2[J_1(\mu_k a)]^2},$$

$$B_{0,k} = 0,$$

$$A_{n,k} = \frac{2}{\pi}\frac{\int_0^{2\pi} d\phi \int_0^a rf(r,\phi)\cos n\phi J_n(\mu_k r)dr}{a^2[J_{n+1}(\mu_k a)]^2},$$

$$B_{n,k} = \frac{2}{\pi}\frac{\int_0^{2\pi} d\phi \int_0^a rf(r,\phi)\sin n\phi J_n(\mu_k r)dr}{a^2[J_{n+1}(\mu_k a)]^2}.$$

4. Obtain the coefficients for the case where the convex surface of the cylinder is impervious to heat.

5. Obtain the coefficients for the case where the convex surface of the cylinder is exposed to air at the temperature zero.

6. Show that if in a drumhead problem of Art. 11 the initial distortion is unsymmetrical, so that we have to solve the equation [XI] Art. 1 subject to the conditions $z = f(r, \phi)$ when $t = 0$, $D_t z = 0$ when $t = 0$, $z = 0$ when $r = a$, the solution is

$$z = \sum_{n=0}^{n=\infty}\sum_{k=1}^{k=\infty} \cos(\mu_k ct)(A_{n,k}\cos n\phi + B_{n,k}\sin n\phi)J_n(\mu_k r)$$

where $A_{0,k}$, $B_{0,k}$, $A_{n,k}$, and $B_{n,k}$ have the values given in Ex. 3.

7. What modifications do the statements made in Ex. 2, Art. 126, need to make them apply to the unsymmetrical case treated in Ex. 6?

Show that any possible nodal system in Ex. 6 is composed of concentric circles and of radii whose outer extremities are equidistant. v. Rayleigh's Sound, Vol. I., Arts. (202-207).

8. Solve the problem of Art. 127 and of Art. 127. Ex. for the unsymmetrical case. *Suggestion:* $AJ_n(x) + BK_n(x)$ is a solution of Bessel's Equation.

9. Solve the problem of Art. 128 and of Art. 128. Ex. 1. for the case where $u = f(z, \phi)$ when $r = a$. *Suggestion:* $u = \sin \mu z(A \cos n\phi + B \sin n\phi) J_n(\mu r i)$ is a solution of Laplace's Equation, and $f(z, \phi)$ can be developed into a double Fourier's Series [v. (15) Art. 71].

10. Show that in dealing with a wedge cut from a cylinder by planes passed through the axis, or with a membrane in the form of a circular sector, it may be necessary to use Bessel's Functions of fractional or incommensurable orders.

11. *Bernouilli's Problem* (v. Chapter IX). In considering small transverse vibrations of a uniform, heavy, flexible, inelastic string fastened at one end and initially distorted into some given curve, we have to solve the equation $D_t^2 y = c^2(x D_x^2 y + D_x y)$, subject to the conditions $D_t y = 0$ when $t = 0$, $y = f(x)$ when $t = 0$, $y = 0$ when $x = a$; the origin being taken at the distance a below the point of suspension and the axis of X taken vertical.

Show that
$$y = \sum_{k=1}^{k=\infty} A_k \cos \mu_k ct\, B_0(\mu_k^2 x),$$

where
$$B_0(x) = 1 - \frac{x}{1^2} + \frac{x^2}{1^2.2^2} - \frac{x^3}{1^2.2^2.3^2} + \cdots$$
$$= J_0(2\sqrt{x})$$

and μ_k is a root of the equation
$$B_0(\mu^2 a) = J_0(2\mu\sqrt{a}) = 0,$$

and
$$A_k = \frac{\int_0^a f(x) B_0(\mu_k^2 x) dx}{\mu^2 a^2 [B_0'(\mu_k^2 a)]^2} = \frac{\int_0^a f(x) J_0(2\mu_k\sqrt{x}) dx}{a[J_1(2\mu_k\sqrt{a})]^2}.$$

12. As a simple case under Example 10 consider the vibrations of a circular membrane fastened at the perimeter and also along a radius and then initially distorted (v. Rayleigh's Sound, Art. 207). In this case we must modify the formula given in Ex. 6 by dropping out the terms involving $\cos n\phi$ and by taking $n = \frac{m}{2}$. The required solution is

$$z = \sum_{m=1}^{m=\infty} \sum_{k=1}^{k=\infty} B_{m,k} \cos \mu_k ct \sin \frac{m\phi}{2} J_{\frac{m}{2}}(\mu_k r)$$

where μ_k is a root of $$\frac{J_{\frac{m}{2}}(\mu a)}{\mu^{\frac{m}{2}} a^{\frac{m}{2}}} = 0$$

and $$B_{m,k} = \frac{2}{\pi} \frac{\int_0^{2\pi} d\phi \int_0^a r f(r,\phi) \sin \frac{m\phi}{2} J_{\frac{m}{2}}(\mu_k r) dr}{a^2 [J_{\frac{m}{2}}'(\mu_k a)]^2}.$$

For the terms in which m is odd, $J_{\frac{m}{2}}(x)$ can be readily obtained from (13) Art. 122, which will become a finite sum.

For example, (13) Art. 122 gives the values

$$J_{\frac{1}{2}}(x) = \sqrt{\frac{2}{\pi x}} \sin x; \quad J_{\frac{3}{2}}(x) = \sqrt{\frac{2}{\pi x}} \left(\frac{1}{x} \sin x - \cos x \right);$$

$$J_{\frac{5}{2}}(x) = -\sqrt{\frac{2}{\pi x}} \left[\left(1 + \frac{3}{x^2} \right) \sin x + \frac{3}{x} \cos x \right]; \quad \text{\&c.}$$

13. The question of the flow of heat in three dimensions involves a problem not unlike the last.

Suppose the initial temperatures of all points in a sphere of radius c given, and let the surface be kept at the temperature zero. Then we have to solve the equation

$$D_t u = \frac{a^2}{r^2} \left[D_r(r^2 D_r u) + \frac{1}{\sin\theta} D_\theta(\sin\theta D_\theta u) + \frac{1}{\sin^2\theta} D_\phi^2 u \right] \tag{1}$$

([IV] Art. 1) subject to the conditions

$$u = 0 \quad \text{when} \quad r = c,$$
$$u = f(r, \theta, \phi) \quad \text{when} \quad t = 0.$$

If we assume $u = T.R.V$ where T is a function of t only, R of r only, and V of θ and ϕ only, (1) can be broken up into

$$\frac{dT}{dt} + a^2\alpha^2 T = 0 \tag{2}$$

$$m(m+1)V + \frac{1}{\sin\theta} D_\theta(\sin\theta D_\theta V) + \frac{1}{\sin^2\theta} D_\phi^2 V = 0 \tag{3}$$

and $$\frac{d^2R}{dr^2} + \frac{2}{r}\frac{dR}{dr} + \left[\alpha^2 - \frac{m(m+1)}{r^2} \right] R = 0. \tag{4}$$

Hence $T = e^{-a^2\alpha^2 t}$, $V = Y_m(\mu, \phi)$ [v. Art. 102 (2)], and R is still to be found. If in (4) we let $x = \alpha r$ and $z = R\sqrt{\alpha r}$ it becomes

$$\frac{d^2z}{dx^2} + \frac{1}{x}\frac{dz}{dx} + \left[1 - \frac{(m + \frac{1}{2})^2}{x^2} \right] z = 0$$

which is satisfied by $z = J_{m+\frac{1}{2}}(x)$. (v. Art. 17.)

Therefore
$$R = \frac{1}{\sqrt{\alpha r}} J_{m+\frac{1}{2}}(\alpha r).$$

$$f(r,\theta,\phi) = \frac{1}{4\pi}\sum_{m=0}^{m=\infty}(2m+1)\int_0^{2\pi} d\phi_1 \int_0^{\pi} f(r,\theta_1,\phi_1)P_m(\cos\gamma)\sin\theta_1 d\theta_1$$

by (3) Art. 114,

$$= \sum_{m=0}^{m=\infty}\sum_{n=0}^{n=m}[A_{m,n}f_{m,n}(r)\cos n\phi + B_{m,n}F_{m,n}(r)\sin n\phi]P_m^n(\mu).$$

$$\sqrt{r}f_{m,n}(r) = \sum_{k=0}^{k=\infty} C_{m,n,k}J_{m+\frac{1}{2}}(\alpha_k r)$$

where α_k is a root of the equation

$$\frac{J_{m+\frac{1}{2}}(\alpha c)}{(\alpha c)^{m+\frac{1}{2}}} = 0$$

and
$$C_{m,n,k} = \frac{2\int_0^c r^{\frac{3}{2}} f_{m,n}(r)J_{m+\frac{1}{2}}(\alpha_k r)dr}{c^2[J'_{m+\frac{1}{2}}(\alpha_k c)]^2}.$$

$$\sqrt{r}F_{m,n}(r) = \sum_{m=0}^{m=\infty} D_{m,n,k}J_{m+\frac{1}{2}}(\alpha_k r)$$

where
$$D_{m,n,k} = \frac{2\int_0^c r^{\frac{3}{2}} F_{m,n}(r)J_{m+\frac{1}{2}}(\alpha_k r)dr}{c^2[J'_{m+\frac{1}{2}}(\alpha_k c)]^2}.$$

The final solution is

$$u = \frac{1}{\sqrt{r}}\sum_{m=0}^{m=\infty}\sum_{n=0}^{n=m}\Bigg[P_m^n(\mu)\sum_{k=1}^{k=\infty}(A_{m,n}C_{m,n,k}\cos n\phi + B_{m,n}D_{m,n,k}\sin n\phi)e^{-a^2\alpha_k^2 t}J_{m+\frac{1}{2}}(\alpha_k r)\Bigg]$$

cf. Riemann, Par. Dif. Gl., §§ 72 and 73.

CHAPTER VIII.

LAPLACE'S EQUATION IN CURVILINEAR COÖRDINATES. ELLIPSOIDAL HARMONICS.

130. *Orthogonal Curvilinear Coördinates.*

If

$$\left.\begin{aligned} F_1(x,y,z) &= \rho_1 \\ F_2(x,y,z) &= \rho_2 \\ F_3(x,y,z) &= \rho_3 \end{aligned}\right\} \tag{1}$$

are the equations in rectangular coördinates of three surfaces that are mutually perpendicular no matter what the values of ρ_1, ρ_2, and ρ_3, the parameters ρ_1, ρ_2, and ρ_3, may be regarded as a set of coördinates for a point of intersection of the three surfaces, in the sense that when ρ_1, ρ_2, ρ_3 are given the point in question is determined, and when the point is given the corresponding values of ρ_1, ρ_2, ρ_3, can be found.

From equations (1) x, y, and z can be expressed in terms of ρ_1, ρ_2, and ρ_3. Suppose this done. If now x, y, z are the rectangular coördinates of the point $\rho_1 = a$, $\rho_2 = b$, $\rho_3 = c$, the rectangular coördinates of the points $\rho_1 = a + d\rho_1$, $\rho_2 = b$, $\rho_3 = c$, are obviously $x + D_{\rho_1}x.d\rho_1 + \epsilon_1$, $y + D_{\rho_1}y.d\rho_1 + \epsilon_2$, $z + D_{\rho_1}z.d\rho_1 + \epsilon_3$, where ϵ_1, ϵ_2, and ϵ_3 are infinitesimals of higher order than $d\rho_1$. Hence the square of the distance between the points will differ by an infinitesimal of higher order than that of $d\rho_1^2$ from dn_1^2 where

$$dn_1^2 = [(D_{\rho_1}x)^2 + (D_{\rho_1}y)^2 + (D_{\rho_1}z)^2]d\rho_1^2.$$

Let

$$\left.\begin{aligned} \frac{1}{h_1^2} &= (D_{\rho_1}x)^2 + (D_{\rho_1}y)^2 + (D_{\rho_1}z)^2 \\ \frac{1}{h_2^2} &= (D_{\rho_2}x)^2 + (D_{\rho_2}y)^2 + (D_{\rho_2}z)^2 \\ \frac{1}{h_3^2} &= (D_{\rho_3}x)^2 + (D_{\rho_3}y)^2 + (D_{\rho_3}z)^2. \end{aligned}\right\} \tag{2}$$

Then if dn_1 is the element of length normal to the surface $\rho_1 = a$, dn_2 normal to $\rho_2 = b$, and dn_3 normal to $\rho_3 = c$

$$dn_1 = \frac{d\rho_1}{h_1}, \quad dn_2 = \frac{d\rho_2}{h_2}, \quad dn_3 = \frac{d\rho_3}{h_3}. \tag{3}$$

The element of surface dS_1 on the surface $\rho_1 = a$ is easily seen to be

$$dS_1 = \frac{d\rho_2 d\rho_3}{h_2 h_3}; \tag{4}$$

and the element of volume dv is

$$dv = \frac{d\rho_1 d\rho_2 d\rho_3}{h_1 h_2 h_3}. \tag{5}$$

EXAMPLE.

Show that
$$h_1^2 = (D_x\rho_1)^2 + (D_y\rho_1)^2 + (D_z\rho_1)^2$$
$$h_2^2 = (D_x\rho_2)^2 + (D_y\rho_2)^2 + (D_z\rho_2)^2$$
$$h_3^2 = (D_x\rho_3)^2 + (D_y\rho_3)^2 + (D_z\rho_3)^2.$$

Suggestion: If h_1 has the value just given $\frac{D_x\rho_1}{h_1}, \frac{D_y\rho_1}{h_1}, \frac{D_z\rho_1}{h_1}$ are the direction cosines of the normal at any given point of $\rho_1 = a$. (v. Int. Cal. page 161.) Then
$$dn_1 = \frac{D_x\rho_1}{h_1}dx + \frac{D_y\rho_1}{h_1}dy + \frac{D_z\rho_1}{h_1}dz = \frac{1}{h_1}d\rho_1.$$

131. *Laplace's Equation in orthogonal curvilinear coördinates.*
If we apply the special form of Green's Theorem
$$\iiint \nabla^2 V dxdydz = \int D_n V dS \qquad \text{(v. Art. 98)}$$
to the space bounded by the surfaces $\rho_1 = a$, $\rho_2 = b$, $\rho_3 = c$, $\rho_1 = a + d\rho_1$, $\rho_2 = b + d\rho_2$, $\rho_3 = c + d\rho_3$, we have
$$\frac{\nabla^2 V d\rho_1 d\rho_2 d\rho_3}{h_1h_2h_3} =$$
$$-h_1D_{\rho_1}V\frac{d\rho_2d\rho_3}{h_2h_3} + h_1D_{\rho_1}V\frac{d\rho_2d\rho_3}{h_2h_3} + D_{\rho_1}\left(\frac{h_1}{h_2h_3}D_{\rho_1}V\right)d\rho_1d\rho_2d\rho_3$$
$$-h_2D_{\rho_2}V\frac{d\rho_3d\rho_1}{h_3h_1} + h_2D_{\rho_2}V\frac{d\rho_3d\rho_1}{h_3h_1} + D_{\rho_2}\left(\frac{h_2}{h_3h_1}D_{\rho_2}V\right)d\rho_1d\rho_2d\rho_3$$
$$-h_3D_{\rho_3}V\frac{d\rho_1d\rho_2}{h_1h_2} + h_3D_{\rho_3}V\frac{d\rho_1d\rho_2}{h_1h_2} + D_{\rho_3}\left(\frac{h_3}{h_1h_2}D_{\rho_3}V\right)d\rho_1d\rho_2d\rho_3;$$
whence
$$\nabla^2 V = h_1h_2h_3\left[D_{\rho_1}\left(\frac{h_1}{h_2h_3}D_{\rho_1}V\right) + D_{\rho_2}\left(\frac{h_2}{h_3h_1}D_{\rho_2}V\right) + D_{\rho_3}\left(\frac{h_3}{h_1h_2}D_{\rho_3}V\right)\right], \tag{6}$$
and Laplace's Equation in our curvilinear system is
$$h_1h_2h_3\left[D_{\rho_1}\left(\frac{h_1}{h_2h_3}D_{\rho_1}V\right) + D_{\rho_2}\left(\frac{h_2}{h_3h_1}D_{\rho_2}V\right) + D_{\rho_3}\left(\frac{h_3}{h_1h_2}D_{\rho_3}V\right)\right] = 0. \tag{7}$$

If it happens that $\nabla^2\rho_1 = 0$, $V = \rho_1$ will satisfy (7) and we shall have $h_1h_2h_3D_{\rho_1}\left(\frac{h_1}{h_2h_3}\right) = 0$. In like manner if $\nabla^2\rho_2 = 0$ we have $D_{\rho_2}\left(\frac{h_2}{h_3h_1}\right) = 0$, and if $\nabla^2\rho_3 = 0$ we have $D_{\rho_3}\left(\frac{h_3}{h_1h_2}\right) = 0$; and therefore (7) reduces to
$$h_1^2D_{\rho_1}^2V + h_2^2D_{\rho_2}^2V + h_3^2D_{\rho_3}^2V = 0 \tag{8}$$

when $\nabla^2\rho_1 = 0$, $\nabla^2\rho_2 = 0$, and $\nabla^2\rho_3 = 0$.

132. If instead of having the value of the Potential Function V given on the surface of a sphere as in our Spherical Harmonic problem, we have it given at all the points on the surface of an *oblate spheroid*, and are required to find its value at any internal or external point, we can easily get a solution by methods in no essential respect different from those already employed, if only we rightly choose our system of coördinates.

If we take an ellipse and an hyperbola having the same foci, and revolve them about the minor axis of the ellipse, we shall get a pair of surfaces which are mutually perpendicular; a plane through the axis of revolution will cut both the *spheroid* and the *hyperboloid* orthogonally.

The equations of the three surfaces can be written:—

$$F_1(x,y,z,\lambda) = \frac{x^2}{\lambda^2} + \frac{y^2}{\lambda^2 - b^2} + \frac{z^2}{\lambda^2} - 1 = 0 \tag{1}$$

$$F_2(x,y,z,\mu) = \frac{x^2}{\mu^2} + \frac{y^2}{\mu^2 - b^2} + \frac{z^2}{\mu^2} - 1 = 0 \tag{2}$$

$$F_3(x,y,z,\nu) = z - \nu x = 0, \tag{3}$$

where $\lambda^2 > b^2 > \mu^2$, $2b$ being the distance between the foci.

For all values of λ, μ, and ν consistent with the inequality above written the surfaces (1), (2), (3) intersect in real points and cut orthogonally.

λ, μ, and ν can be so chosen that the surfaces will intersect in any given point, and therefore can be taken as a set of curvilinear coördinates, and Laplace's Equation can be expressed in terms of them by the aid of Formula [XV] Art. 1.

From (1), (2), and (3) we readily get

$$\left.\begin{aligned} x^2 &= \frac{\lambda^2\mu^2}{b^2(1+\nu^2)} \\ y^2 &= \frac{(\lambda^2 - b^2)(b^2 - \mu^2)}{b^2} \\ z^2 &= \frac{\lambda^2\mu^2\nu^2}{b^2(1+\nu^2)}; \end{aligned}\right\} \tag{4}$$

whence $\quad D_\lambda x = \dfrac{\mu}{b\sqrt{1+\nu^2}}, \quad D_\lambda y = \dfrac{\lambda}{b}\sqrt{\dfrac{b^2-\mu^2}{\lambda^2-b^2}}, \quad D_\lambda z = \dfrac{\mu\nu}{b\sqrt{1+\nu^2}};$

and

$$\frac{1}{h_1^2} = \frac{\lambda^2 - \mu^2}{\lambda^2 - b^2} \tag{5}$$

[v. 130 (2)]. In like manner we get

$$\frac{1}{h_2^2} = \frac{\lambda^2 - \mu^2}{b^2 - \mu^2} \tag{6}$$

and
$$\frac{1}{h_3^2} = \frac{\lambda^2\mu^2}{b^2(1+\nu^2)^2}, \tag{7}$$

and [xv] Art. 1 becomes

$$\begin{aligned}&\frac{\mu}{b(1+\nu^2)\sqrt{b^2-\mu^2}}D_\lambda[\lambda\sqrt{\lambda^2-b^2}.D_\lambda V]\\ &+\frac{\lambda}{b(1+\nu^2)\sqrt{\lambda^2-b^2}}D_\mu[\mu\sqrt{b^2-\mu^2}.D_\mu V]\\ &+\frac{b(\lambda^2-\mu^2)}{\lambda\mu\sqrt{(\lambda^2-b^2)(b^2-\mu^2)}}D_\nu[(1+\nu^2)D_\nu V]=0,\end{aligned} \tag{8}$$

which is Laplace's Equation in terms of our *Spheroidal Coördinates* λ, μ, and ν.

If now in place of λ, μ, and ν we can introduce some function of λ, some function of μ, and some function of ν which, therefore, will represent the same set of orthogonal surfaces, and if we can choose these functions α, β, and γ, which of course are functions of x, y, and z, so that $\nabla^2\alpha = 0$, $\nabla^2\beta = 0$, and $\nabla^2\gamma = 0$, equation (8) must reduce to the simple and symmetrical form given in [xvi] Art. 1.

These functions α, β, and γ are easily found. Equation (8) is $\nabla^2 V = 0$ expressed in terms of λ, μ, and ν. Assume that V is a function of λ only; then $D_\mu V = 0$, and $D_\nu V = 0$, and (8) reduces to

$$D_\lambda[\lambda\sqrt{\lambda^2-b^2}.D_\lambda V] = 0$$

whence
$$\lambda\sqrt{\lambda^2-b^2}\,\frac{dV}{d\lambda} = c_1,$$

$$dV = \frac{c_1 d\lambda}{\lambda\sqrt{\lambda^2-b^2}},$$

and
$$V = \frac{c_1}{b}\sec^{-1}\frac{\lambda}{b},$$

and is a function of λ which satisfies Laplace's Equation.

Take this as α leaving c_1 at present undetermined, so that

$$d\alpha = \frac{c_1 d\lambda}{\lambda\sqrt{\lambda^2-b^2}} \quad\text{and}\quad \alpha = \frac{c_1}{b}\sec^{-1}\frac{\lambda}{b}.$$

In the same way we get

$$d\beta = \frac{c_2 d\mu}{\mu\sqrt{b^2-\mu^2}} \quad\text{and}\quad \beta = \frac{c_2}{b}\operatorname{sech}^{-1}\frac{\mu}{b},$$

(v. Int. Cal. Art. 46, Ex.)

$$d\gamma = \frac{c_3 d\nu}{1+\nu^2}, \quad\text{and}\quad \gamma = c_3\tan^{-1}\nu.$$

Substituting these values in (8) and taking $c_1 = -c_2 = b$, and $c_3 = 1$, (8) reduces at once to

$$\frac{D_\alpha^2 V}{\lambda^2} + \frac{D_\beta^2 V}{\mu^2} + \frac{\lambda^2 - \mu^2}{\lambda^2 \mu^2} D_\gamma^2 V = 0, \tag{9}$$

or since $\lambda = b \sec\alpha, \quad \mu = b \operatorname{sech}\beta, \quad \text{and} \quad \nu = \tan\gamma,$ (10)

to

$$\cos^2\alpha D_\alpha^2 V + \cosh^2\beta D_\beta^2 V + (\cosh^2\beta - \cos^2\alpha) D_\gamma^2 V = 0 \tag{11}$$

which is Laplace's Equation in terms of what we may call *Normal Oblate Spheroidal Coördinates.*

In using (11) it is to be noted that the point whose coördinates are (α, β, γ) is the point of intersection of an oblate spheroid whose semi-axes are $b \sec\alpha$ and $b \tan\alpha$, an unparted hyperboloid of revolution whose semi-axes are $b \operatorname{sech}\beta$ and $b \tanh\beta$, and a plane containing the axis of the system and making the angle γ with a fixed plane; and that if the axis of revolution is the axis of Y and the fixed plane is the plane of XY, the rectangular coördinates of (α, β, γ) are

$$x = b \sec\alpha \operatorname{sech}\beta \cos\gamma, \quad y = b \tan\alpha \tanh\beta, \quad z = b \sec\alpha \operatorname{sech}\beta \sin\gamma \tag{12}$$

[v. (4)].

If now we let α range from 0 to $\frac{\pi}{2}$, β from $-\infty$ to ∞, and γ from 0 to 2π, we shall be able to represent all points in space; and if we agree that negative values of β shall belong to points below a plane through the origin and perpendicular to the axis of revolution and positive values of β to points above that plane, not only shall we have no ambiguity, but also the rectangular coördinates of any point as given in (12) will have their proper signs.

EXAMPLES.

1. If the spheroid is a *prolate* spheroid, the ellipse and confocal hyperbola must be revolved about the major axis of the ellipse, and the plane must contain that axis. In place of equations (1), (2), and (3) of Art. 132 we have, then,

$$\frac{x^2}{\lambda^2} + \frac{y^2}{\lambda^2 - b^2} + \frac{z^2}{\lambda^2 - b^2} - 1 = 0$$

$$\frac{x^2}{\mu^2} + \frac{y^2}{\mu^2 - b^2} + \frac{z^2}{\mu^2 - b^2} - 1 = 0$$

$$z - \nu y = 0$$

where $\lambda^2 > b^2 > \mu^2.$

$$h_1^2 = \frac{\lambda^2 - b^2}{\lambda^2 - \mu^2}, \quad h_2^2 = \frac{b^2 - \mu^2}{\lambda^2 - \mu^2}, \quad h_3^2 = \frac{b^2(1 + \nu^2)^2}{(\lambda^2 - b^2)(b^2 - \mu^2)}.$$

Laplace's Equation becomes

$$\frac{1}{b^2(1+\nu^2)}D_\lambda[(\lambda^2-b^2)D_\lambda V]+\frac{1}{b^2(1+\nu^2)}D_\mu[(b^2-\mu^2)D_\mu V]$$
$$+\frac{\lambda^2-\mu^2}{(\lambda^2-b^2)(b^2-\mu^2)}D_\nu[(1+\nu^2)D_\nu V]=0. \quad (1)$$

(1) reduces to $$\frac{D_\alpha^2 V}{\lambda^2-b^2}+\frac{D_\beta^2 V}{b^2-\mu^2}+\frac{\lambda^2-\mu^2}{(\lambda^2-b^2)(b^2-\mu^2)}D_\gamma^2 V=0, \quad (2)$$

where $$d\alpha=-\frac{bd\lambda}{\lambda^2-b^2},\quad d\beta=\frac{bd\mu}{b^2-\mu^2},\quad d\gamma=\frac{d\nu}{1+\nu^2},$$

$$\alpha=\text{ctnh}^{-1}\frac{\lambda}{b},\quad \beta=\tanh^{-1}\frac{\mu}{b},\quad \text{and}\quad \gamma=\tan^{-1}\nu.$$

Since $\lambda=b\,\text{ctnh}\,\alpha,\quad \mu=b\tanh\beta,\quad$ and $\quad \nu=\tan\gamma$

(2) can be reduced to

$$\sinh^2\alpha D_\alpha^2 V+\cosh^2\beta D_\beta^2 V+(\sinh^2\alpha+\cosh^2\beta)D_\gamma^2 V=0. \quad (3)$$

In using (3) it is to be noted that the point (α,β,γ) is the point of intersection of a prolate spheroid whose semi-axes are $b\,\text{ctnh}\,\alpha$ and $b\,\text{csch}\,\alpha$, a biparted hyperboloid of revolution whose semi-axes are $b\tanh\beta$ and $b\,\text{sech}\,\beta$, and a plane containing the axis of revolution and making the angle γ with a fixed plane.

If the fixed plane is that of (XY) the rectangular coördinates of any point (α,β,γ) are

$$x=b\,\text{ctnh}\,\alpha\tanh\beta,\quad y=b\,\text{csch}\,\alpha\,\text{sech}\,\beta\cos\gamma,\quad z=b\,\text{csch}\,\alpha\,\text{sech}\,\beta\sin\gamma,$$

and α may range from ∞ to 0, β from $-\infty$ to ∞, and γ from 0 to 2π. Negative values of β are to be taken for points lying to the left of a plane through the origin perpendicular to the axis of revolution.

2. Transform Laplace's Equation in Spherical Coördinates [XIII] Art. 1 to the symmetrical form

$$\alpha^2 D_\alpha^2 V+\cosh^2\beta D_\beta^2 V+\cosh^2\beta D_\gamma^2 V=0$$

where $$\alpha=\frac{1}{r},\quad \beta=\log\tan\frac{\theta}{2},\quad \text{and}\quad \gamma=\phi.$$

3. Transform Laplace's Equation in Cylindrical Coördinates [XIV] Art. 1 to the symmetrical form

$$D_\alpha^2 V+D_\beta^2 V+e^{2\alpha}D_\gamma^2 V=0$$

where $$\alpha=\log r,\quad \beta=\phi,\quad \text{and}\quad \gamma=z.$$

133. In each of the cases we have considered, it has been easy to pass from Laplace's Equation in terms of the chosen coördinates representing an orthogonal system of surfaces to the symmetrical form [XVI] Art. 1; and it is evident that our new coördinate α is a value of V corresponding to such a distribution that the surfaces obtained by giving particular values to ρ_1 are *equipotential* surfaces; that β is a value of V corresponding to such a distribution that the surfaces obtained by giving particular values to ρ_2 are equipotential surfaces; and that γ is a value of V corresponding to such a distribution that the surfaces obtained by giving particular values to ρ_3 are equipotential surfaces. α, β, and γ are called by Lamé "*thermometric parameters.*"

The condition that these values should exist, for a given system of surfaces, that is, that the distribution described above should be possible, is readily obtained. We shall work it out for α. It is merely the condition that V in Laplace's Equation may be a function of ρ_1 alone.

If V is a function of ρ_1 alone

$$D_xV = \frac{dV}{d\rho_1}D_x\rho_1, \quad D_yV = \frac{dV}{d\rho_1}D_y\rho_1, \quad D_zV = \frac{dV}{d\rho_1}D_z\rho_1,$$

$$D_x^2V = \frac{d^2V}{d\rho_1{}^2}(D_x\rho_1)^2 + \frac{dV}{d\rho_1}D_x^2\rho_1$$

$$D_y^2V = \frac{d^2V}{d\rho_1{}^2}(D_y\rho_1)^2 + \frac{dV}{d\rho_1}D_y^2\rho_1$$

$$D_z^2V = \frac{d^2V}{d\rho_1{}^2}(D_z\rho_1)^2 + \frac{dV}{d\rho_1}D_z^2\rho_1.$$

Therefore $[(D_x\rho_1)^2 + (D_y\rho_1)^2 + (D_z\rho_1)^2]\dfrac{d^2V}{d\rho_1{}^2} + [D_x^2\rho_1 + D_y^2\rho_1 + D_z^2\rho_1]\dfrac{dV}{d\rho_1} = 0$

whence
$$\frac{D_x^2\rho_1 + D_y^2\rho_1 + D_z^2\rho_1}{(D_x\rho_1)^2 + (D_y\rho_1)^2 + (D_z\rho_1)^2} = -\frac{d^2V}{d\rho_1{}^2} \div \frac{dV}{d\rho_1}.$$

or
$$\frac{\nabla^2\rho_1}{h_1^2} = F_1(\rho_1)$$

where $F_1(\rho_1)$ may be any function of ρ_1 alone. Our required conditions are then

$$\left.\begin{aligned} \frac{\nabla^2\rho_1}{h_1^2} &= F_1(\rho_1) \\ \frac{\nabla^2\rho_2}{h_2^2} &= F_2(\rho_2) \\ \frac{\nabla^2\rho_3}{h_3^2} &= F_3(\rho_3) \end{aligned}\right\} \qquad (1)$$

and when they are fulfilled the original curvilinear coördinates ρ_1, ρ_2, ρ_3, correspond to possible *equipotential* or *isothermal* surfaces, *thermometric parameters*

α, β, and γ exist, and the reduction of Laplace's Equation to the symmetrical form [xvi] Art. 1 is possible.

134. Returning to our Oblate Spheroid problem of Art. 132 we can proceed as usual to break up our equation (**11**) Art. 132.

Assume that $V = L.M.N$, where L is a function of α only, M of β only, and N of γ only. (11) Art. 132 becomes

$$\frac{\cos^2\alpha}{L}\frac{d^2L}{d\alpha^2} + \frac{\cosh^2\beta}{M}\frac{d^2M}{d\beta^2} + \frac{[\cosh^2\beta - \cos^2\alpha]}{N}\frac{d^2N}{d\gamma^2} = 0$$

or
$$\frac{1}{L}\frac{\cos^2\alpha}{\cosh^2\beta - \cos^2\alpha}\frac{d^2L}{d\alpha^2} + \frac{1}{M}\frac{\cosh^2\beta}{\cosh^2\beta - \cos^2\alpha}\frac{d^2M}{d\beta^2} = -\frac{1}{N}\frac{d^2N}{d\gamma^2}.$$

The first member is independent of γ, and the second member is independent of α and β, and the two members are identically equal. The second member is then independent of α, β, and γ and must be constant; call it n^2. We have, then,

$$\frac{d^2N}{d\gamma^2} + n^2N = 0 \tag{1}$$

and
$$\frac{\cos^2\alpha}{L}\frac{d^2L}{d\alpha^2} + \frac{\cosh^2\beta}{M}\frac{d^2M}{d\beta^2} - n^2(\cosh^2\beta - \cos^2\alpha) = 0. \tag{2}$$

(1) gives us
$$N = A\cos n\gamma + B\sin n\gamma. \tag{3}$$

(2) can be written

$$\frac{\cos^2\alpha}{L}\frac{d^2L}{d\alpha^2} + n^2\cos^2\alpha = n^2\cosh^2\beta - \frac{\cosh^2\beta}{M}\frac{d^2M}{d\beta^2} = m(m+1),$$

whence
$$\cos^2\alpha\frac{d^2L}{d\alpha^2} + [n^2\cos^2\alpha - m(m+1)]L = 0 \tag{4}$$

and
$$\cosh^2\beta\frac{d^2M}{d\beta^2} + [m(m+1) - n^2\cosh^2\beta]M = 0. \tag{5}$$

If we introduce $x = \tanh\beta$ in (5) it becomes

$$(1-x^2)\frac{d^2M}{dx^2} - 2x\frac{dM}{dx} + \left[m(m+1) - \frac{n^2}{1-x^2}\right]M = 0 \tag{6}$$

where since $x = \tanh\beta$ and β may have any value from $-\infty$ to ∞, x may have any value between -1 and 1. (6) is a familiar equation having for a particular solution

$$M = (1-x^2)^{\frac{n}{2}}\frac{d^nP_m(x)}{dx^n} = P_m^n(x) = P_m^n(\tanh\beta). \tag{7}$$

(v. Arts. 101 and 102).

If we introduce in (4) $x = \tan\alpha$ it reduces to

$$(1+x^2)\frac{d^2L}{dx^2} + 2x\frac{dL}{dx} + \left[\frac{n^2}{1+x^2} - m(m+1)\right]L = 0. \tag{8}$$

(8) is an unfamiliar equation, but it can be treated as (6) was treated if we take the pains to go back to the beginning and follow the steps of the treatment of Legendre's Equation.

This labor can be saved, however, by noting that if we let $x = \frac{y}{i}$ (8) becomes

$$(1-y^2)\frac{d^2L}{dy^2} - 2y\frac{dL}{dy} + \left[m(m+1) - \frac{n^2}{1-y^2}\right]L = 0$$

and is identical in form with (6). Hence

$$L = P_m^n(y) \quad \text{and} \quad L = (1-y^2)^{\frac{n}{2}}\frac{d^nQ_m(y)}{dy^n} \qquad \text{(v. Art. 101)},$$

where $y = i\tan\alpha$, are particular solutions of (4).

We can avoid imaginaries if we use the values

$$L = (-i)^{m-n}P_m^n(y) \quad \text{and} \quad L = i^{m+n+1}(1-y^2)^{\frac{n}{2}}\frac{d^nQ_m(y)}{dy^n}. \tag{9}$$

Since we assumed $V = L.M.N$ we have

$$\left.\begin{aligned} V &= (A\cos n\gamma + B\sin n\gamma)P_m^n(\tanh\beta)(-i)^{m-n}P_m^n(i\tan\alpha) \\ \text{and} \quad V &= (A\cos n\gamma + B\sin n\gamma)P_m^n(\tanh\beta)i^{m+n+1}\sec^n\alpha\frac{d^nQ_m(i\tan\alpha)}{(d(i\tan\alpha))^n}\end{aligned}\right\} \tag{10}$$

as particular solutions of (**11**) Art. 132.

If the problem is symmetrical with respect to the axis of the spheroid $D_\gamma^2 V = 0$, $n^2 = 0$ and our particular solutions (10) reduce to

$$\left.\begin{aligned} V &= (-i)^m P_m(i\tan\alpha)P_m(\tanh\beta) \\ \text{and} \quad V &= i^{m+1}Q_m(i\tan\alpha)P_m(\tanh\beta).\end{aligned}\right\} \tag{11}$$

If, then, V is given on the surface of a spheroid as a function of β and γ, we must express it as a function of $\tanh\beta$ and γ, and shall be obliged to develop it in terms of *Spherical Harmonics* of $\tanh\beta$ and γ by the formulas of Chapter VII, using the first equation in (10) for the value of V at an internal point, and the second for the value of V at an external point. If the problem is symmetrical, we must develop in Zonal Harmonics of $\tanh\beta$ by the formulas of Chapter VI.

A convenient form for $Q_m(i\tan\alpha)$ is obtained from (2) Art. 100; it is

$$Q_m(i\tan\alpha) = -iP_m(i\tan\alpha)\int_{\tan\alpha}^{\infty}\frac{dx}{(1+x^2)[P_m(xi)]^2}. \tag{12}$$

Hence

$$Q_0(i\tan\alpha) = -i\int_{\tan\alpha}^{\infty}\frac{dx}{1+x^2} = -i\left(\frac{\pi}{2} - \alpha\right). \tag{13}$$

EXAMPLES.

1. A conductor in the form of an oblate spheroid whose semi-axes are $b\sec\alpha_0$ and $b\tan\alpha_0$ is charged with electricity and is found to be at potential V_0; find the value of the potential function at any internal or external point.

Here $V_0 = V_0 P_0(\tanh\beta)$. Hence at an internal point

$$V = V_0 \frac{P_0(i\tan\alpha)}{P_0(i\tan\alpha_0)} P_0(\tanh\beta) = V_0, \tag{1}$$

and at an external point

$$V = V_0 \frac{Q_0(i\tan\alpha)}{Q_0(i\tan\alpha_0)} P_0(\tanh\beta) = V_0 \frac{\left(\frac{\pi}{2} - \alpha\right)}{\left(\frac{\pi}{2} - \alpha_0\right)}. \tag{2}$$

Since V in (2) involves α only, the equipotential surfaces are all spheroids confocal with the conductor.

2. The upper half of an oblate spheroid whose semi-axes are $b\sec\alpha_0$ and $b\tan\alpha_0$ is kept at the temperature unity, and the lower half at the temperature zero. Find the permanent temperature at any internal point.

Ans. $$u = \frac{1}{2} + \frac{3}{4}\frac{P_1(i\tan\alpha)}{P_1(i\tan\alpha_0)} P_1(\tanh\beta) - \frac{7}{8}\cdot\frac{1}{2}\frac{P_3(i\tan\alpha)}{P_3(i\tan\alpha_0)} P_3(\tanh\beta) + \cdots$$

(v. Art. 93). u may be expressed in terms of x, y, and z without serious difficulty [v. (12) Art. 132].

$$u = \frac{1}{2} + \frac{3}{4}\frac{y}{c} - \frac{7}{8}\cdot\frac{1}{2}\cdot\frac{1}{2}\frac{[25y^3 - 15y(x^2+y^2+z^2-b^2) - 9b^2y]}{5c^3+3b^2c} + \cdots$$

if $2c = 2b\tan\alpha_0$ = minor axis of spheroid.

135. Let us now find the potential function at an external point due to the attraction of a solid homogeneous oblate spheroid, using the method employed in Arts. 98 and 99.

Consider first the potential function due to a shell bounded by the spheroids for which $\alpha = \phi$ and $\alpha = \phi + d\phi$.

By (1) Art. 98 we have

$$4\pi\rho\kappa = [D_n V_1 - D_n V_2]_{\alpha=\phi}, \tag{1}$$

where ρ is the density and κ the thickness of the shell, V_1 the value of the potential function at an internal point, and V_2 the value of the potential function at an external point.

Let $$V_1 = \sum A_m (-i)^m P_m(i\tan\alpha) P_m(\tanh\beta)$$

and $$V_2 = \sum B_m i^{m+1} Q_m(i\tan\alpha) P_m(\tanh\beta) \qquad \text{[v. (11) Art. 134].}$$

Since V_1 and V_2 must have the same value when $\alpha = \phi$

$$A_m = B_m i^{2m+1}\frac{Q_m(i\tan\phi)}{P_m(i\tan\phi)} = (-1)^m B_m \int\limits_{\tan\phi}^{\infty}\frac{dx}{(1+x^2)[P_m(xi)]^2} \tag{2}$$

[v. (12) Art. 134].

$$\left.\begin{aligned} \text{Hence}\quad V_1 &= \sum i^m B_m P_m(\tanh\beta)P_m(i\tan\alpha)\int\limits_{\tan\phi}^{\infty}\frac{dx}{(1+x^2)[P_m(xi)]^2} \\ \text{and}\quad V_2 &= \sum i^m B_m P_m(\tanh\beta)P_m(i\tan\alpha)\int\limits_{\tan\alpha}^{\infty}\frac{dx}{(1+x^2)[P_m(xi)]^2}. \end{aligned}\right\} \tag{3}$$

$$D_nV_1 = D_\alpha V_1 . D_n\alpha. \qquad D_nV_2 = D_\alpha V_2 . D_n\alpha$$

$$\begin{aligned}[D_nV_1 - D_nV_2]_{\alpha=\phi} &= [D_\alpha V_1 - D_\alpha V_2]_{\alpha=\phi}(D_n\alpha)_{\alpha=\phi} \\ &= [D_\alpha(V_1 - V_2)]_{\alpha=\phi}[D_n\alpha]_{\alpha=\phi}.\end{aligned}$$

$$V_1 - V_2 = \sum i^m B_m P_m(\tanh\beta)P_m(i\tan\alpha)\int\limits_{\tan\phi}^{\tan\alpha}\frac{dx}{(1+x^2)[P_m(xi)]^2}.$$

$$\begin{aligned}D_\alpha(V_1 - V_2) = \sum i^m B_m P_m(\tanh\beta)\bigg[&P_m(i\tan\alpha)\frac{\sec^2\alpha}{(1+\tan^2\alpha)[P_m(i\tan\alpha)]^2} \\ &+ \frac{dP_m(i\tan\alpha)}{d\alpha}\int\limits_{\tan\phi}^{\tan\alpha}\frac{dx}{(1+x^2)[P_m(xi)]^2}\bigg].\end{aligned}$$

$$D_\alpha[V_1 - V_2]_{\alpha=\phi} = \sum i^m B_m \frac{P_m(\tanh\beta)}{P_m(i\tan\phi)}.$$

$$D_n\alpha = \frac{d\alpha}{dn}$$

$$dn = \frac{d\rho_1}{h_1} = \frac{d\lambda}{h_1} = \frac{\sqrt{\lambda^2-\mu^2}}{\sqrt{\lambda^2-b^2}}d\lambda = b\sec\alpha\sqrt{\tan^2\alpha + \tanh^2\beta}.d\alpha \tag{4}$$

v. Art. 130 (3), and Art. 132 (5) and (10).

$$[D_n\alpha]_{\alpha=\phi} = \frac{1}{b\sec\phi\sqrt{\tan^2\phi+\tanh^2\beta}}.$$

Hence

$$[D_nV_1 - D_nV_2]_{\alpha=\phi} = \frac{1}{b\sec\phi\sqrt{\tan^2\phi+\tanh^2\beta}}\sum i^m B_m\frac{P_m(\tanh\beta)}{P_m(i\tan\phi)}.$$

$$\kappa = [dn]_{\alpha=\phi} = b\sec\phi\sqrt{\tan^2\phi+\tanh^2\beta}.d\phi$$

by (4), and (1) may be written

$$4\pi\rho b^2 \sec^2\phi(\tan^2\phi + \tanh^2\beta)d\phi = \sum i^m B_m \frac{P_m(\tanh\beta)}{P_m(i\tan\phi)}. \quad (5)$$

Since $\tanh^2\beta = \frac{1}{3}P_0(\tanh\beta) + \frac{2}{3}P_2(\tanh\beta)$

by (5) Art. 95, to satisfy (5) we must give m the values 0 and 2 and

$$B_0 = \tfrac{4}{3}\pi\rho b^2 \sec^2\phi(3\tan^2\phi + 1)d\phi$$

and

$$B_2 = \tfrac{4}{3}\pi\rho b^2 \sec^2\phi(3\tan^2\phi + 1)d\phi.$$

So that by (3)

$$V_1 = \tfrac{4}{3}\pi\rho b^2 \sec^2\phi(3\tan^2\phi + 1)d\phi\left[\int_{\tan\phi}^{\infty} \frac{dx}{1+x^2} \right.$$
$$\left. - P_2(\tanh\beta)P_2(i\tan\alpha)\int_{\tan\phi}^{\infty} \frac{dx}{(1+x^2)[P_2(xi)]^2}\right] \quad (6)$$

and $V_2 = \frac{4}{3}\pi\rho b^2 \sec^2\phi(3\tan^2\phi + 1)d\phi[iQ_0(i\tan\alpha)$
$+ i^3 P_2(\tanh\beta)Q_2(i\tan\alpha)]. \quad (7)$

The potential function at an external point due to the solid spheroid for which $\alpha = \alpha_0$ is

$$V = \int_{\phi=0}^{\phi=\alpha_0} V_2 = \tfrac{4}{3}\pi\rho b^2 \sec^2\alpha_0 \tan\alpha_0[iQ_0(i\tan\alpha) + i^3 P_2(\tanh\beta)Q_2(i\tan\alpha)]. \quad (8)$$

If $2a$ is the major axis and $2c$ the minor axis of the spheroid

$$\tfrac{4}{3}\pi\rho b^2 \sec^2\alpha_0 \tan\alpha_0 = \tfrac{4}{3}\frac{\pi\rho a^2 c}{b} = \frac{M}{b}$$

where M is the mass of the spheroid. Therefore

$$V = \frac{M}{b}[iQ_0(i\tan\alpha) + i^3 P_2(\tanh\beta)Q_2(i\tan\alpha)] \quad (9)$$

is the required value. (9) can be reduced to

$$V = \frac{M}{b}\left\{\frac{\pi}{2} - \alpha + \frac{1}{4}\left[\left(\frac{\pi}{2} - \alpha\right)(3\tan^2\alpha + 1) - 3\tan\alpha\right][3\tanh^2\beta - 1]\right\}. \quad (10)$$

EXAMPLES.

1. Break up the equation (3) Ex. 1, Art. 132, for the prolate spheroid, and obtain particular solutions of the term

$$V = (A\cos n\gamma + B\sin n\gamma)P_m^n(\tanh\beta)P_m^n(\operatorname{ctnh}\alpha),$$

$$V = (A\cos n\gamma + B\sin n\gamma)P_m^n(\tanh\beta)(-1)^{\frac{n}{2}}\operatorname{csch}^n\alpha\frac{d^nQ_m(\operatorname{ctnh}\alpha)}{(d\operatorname{ctnh}\alpha)^n}.$$

2. Break up and solve the equations of Exs. 2 and 3, Art. 132, and show that they lead to familiar forms.

3. If in Ex. 1, Art. 132, the conductor is a prolate spheroid whose semi-axes are $b\operatorname{ctnh}\alpha_0$ and $b\operatorname{csch}\alpha_0$ show that

$$V = V_0 \text{ at an internal point.} \qquad V = V_0\frac{\alpha}{\alpha_0} \text{ at an external point.}$$

4. Show that the potential function at an external point due to the attraction of a homogeneous solid prolate spheroid is

$$V = \frac{M}{b}[Q_0(\operatorname{ctnh}\alpha) - P_2(\tanh\beta)Q_2(\operatorname{ctnh}\alpha)].$$

Ellipsoidal Harmonics.

136. If we are dealing with an *ellipsoid* instead of a spheroid, we can take as our orthogonal system of surfaces a set of *confocal quadrics*;

$$\left.\begin{aligned}\frac{x^2}{\lambda^2}+\frac{y^2}{\lambda^2-b^2}+\frac{z^2}{\lambda^2-c^2}-1=0\\ \frac{x^2}{\mu^2}+\frac{y^2}{\mu^2-b^2}+\frac{z^2}{\mu^2-c^2}-1=0\\ \frac{x^2}{\nu^2}+\frac{y^2}{\nu^2-b^2}+\frac{z^2}{\nu^2-c^2}-1=0\end{aligned}\right\} \tag{1}$$

where $\lambda^2 > c^2 > \mu^2 > b^2 > \nu^2$. Here the first surface is an ellipsoid, the second an unparted hyperboloid, and the third a biparted hyperboloid. Each of the three principal sections of the system consists of confocal conics, and it is well known and is easily shown that the surfaces cut orthogonally. λ, μ, and ν will be our curvilinear coördinates, and are known as Ellipsoidal Coördinates.

We find without difficulty that

$$x^2 = \frac{\lambda^2\mu^2\nu^2}{b^2c^2}, \quad y^2 = \frac{(\lambda^2-b^2)(\mu^2-b^2)(b^2-\nu^2)}{b^2(c^2-b^2)},$$

$$z^2 = \frac{(\lambda^2-c^2)(c^2-\mu^2)(c^2-\nu^2)}{c^2(c^2-b^2)}, \tag{2}$$

$$h_1^2 = \frac{(\lambda^2 - b^2)(\lambda^2 - c^2)}{(\lambda^2 - \mu^2)(\lambda^2 - \nu^2)}, \quad h_2^2 = \frac{(\mu^2 - b^2)(c^2 - \mu^2)}{(\mu^2 - \nu^2)(\lambda^2 - \mu^2)},$$

$$h_3^2 = \frac{(b^2 - \nu^2)(c^2 - \nu^2)}{(\lambda^2 - \nu^2)(\mu^2 - \nu^2)}. \quad (3)$$

To avoid ambiguity, we shall suppose that of the nine semi-axes in (1) $\sqrt{c^2 - \mu^2}$ is to be taken with the positive sign for a point on the half of the unparted hyperboloid on which z is positive, and with the negative sign for a point on the half on which z is negative; $\sqrt{b^2 - \nu^2}$ is to be taken with the positive sign for a point on the half of the biparted hyperboloid on which y is positive, and with the negative sign for a point on the half on which y is negative; ν is to be taken positive for a point on the half of the biparted hyperboloid on which x is positive, and negative for a point on the half on which x is negative, and that the remaining six are to be always positive. It follows that our Ellipsoidal Coördinates have the disadvantage that to fully fix a point we need to know not merely the values of its coördinates λ, μ, and ν, but the signs of $\sqrt{c^2 - \mu^2}$, and $\sqrt{b^2 - \nu^2}$ as well.

We shall see later, Art. 139, when we come to introduce what we may call the *Normal Ellipsoidal Coördinates* α, β, and γ that they are free from this disadvantage.

It is to be observed that λ may range from c to ∞, μ from b to c, and ν from $-b$ to b.

The element of length perpendicular to the Ellipsoid is

$$dn = \frac{d\lambda}{h_1} = \sqrt{\frac{(\lambda^2 - \mu^2)(\lambda^2 - \nu^2)}{(\lambda^2 - b^2)(\lambda^2 - c^2)}}.d\lambda. \quad (4)$$

The element of Ellipsoidal surface is

$$dS = \frac{d\mu d\nu}{h_2 h_3} = (\mu^2 - \nu^2)\sqrt{\frac{(\lambda^2 - \mu^2)(\lambda^2 - \nu^2)}{(\mu^2 - b^2)(c^2 - \mu^2)(b^2 - \nu^2)(c^2 - \nu^2)}}.d\mu d\nu, \quad (5)$$

and the element of volume is

$$dv = \frac{d\lambda d\mu d\nu}{h_1 h_2 h_3}$$

$$= \frac{(\lambda^2 - \mu^2)(\lambda^2 - \nu^2)(\mu^2 - \nu^2)}{\sqrt{(\lambda^2 - b^2)(\lambda^2 - c^2)(\mu^2 - b^2)(c^2 - \mu^2)(b^2 - \nu^2)(c^2 - \nu^2)}} d\lambda d\mu d\nu. \quad (6)$$

The surface integral of any given function of μ and ν taken over the ellipsoid is

$$\int f(\mu, \nu) dS = \int_{-b}^{b} d\nu \int_{b}^{c} [f_1(\mu, \nu) + f_2(\mu, \nu) + f_3(\mu, \nu)$$

$$+ f_4(\mu, \nu)](\mu^2 - \nu^2)\sqrt{\frac{(\lambda^2 - \mu^2)(\lambda^2 - \nu^2)}{(\mu^2 - b^2)(c^2 - \mu^2)(b^2 - \nu^2)(c^2 - \nu^2)}}.d\mu, \quad (7)$$

where $f_1(\mu,\nu)$, $f_2(\mu,\nu)$, $f_3(\mu,\nu)$ and $f_4(\mu,\nu)$ are the values of the given function on the four quarters of the ellipsoid into which it is divided by the planes of (XY) and (XZ).

Laplace's Equation proves reducible to

$$(\mu^2-\nu^2)D_\alpha^2V+(\lambda^2-\nu^2)D_\beta^2V+(\lambda^2-\mu^2)D_\gamma^2V=0 \tag{8}$$

where $$\alpha=c\int_c^\lambda\frac{d\lambda}{\sqrt{(\lambda^2-b^2)(\lambda^2-c^2)}},\quad \beta=c\int_b^\mu\frac{d\mu}{\sqrt{(c^2-\mu^2)(\mu^2-b^2)}},$$
$$\gamma=c\int_0^\nu\frac{d\nu}{\sqrt{(b^2-\nu^2)(c^2-\nu^2)}}. \tag{9}$$

α, β, and γ can be expressed as Elliptic Integrals of the first class and are

$$\alpha=F\Big(\frac{b}{c},\frac{\pi}{2}\Big)-F\Big(\frac{b}{c},\sin^{-1}\frac{c}{\lambda}\Big),\quad \beta=F\Bigg(\sqrt{1-\frac{b^2}{c^2}},\sin^{-1}\sqrt{\frac{1-\frac{b^2}{\mu^2}}{1-\frac{b^2}{c^2}}}\Bigg),$$
$$\gamma=F\Big(\frac{b}{c},\sin^{-1}\frac{\nu}{b}\Big); \tag{10}$$

whence $$\lambda=\frac{c}{\operatorname{sn}(K-\alpha)}\Big(\operatorname{mod}\frac{b}{c}\Big)=c\frac{\operatorname{dn}\alpha}{\operatorname{cn}\alpha}\Big(\operatorname{mod}\frac{b}{c}\Big),$$
$$\mu=\frac{b}{\operatorname{dn}\beta}\Big(\operatorname{mod}\Big(1-\frac{b^2}{c^2}\Big)^{\frac12}\Big),\quad \nu=b\operatorname{sn}\gamma\Big(\operatorname{mod}\frac{b}{c}\Big) \tag{11}$$

(v. Int. Cal. Arts. 179, 192, and 196).

137. If in (8) Art. 136 we assume $V=L.M.N$ where L involves α only, M involves β only, and N involves γ only, (8) can be written

$$\frac{\mu^2-\nu^2}{L}\frac{d^2L}{d\alpha^2}+\frac{\lambda^2-\nu^2}{M}\frac{d^2M}{d\beta^2}+\frac{\lambda^2-\mu^2}{N}\frac{d^2N}{d\gamma^2}=0. \tag{1}$$

(1) is too complicated to be broken up by our usual method.

If, however, we let

$$\frac{1}{L}\frac{d^2L}{d\alpha^2}=\sum a_k\lambda^k,\quad \frac{1}{M}\frac{d^2M}{d\beta^2}=\sum b_k\mu^k,\quad \frac{1}{N}\frac{d^2N}{d\gamma^2}=\sum c_k\nu^k,$$

substitute in (1) and make use of the fact that the result must be identically zero, we find that the coefficients are zero for all values of k except $k=0$ and $k=2$, and that $a_0=-b_0=c_0$, and $a_2=-b_2=c_2$.

Therefore (1) can be broken up into the three equations

$$\frac{d^2L}{d\alpha^2} = (a_0 + a_2\lambda^2)L$$
$$\frac{d^2M}{d\beta^2} = -(a_0 + a_2\mu^2)M$$
$$\frac{d^2N}{d\gamma^2} = (a_0 + a_2\nu^2)N.$$

We shall find it convenient to take a_2 as $m(m+1)$ and a_0 as $-(b^2+c^2)p$; whence

$$\left.\begin{aligned}&\frac{d^2L}{d\alpha^2} - [m(m+1)\lambda^2 - (b^2+c^2)p]L = 0\\ &\frac{d^2M}{d\beta^2} + [m(m+1)\mu^2 - (b^2+c^2)p]M = 0\\ &\frac{d^2N}{d\gamma^2} - [m(m+1)\nu^2 - (b^2+c^2)p]N = 0.\end{aligned}\right\} \tag{2}$$

If now in (2) we replace α, β, and γ by their values in terms of λ, μ, and ν, we get

$$\left.\begin{aligned}&(\lambda^2-b^2)(\lambda^2-c^2)\frac{d^2L}{d\lambda^2} + \lambda(\lambda^2-b^2+\lambda^2-c^2)\frac{dL}{d\lambda}\\ &\qquad\qquad -[m(m+1)\lambda^2-(b^2+c^2)p]L = 0\\ &(\mu^2-b^2)(\mu^2-c^2)\frac{d^2M}{d\mu^2} + \mu(\mu^2-b^2+\mu^2-c^2)\frac{dM}{d\mu}\\ &\qquad\qquad -[m(m+1)\mu^2-(b^2+c^2)p]M = 0\\ &(\nu^2-b^2)(\nu^2-c^2)\frac{d^2N}{d\nu^2} + \nu(\nu^2-b^2+\nu^2-c^2)\frac{dN}{d\nu}\\ &\qquad\qquad -[m(m+1)\nu^2-(b^2+c^2)p]N = 0.\end{aligned}\right\} \tag{3}$$

Whence if $L = E_m^p(\lambda)$, it follows that $M = E_m^p(\mu)$ and $N = E_m^p(\nu)$, and that

$$V = E_m^p(\lambda)E_m^p(\mu)E_m^p(\nu) \tag{4}$$

is a solution of Laplace's Equation, (8) Art. 136.

The equation

$$(x^2-b^2)(x^2-c^2)\frac{d^2z}{dx^2} + x(x^2-b^2+x^2-c^2)\frac{dz}{dx} \\ - [m(m+1)x^2-(b^2+c^2)p]z = 0 \tag{5}$$

is known as Lamé's Equation, and $E_m^p(x)$ as a *Lamé's Function* or an *Ellipsoidal Harmonic*. We shall suppose m a positive integer.

To get a particular solution of (5) let $z = \sum a_k x^k$. Substitute in (5) and reduce and we get

$$[k(k+1) - m(m+1)]a_k - (b^2+c^2)[(k+2)^2 - p]a_{k+2} + b^2c^2(k+3)(k+4)a_{k+4} = 0. \quad (6)$$

We have now only to choose a sequence of coefficients satisfying (6), and we may take any two consecutive coefficients arbitrarily.

(6) which is ordinarily a relation connecting three consecutive coefficients reduces to a relation between two when $k = m$, when $k = -3$, and when $k = -4$. If we take $a_{m+2} = 0$, a_{m+4}, a_{m+6}, &c., will vanish. Let $a_m = 1$. If m is even the coefficient of a_0 in (6) will be zero; if p has such a value that a_{-2} is zero, a_{-4}, a_{-6}, &c., will be zero, and there will be no terms in the solution involving negative powers of x.

If we write the values of a_{m-2}, a_{m-4}, &c., by the aid of (6) we see that a_{m-2} is of the first degree in p, a_{m-4} of the second degree in p, &c., and a_{-2} of the degree $\frac{m}{2}+1$ in p. There are then $\frac{m}{2}+1$ values of p which we shall call p_1, p_2, p_3, &c., for which a_{-2} will vanish, and for which our solutions will be of the form

$$E_m^p(x) = x^m + a_{m-2}x^{m-2} + a_{m-4}x^{m-4} + \cdots + a_0$$

if m is even.

If m is odd, the coefficient of a_1 in (6) will vanish and we can choose p so that a_{-1} shall be zero, and then all coefficients of lower order will vanish. a_{-1} is of the degree $\frac{m+1}{2}$ in p, and there will be $\frac{m+1}{2}$ values p_1, p_2, p_3, &c., of p for which

$$E_m^p(x) = x^m + a_{m-2}x^{m-2} + a_{m-4}x^{m-4} + \cdots + a_1 x.$$

Following Heine we shall call the solution just obtained $K_m^p(x)$ so that

$$K_m^p(x) = x^m + a_{m-2}x^{m-2} + a_{m-4}x^{m-4} + \cdots \quad (7)$$

terminating with a_0 if m is even, and with $a_1 x$ if m is odd. If m is even, there are $\frac{m}{2}+1$ of these functions $K_m^{p_1}(x)$, $K_m^{p_2}(x)$, &c., and there are $\frac{m+1}{2}$ of them if m is odd. The coefficients can be computed by the aid of (6).

If in Lamé's Equation (5) we let $z = v\sqrt{x^2-b^2}$ we get the equation

$$(x^2-b^2)(x^2-c^2)\frac{d^2v}{dx^2} + x[x^2-b^2+3(x^2-c^2)]\frac{dv}{dx} - [(m+2)(m-1)x^2 + c^2 - (b^2+c^2)p]v = 0. \quad (8)$$

Letting $v = \sum a_k x^k$ we obtain the relation

$$[k(k+3) - (m+2)(m-1)]a_k - \{(b^2+c^2)[(k+2)^2-p] + c^2(2k+5)\}a_{k+2} + b^2c^2(k+3)(k+4)a_{k+4} = 0. \quad (9)$$

Proceeding exactly as before, we find that there are $\frac{m}{2}$ values q_1, q_2, q_3, &c., of p for which $v = x^{m-1} + a_{m-3}x^{m-3} + \cdots + a_1 x$ if m is even, and $\frac{m+1}{2}$ values for which $v = x^{m-1} + a_{m-3}x^{m-3} + \cdots + a_0$ if m is odd.

Calling $v\sqrt{x^2 - b^2}$ $L_m^p(x)$ so that

$$L_m^p(x) = \sqrt{x^2 - b^2}[x^{m-1} + a_{m-3}x^{m-3} + a_{m-5}x^{m-5} + \cdots], \tag{10}$$

terminating with $a_1 x$ if m is even and with a_0 if m is odd, we have $\frac{m}{2}$ values of $E_m^p(x)$, namely $L_m^{q_1}(x)$, $L_m^{q_2}(x)$, &c., of the form (10) if m is even and $\frac{m+1}{2}$ values if m is odd.

By interchanging b and c in (8), (9), and (10) we may show that if

$$M_m^p(x) = \sqrt{x^2 - c^2}[x^{m-1} + a_{m-3}x^{m-3} + a_{m-5}x^{m-5} + \cdots] \tag{11}$$

there are $\frac{m}{2}$ values of $E_m^p(x)$, namely $M_m^{r_1}(x)$, $M_m^{r_2}(x)$, $M_m^{r_3}(x)$, &c., of the form (11) if m is even and $\frac{m+1}{2}$ values if m is odd.

Finally if in Lamé's Equation (5) we let $z = v\sqrt{(x^2 - b^2)(x^2 - c^2)}$ we get

$$\begin{aligned}(x^2 - b^2)(x^2 - c^2)\frac{d^2v}{dx^2} + 3x(x^2 - b^2 + x^2 - c^2)\frac{dv}{dx}& \\ - [(m+3)(m-2)x^2 - (b^2 + c^2)(p-1)]v = 0.&\end{aligned} \tag{12}$$

If now we let $v = \sum a_k x^k$ we obtain the relation

$$\begin{aligned}&[k(k+5) - (m-2)(m+3)]a_k \\ &\quad - (b^2 + c^2)[(k+2)(k+4) + 1 - p]a_{k+2} + b^2c^2(k+3)(k+4)a_{k+4} = 0.\end{aligned} \tag{13}$$

Proceeding as before we find that there are $\frac{m}{2}$ values s_1, s_2, s_3, &c., of p for which $v = x^{m-2} + a_{m-4}x^{m-4} + a_{m-6}x^{m-6} + \cdots + a_0$ if m is even, and $\frac{m+1}{2}$ values for which $v = x^{m-2} + a_{m-4}x^{m-4} + \cdots + a_1 x$ if m is odd.

Calling $v\sqrt{(x^2 - b^2)(x^2 - c^2)}$ $N_m^p(x)$ so that

$$N_m^p(x) = \sqrt{(x^2 - b^2)(x^2 - c^2)}[x^{m-2} + a_{m-4}x^{m-4} + a_{m-6}x^{m-6} + \cdots] \tag{14}$$

terminating with a_0 if m is even and with $a_1 x$ if m is odd, we have $\frac{m}{2}$ values of $E_m^p(x)$, namely $N_m^{s_1}(x)$, $N_m^{s_2}(x)$, $N_m^{s_3}(x)$, &c., of the form (14) if m is even and $\frac{m-1}{2}$ values if m is odd.

Summing up our results we see that there are $2m + 1$ Ellipsoidal Harmonics $E_m^p(x)$ each of which is a finite sum of the mth degree in x, or in x and $\sqrt{x^2 - b^2}$, or in x and $\sqrt{x^2 - c^2}$, or in x and $\sqrt{x^2 - b^2}$ and $\sqrt{x^2 - c^2}$.

It was proved by Lamé that the $2m+1$ values of p, namely p_1, p_2, p_3, &c., q_1, q_2, q_3, &c., r_1, r_2, r_3, &c., s_1, s_2, s_3, &c., were all real, and by Liouville that they were all different.

We give tables of the Ellipsoidal Harmonics for $m=0$, $m=1$, $m=2$, and $m=3$. The coefficients were obtained by the aid of formulas (6), (9), and (13).

$E_0(x)$

$$\begin{aligned} K_0(x) &= 1 \\ L_0(x) &= 0 \\ M_0(x) &= 0 \\ N_0(x) &= 0 \end{aligned}$$

$E_1(x)$

$$\begin{aligned} K_1(x) &= x \\ L_1(x) &= \sqrt{x^2-b^2} \\ M_1(x) &= \sqrt{x^2-c^2} \\ N_1(x) &= 0 \end{aligned}$$

$E_2(x)$

$$\begin{aligned} K_2^{p_1}(x) &= x^2 - \tfrac{1}{3}[b^2+c^2-\sqrt{(b^2+c^2)^2-3b^2c^2}] \\ K_2^{p_2}(x) &= x^2 - \tfrac{1}{3}[b^2+c^2+\sqrt{(b^2+c^2)^2-3b^2c^2}] \\ L_2(x) &= x\sqrt{x^2-b^2} \\ M_2(x) &= x\sqrt{x^2-c^2} \\ N_2(x) &= \sqrt{(x^2-b^2)(x^2-c^2)} \end{aligned}$$

$E_3(x)$

$$\begin{aligned} K_3^{p_1}(x) &= x^3 - \frac{x}{5}[2(b^2+c^2)-\sqrt{4(b^2+c^2)^2-15b^2c^2}] \\ K_3^{p_2}(x) &= x^3 - \frac{x}{5}[2(b^2+c^2)+\sqrt{4(b^2+c^2)^2-15b^2c^2}] \\ L_3^{q_1}(x) &= \sqrt{x^2-b^2}[x^2-\tfrac{1}{5}(b^2+2c^2-\sqrt{(b^2+2c^2)^2-5b^2c^2})] \\ L_3^{q_2}(x) &= \sqrt{x^2-b^2}[x^2-\tfrac{1}{5}(b^2+2c^2+\sqrt{(b^2+2c^2)^2-5b^2c^2})] \\ M_3^{r_1}(x) &= \sqrt{x^2-c^2}[x^2-\tfrac{1}{5}(2b^2+c^2-\sqrt{(2b^2+c^2)^2-5b^2c^2})] \\ M_3^{r_2}(x) &= \sqrt{x^2-c^2}[x^2-\tfrac{1}{5}(2b^2+c^2+\sqrt{(2b^2+c^2)^2-5b^2c^2})] \\ N_3(x) &= x\sqrt{(x^2-b^2)(x^2-c^2)} \end{aligned}$$

It is to be noted that since in the solution (4) of Laplace's Equation,

$$V = E_m^p(\lambda)E_m^p(\mu)E_m^p(\nu),$$

we have the same m and p in each of the three factors, we shall have to deal merely with products made up of factors of the same form, for example,

$$K_m^{p_k}(\lambda)K_m^{p_k}(\mu)K_m^{p_k}(\nu), \quad L_m^{q_k}(\lambda)L_m^{q_k}(\mu)L_m^{q_k}(\nu), \quad \&c.;$$

and that in a solution of the form

$$V = \sum A_{m,p}E_m^p(\lambda)E_m^p(\mu)E_m^p(\nu)$$

we shall have for a given m just $2m+1$ terms.

138. From the particular solution of Lamé's Equation [(5) Art. 137] $z = E_m^p(x)$, we can get by formula (5), Art. 18, the general solution.

It is $$z = AE_m^p(x) + BE_m^p(x)\int_x^\infty \frac{dx}{\sqrt{(x^2-b^2)(x^2-c^2)}[E_m^p(x)]^2}. \tag{1}$$

Making $A = 0$ and $B = 2m+1$ we get a second form of particular solution of Lamé's Equation, $z = F_m^p(x)$ where

$$F_m^p(x) = (2m+1)E_m^p(x)\int_x^\infty \frac{dx}{\sqrt{(x^2-b^2)(x^2-c^2)}[E_m^p(x)]^2}. \tag{2}$$

We shall call $F_m^p(x)$ a *Lamé's Function of the second kind.*

It is easily seen to approach the value zero as x is indefinitely increased.

EXAMPLES.

1. If an ellipsoidal conductor is charged with electricity, and is found to be at potential V_0, show that since $V_0 = V_0K_0(\lambda)$,

$$V = V_0K_0(\lambda)K_0(\mu)K_0(\nu) = V_0$$

at an internal point, and

$$V = V_0K_0(\mu)K_0(\nu)\left[K_0(\lambda)\int_\lambda^\infty \frac{dx}{\sqrt{(x^2-b^2)(x^2-c^2)}[K_0(x)]^2}\right.$$
$$\left.\div K_0(\lambda_0)\int_{\lambda_0}^\infty \frac{dx}{\sqrt{(x^2-b^2)(x^2-c^2)}[K_0(x)]^2}\right]$$

$$= V_0\left[\int_\lambda^\infty \frac{dx}{\sqrt{(x^2-b^2)(x^2-c^2)}} \div \int_{\lambda_0}^\infty \frac{dx}{\sqrt{(x^2-b^2)(x^2-c^2)}}\right] = V_0\frac{F\left(\frac{b}{c}, \sin^{-1}\frac{c}{\lambda}\right)}{F\left(\frac{b}{c}, \sin^{-1}\frac{c}{\lambda_0}\right)},$$

whence $$V = V_0\frac{\left(F\left(\frac{b}{c}, \frac{\pi}{2}\right) - \alpha\right)}{F\left(\frac{b}{c}, \frac{\pi}{2}\right) - \alpha_0}$$ v. (10) Art. 136.

2. Find the value of the potential function at an external point due to the attraction of a solid homogeneous ellipsoid (v. Art. 135).

Observe that

$$(l^2-\mu^2)(l^2-\nu^2)=\tfrac{1}{3}[3l^4-2(b^2+c^2)l^2+b^2c^2]K_0(\mu)K_0(\nu)$$
$$+\tfrac{1}{2}\left[1+\frac{b^2+c^2-3l^2}{\sqrt{(b^2+c^2)^2-3b^2c^2}}\right]K_2^{p_1}(\mu)K_2^{p_1}(\nu)$$
$$+\tfrac{1}{2}\left[1-\frac{b^2+c^2-3l^2}{\sqrt{(b^2+c^2)^2-3b^2c^2}}\right]K_2^{p_2}(\mu)K_2^{p_2}(\nu);$$

and that

$$\int_0^{\lambda_0}\tfrac{4}{3}\pi\rho\frac{3l^4-2(b^2+c^2)l^2+b^2c^2}{\sqrt{(l^2-b^2)(l^2-c^2)}}dl=\tfrac{4}{3}\pi\rho\lambda_0\sqrt{(\lambda_0^2-b^2)(\lambda_0^2-c^2)}=M$$

where M is the mass of the ellipsoid.

Ans. $$V=M\left\{\int_\lambda^\infty\frac{dx}{\sqrt{(x^2-b^2)(x^2-c^2)}}-\frac{3}{2\sqrt{(b^2+c^2)^2-3b^2c^2}}\right.$$
$$\left[K_2^{p_1}(\mu)K_2^{p_1}(\nu)K_2^{p_1}(\lambda)\int_\lambda^\infty\frac{dx}{\sqrt{(x^2-b^2)(x^2-c^2)}.(K_2^{p_1}(x))^2}\right.$$
$$\left.\left.-K_2^{p_2}(\mu)K_2^{p_2}(\nu)K_2^{p_2}(\lambda)\int_\lambda^\infty\frac{dx}{\sqrt{(x^2-b^2)(x^2-c^2)}.(K_2^{p_2}(x))^2}\right]\right\}.$$

139. If for the sake of brevity we represent $\frac{b}{c}$ by k, and $\left(1-\frac{b^2}{c^2}\right)^{\frac{1}{2}}$ by k' in the formulas (11) Art. 136 we have

$$\lambda=c\frac{\operatorname{dn}\alpha}{\operatorname{cn}\alpha}(\operatorname{mod}k),\quad \mu=\frac{b}{\operatorname{dn}\beta(\operatorname{mod}k')},\quad \nu=b\operatorname{sn}\gamma(\operatorname{mod}k) \tag{1}$$

and from these we get without difficulty (v. Int. Cal. Art. 192)

$$\left.\begin{aligned}
\sqrt{\lambda^2-b^2}&=\frac{ck'}{\operatorname{cn}\alpha(\operatorname{mod}k)}, & \sqrt{\mu^2-b^2}&=\frac{bk'\operatorname{sn}\beta}{\operatorname{dn}\beta}(\operatorname{mod}k'),\\
\sqrt{b^2-\nu^2}&=b\operatorname{cn}\gamma(\operatorname{mod}k), & \sqrt{\lambda^2-c^2}&=\frac{ck'\operatorname{sn}\alpha}{\operatorname{cn}\alpha}(\operatorname{mod}k),\\
\sqrt{c^2-\mu^2}&=\frac{ck'\operatorname{cn}\beta}{\operatorname{dn}\beta}(\operatorname{mod}k'), & \sqrt{c^2-\nu^2}&=c\operatorname{dn}\gamma(\operatorname{mod}k).
\end{aligned}\right\} \tag{2}$$

If we let α range from 0 to K, and β from 0 to $2K'$, and γ from 0 to $4K$, where K and K' are the complete Elliptic Integrals $F\left(k,\frac{\pi}{2}\right)$ and $F\left(k',\frac{\pi}{2}\right)$

respectively, (α, β, γ) may represent any point in space, and there will be no ambiguity in sign (v. Art. 136).

We may note that if $0 < \beta < K'$, z is positive; if $K' < \beta < 2K'$, z is negative; if $0 < \gamma < K$, x and y are both positive; if $K < \gamma < 2K$, x is positive and y negative; if $2K < \gamma < 3K$, x and y are both negative; and if $3K < \gamma < 4K$, x is negative and y positive (v. Art. 136).

We can write the values in (4), (5), (6), and (7), Art. 136, more neatly by bringing in α, β, and γ. We get

$$dn = \frac{1}{c}\sqrt{(\lambda^2-\mu^2)(\lambda^2-\nu^2)}d\alpha, \tag{3}$$

$$dS = \frac{1}{c^2}(\mu^2-\nu^2)\sqrt{(\lambda^2-\mu^2)(\lambda^2-\nu^2)}d\beta d\gamma, \tag{4}$$

$$dv = \frac{1}{c^3}(\lambda^2-\mu^2)(\lambda^2-\nu^2)(\mu^2-\nu^2)d\alpha d\beta d\gamma. \tag{5}$$

For the integral of any function of α, β, and γ over the ellipsoid $\alpha = \alpha_0$, we shall have

$$\int F(\alpha,\beta,\gamma)dS = \frac{1}{c^2}\int_0^{2K'} d\beta \int_0^{4K} F(\alpha_0,\beta,\gamma)(\mu^2-\nu^2)\sqrt{(\lambda^2-\mu^2)(\lambda^2-\nu^2)}d\gamma. \tag{6}$$

140. If we make use of the formula (2) Art. 92

$$\int (UD_nV - VD_nU)dS = 0 \tag{1}$$

and take as our closed surface any given ellipsoid, we can get a very important result.

If $\quad U = E_m^p(\lambda)E_m^p(\mu)E_m^p(\nu) \quad$ and $\quad V = E_n^q(\lambda)E_n^q(\mu)E_n^q(\nu)$

then $$\nabla^2 U = \nabla^2 V = 0.$$

$$D_nU = D_\alpha U D_n\alpha = E_m^p(\mu)E_m^p(\nu)\frac{dE_m^p(\lambda)}{d\alpha}\frac{c}{\sqrt{(\lambda^2-\mu^2)(\lambda^2-\nu^2)}},$$

and $$D_nV = D_\alpha V D_n\alpha = E_n^q(\mu)E_n^q(\nu)\frac{dE_n^q(\lambda)}{d\alpha}\frac{c}{\sqrt{(\lambda^2-\mu^2)(\lambda^2-\nu^2)}},$$

$$UD_nV - VD_nU = E_m^p(\mu)E_m^p(\nu)E_n^q(\mu)E_n^q(\nu)$$
$$\left(E_m^p(\lambda)\frac{dE_n^q(\lambda)}{d\alpha} - E_n^q(\lambda)\frac{dE_m^p(\lambda)}{d\alpha}\right)\frac{c}{\sqrt{(\lambda^2-\mu^2)(\lambda^2-\nu^2)}}.$$

Integrating $UD_nV - VD_nU$ over the whole ellipsoid, and writing the result equal to zero, we have

$$\frac{1}{c}\int_0^{2K'} d\beta \int_0^{4K} E_m^p(\mu)E_m^p(\nu)E_n^q(\mu)E_n^q(\nu)$$
$$\left[E_m^p(\lambda)\frac{dE_n^q(\lambda)}{d\alpha} - E_n^q(\lambda)\frac{dE_m^p(\lambda)}{d\alpha}\right](\mu^2-\nu^2)d\gamma = 0.$$

Hence $$\int_0^{2K'} d\beta \int_0^{4K} E_m^p(\mu)E_m^p(\nu)E_n^q(\mu)E_n^q(\nu)(\mu^2-\nu^2)d\gamma = 0 \tag{2}$$

unless $$E_m^p(\lambda)\frac{dE_n^q(\lambda)}{d\alpha} - E_n^q(\lambda)\frac{dE_m^p(\lambda)}{d\alpha} = 0. \tag{3}$$

But as our ellipsoid may be taken at pleasure, λ and α are unrestricted, and if (3) is true it must be true identically.

If we divide (3) by $[E_m^p(\lambda)]^2$ it becomes

$$\frac{d}{d\alpha}\left[\frac{E_n^q(\lambda)}{E_m^p(\lambda)}\right] = 0 \quad \text{and} \quad \frac{E_n^q(\lambda)}{E_m^p(\lambda)} = \text{a constant;}$$

and this obviously cannot be true unless $n = m$ and $q = p$.

EXAMPLES.

1. Show that it follows from (2) Art. 140 that

$$\int_{-K'}^{K'} d\beta \int_{-K}^{K} E_m^p(\mu)E_m^p(\nu)E_n^q(\mu)E_n^q(\nu)(\mu^2-\nu^2)d\gamma = 0.$$

Suggestion:

$$\int_0^{2K'} E_m^p(\mu)E_n^q(\mu)(\mu^2-\nu^2)d\beta = \int_0^{K'} E_m^p(\mu)E_n^q(\mu)(\mu^2-\nu^2)d\beta$$
$$+ \int_{K'}^{2K'} E_m^p(\mu)E_n^q(\mu)(\mu^2-\nu^2)d\beta.$$

If in the last integral we replace β by $\beta + 2K'$ it becomes

$$\pm \int_{-K'}^{0} E_m^p(\mu)E_n^q(\mu)(\mu^2-\nu^2)d\beta$$

v. Arts. 136 and 139 and Int. Cal. Art. 196.

2. Show that

$$\int_0^{2K'} d\beta \int_0^{4K} [E_m^p(\mu)E_m^p(\nu)]^2(\mu^2-\nu^2)d\gamma = 8\int_0^{K'} d\beta \int_0^{K} [E_m^p(\mu)E_m^p(\nu)]^2(\mu^2-\nu^2)d\gamma.$$

141. We can now solve the problem of finding the value of V at any point in space when it is given at all the points on the surface of the ellipsoid $\alpha = \alpha_0$.

We have first to develop in Ellipsoidal Harmonics a function of μ and ν or rather of α and β given at all points on the surface of the ellipsoid in question; and this is now easily accomplished by our usual method, which leads us to the result

$$f(\alpha_0,\beta,\gamma) = \sum_{m=0}^{m=\infty}\sum_{k=1}^{k=2m+1} A_{m,p_k} E_m^{p_k}(\mu)E_m^{p_k}(\nu), \tag{1}$$

where

$$A_{m,p_k} = \frac{\int_0^{2K'} d\beta \int_0^{4K} f(\alpha_0,\beta,\gamma)E_m^{p_k}(\mu)E_m^{p_k}(\nu)(\mu^2-\nu^2)d\gamma}{8\int_0^{K'} d\beta \int_0^{K} [E_m^{p_k}(\mu)E_m^{p_k}(\nu)]^2(\mu^2-\nu^2)d\gamma}. \tag{2}$$

Our final solution is

$$V = \sum_{m=0}^{m=\infty}\sum_{k=1}^{k=2m+1} A_{m,p_k} \frac{E_m^{p_k}(\lambda)}{E_m^{p_k}(\lambda_0)} E_m^{p_k}(\mu)E_m^{p_k}(\nu) \tag{3}$$

at an internal point;

$$V = \sum_{m=0}^{m=\infty}\sum_{k=1}^{k=2m+1} A_{m,p_k} \frac{F_m^{p_k}(\lambda)}{F_m^{p_k}(\lambda_0)} E_m^{p_k}(\mu)E_m^{p_k}(\nu) \tag{4}$$

at an external point.

Lamé has proved rather ingeniously that

$$\int_0^{K'} d\beta \int_0^{K} [E_m^{p_k}(\mu)E_m^{p_k}(\nu)]^2(\mu^2-\nu^2)d\gamma$$

can always be found and that it is equal to $\frac{\pi}{2}$ multiplied by a rational integral function of the coefficients of $E_m^{p_k}(x)$ and of c^2 and $\left(\frac{b}{c}\right)^2$.

Of course the labor of obtaining even a few terms of the development of a function that is in the least complicated is enormous.

142. If in Laplace's Equation (8) Art. 136 we let $V = E_m^p(\lambda)U$ supposing U to be a function of β and γ only, we get after replacing $\dfrac{1}{E_m^p(\lambda)}\dfrac{d^2E_m^p(\lambda)}{d\alpha^2}$ by its value $m(m+1)\lambda^2 - (b^2+c^2)p$ [v. (2) Art. 137]

$$(\lambda^2-\nu^2)D_\beta^2U + (\lambda^2-\mu^2)D_\gamma^2U + (\mu^2-\nu^2)[m(m+1)\lambda^2-(b^2+c^2)p]U = 0; \quad (1)$$

and since by hypothesis U is independent of λ, the coefficient of λ^2 in (1) must vanish. Hence

$$D_\beta^2U + D_\gamma^2U + (\mu^2-\nu^2)m(m+1)U = 0. \quad (2)$$

Of course $U = E_m^p(\mu)E_m^p(\nu)$ will satisfy (2).

EXAMPLES.

1. Substitute $U = E_m^p(\mu)E_m^p(\nu)$ in (2) Art. 142 and by the aid of (2) Art. 137 show that the equation (2) Art. 142 is satisfied.

2. Obtain (2) Art. 140 directly from (2) Art. 142.

3. *Conical Coördinates.* Consider the system of coördinates defined by the equations

$$\left.\begin{aligned} x^2+y^2+z^2 &= r^2 \\ \frac{x^2}{\mu^2}+\frac{y^2}{\mu^2-b^2}+\frac{z^2}{\mu^2-c^2} &= 0 \\ \frac{x^2}{\nu^2}+\frac{y^2}{\nu^2-b^2}+\frac{z^2}{\nu^2-c^2} &= 0 \end{aligned}\right\} \quad (1)$$

where $c^2 > \mu^2 > b^2 > \nu^2$.

Show that

$$x^2 = \frac{r^2\mu^2\nu^2}{b^2c^2}, \quad y^2 = \frac{r^2(\mu^2-b^2)(\nu^2-b^2)}{b^2(b^2-c^2)}, \quad z^2 = \frac{r^2(\mu^2-c^2)(\nu^2-c^2)}{c^2(c^2-b^2)};$$

$$h_1^2 = \frac{(\mu^2-b^2)(c^2-\mu^2)}{r^2(\mu^2-\nu^2)}, \quad h_2^2 = \frac{(b^2-\nu^2)(c^2-\nu^2)}{r^2(\mu^2-\nu^2)}, \quad h_3^2 = 1.$$

Laplace's Equation is

$$D_\alpha^2V + D_\beta^2V + (\mu^2-\nu^2)D_r(r^2D_rV) = 0 \quad (2)$$

where $\alpha = \displaystyle\int_b^\mu \frac{d\mu}{\sqrt{(\mu^2-b^2)(c^2-\mu^2)}}$ and $\beta = \displaystyle\int_0^\nu \frac{d\nu}{\sqrt{(b^2-\nu^2)(c^2-\nu^2)}}$,

If $V = U.R$ (2) breaks up into

$$\frac{d}{dr}\left(r^2\frac{dR}{dr}\right) = m(m+1)R, \tag{3}$$

$$D_\alpha^2 U + D_\beta^2 U + m(m+1)(\mu^2 - \nu^2)U = 0. \tag{4}$$

(3) gives $$R = Ar^m + Br^{-m-1}.$$

(4) gives $$U = E_m^p(\mu)E_m^p(\nu) \qquad \text{(v. Art. 142).}$$

So that a solution of (2) is

$$V = Ar^m E_m^p(\mu)E_m^p(\nu).$$

But since (2) is Laplace's Equation, $V = Ar^m Y_m(\mu, \phi)$, if expressed in Conical Coördinates, must satisfy it, consequently $E_m^p(\mu)E_m^p(\nu)$ must be simply a Spherical Harmonic of the mth degree.

Toroidal Coördinates.

143. Any pair of circles belonging to the orthogonal system obtained and figured in Art. 46 can be represented by the equations

$$\left.\begin{aligned} \frac{2ax}{\sinh\alpha} &= \frac{x^2+y^2+a^2}{\cosh\alpha} \\ \frac{2ay}{\sin\beta} &= \frac{x^2+y^2-a^2}{\cos\beta} \end{aligned}\right\} \tag{1}$$

if we take $2a$ instead of 2 as the distance between the points common to the second set of circles.

If we rotate the system about the axis of y we get a set of spheres and a set of anchor rings which cut orthogonally. These and a set of planes through the axis of revolution will form an orthogonal system of surfaces, and the parameters corresponding to them may be taken as a set of curvilinear coördinates and may be called *Toroidal Coördinates.*

If we take the axis of the system as the axis of Z, the equations of a set of the surfaces may be written

$$\left.\begin{aligned} \frac{4a^2(x^2+y^2)}{\sinh^2\alpha} &= \frac{[x^2+y^2+z^2+a^2]^2}{\cosh^2\alpha} \\ \frac{2az}{\sin\beta} &= \frac{x^2+y^2+z^2-a^2}{\cos\beta} \\ y &= x\tan\gamma \end{aligned}\right\} \tag{2}$$

α, β, and γ being regarded as the coördinates of a point of intersection of the three surfaces.

Finding Laplace's Equation in the usual manner we get

$$x = \frac{a \sinh \alpha \cos \gamma}{\cosh \alpha \mp \cos \beta}, \quad y = \frac{a \sinh \alpha \sin \gamma}{\cosh \alpha \mp \cos \beta}, \quad z = \frac{a \sin \beta}{\cosh \alpha \mp \cos \beta},$$

$$r = \sqrt{x^2 + y^2} = \frac{a \sinh \alpha}{\cosh \alpha \mp \cos \beta}, \quad a + z \operatorname{ctn} \beta = \frac{a \cosh \alpha}{\cosh \alpha \mp \cos \beta};$$

$$h_1 = \frac{\cosh \alpha \mp \cos \beta}{a}, \quad h_2 = \frac{\cosh \alpha \mp \cos \beta}{a}, \quad h_3 = \frac{\cosh \alpha \mp \cos \beta}{a \sinh \alpha};$$

and Laplace's Equation becomes

$$D_\alpha \left[\frac{a \sinh \alpha}{\cosh \alpha \mp \cos \beta} D_\alpha V \right] + D_\beta \left[\frac{a \sinh \alpha}{\cosh \alpha \mp \cos \beta} D_\beta V \right] + D_\gamma \left[\frac{a}{\sinh \alpha (\cosh \alpha \mp \cos \beta)} D_\gamma V \right] = 0, \tag{1}$$

or

$$D_\alpha (r D_\alpha V) + D_\beta (r D_\beta V) + \frac{1}{\sinh^2 \alpha} r D_\gamma^2 V = 0. \tag{2}$$

We cannot proceed further by our usual method, for the assumption that V is a function of α alone, or that V is a function of β alone, proves to be inadmissible. Indeed, not only are α, β, and γ not *thermometric parameters* (v. Art. 133), but no thermometric parameters exist, and no possible distribution can make our anchor rings or our spheres a set of equipotential surfaces.

We can, however, simplify (2). It can be written

$$D_\alpha^2 (V \sqrt{r}) + D_\beta^2 (V \sqrt{r}) + \frac{1}{\sinh^2 \alpha} D_\gamma^2 (V \sqrt{r}) - V (D_\alpha^2 \sqrt{r} + D_\beta^2 \sqrt{r}) = 0. \tag{3}$$

$D_\alpha^2 \sqrt{r} + D_\beta^2 \sqrt{r}$ proves equal to $-\dfrac{\sqrt{r}}{4 \sinh^2 \alpha}$; hence if $U = V\sqrt{r}$ (3) becomes

$$\sinh^2 \alpha (D_\alpha^2 U + D_\beta^2 U) + D_\gamma^2 U + \tfrac{1}{4} U = 0, \tag{4}$$

for which particular solutions can readily be found by our usual process.

(4) can be broken up into the three equations

$$\frac{d^2 N}{d\gamma^2} + (m + \tfrac{1}{2})^2 N = 0 \tag{5}$$

$$\frac{d^2 M}{d\beta^2} + n^2 M = 0 \tag{6}$$

$$\sinh^2 \alpha \frac{d^2 L}{d\alpha^2} - [m(m+1) + n^2 \sinh^2 \alpha] L = 0. \tag{7}$$

$$N = A \cos(m + \tfrac{1}{2})\gamma + B \sin(m + \tfrac{1}{2})\gamma$$

$$M = A_1 \cos n\beta + B_1 \sin n\beta.$$

If we introduce into (7) $x = \operatorname{ctnh}\alpha$ it becomes

$$(1-x^2)\frac{d^2L}{dx^2} - 2x\frac{dL}{dx} + \left[m(m+1) - \frac{n^2}{1-x^2}\right]L = 0,$$

a solution of which is

$$L = P_m^n(x) = (1-x^2)^{\frac{n}{2}}\frac{d^nP_m(x)}{dx^n} \qquad \text{(v. Art. 102).}$$

It is to be noted that since $\operatorname{ctnh}\alpha$ is greater than 1

$$P_m^n(\operatorname{ctnh}\alpha) = i^{\frac{n}{2}}\operatorname{csch}^n\alpha\frac{d^nP_m(\operatorname{ctnh}\alpha)}{(d\operatorname{ctnh}\alpha)^n}.$$

The constant coefficient $i^{\frac{n}{2}}$ can be rejected and we get

$$U = [A\cos(m+\tfrac{1}{2})\gamma + B\sin(m+\tfrac{1}{2})\gamma](A_1\cos n\beta + B_1\sin n\beta)\operatorname{csch}^n\alpha\frac{d^nP_m(\operatorname{ctnh}\alpha)}{(d\operatorname{ctnh}\alpha)^n}$$

as a particular solution of (4).

$$\frac{1}{i^{\frac{n}{2}}}P_m^n(\operatorname{ctnh}\alpha) = \operatorname{csch}^n\alpha\frac{d^nP_m(\operatorname{ctnh}\alpha)}{(d\operatorname{ctnh}\alpha)^n}$$

has been called a Toroidal Harmonic.

EXAMPLES.

1. Given the value of the potential function at all points on the surface of an anchor ring; find its value at any point within the ring.

Suggestion: If $V = f(\beta,\gamma)$ when $\alpha = \alpha_0$, the function to be developed is

$$\sqrt{r}.f(\beta,\gamma) \quad i.e. \quad \left[\frac{a\sinh\alpha_0}{\cosh\alpha_0 \mp \cos\beta}\right]^{\frac{1}{2}} f(\beta,\gamma)$$

and the development will be in a double Fourier's Series (v. Art. 71).

2. Show that if we let α range from 0 to ∞, β from $-\pi$ to π, and γ from 0 to 2π, each of the double signs on page 264 may be replaced by the minus sign without loss of generality.

CHAPTER IX.[1]

HISTORICAL SUMMARY.

The method of development in series which has enabled us in the preceding chapters to solve problems in various branches of mathematical physics, had its origin, as might have been expected, in the theory of the musical vibrations of a stretched string. It was in the year 1753[2] that Daniel Bernoulli enunciated the principle of the coexistence of small oscillations, which, in connection with Taylor's and John Bernoulli's theory of the vibrating string, led him to believe that the general solution of this problem could be put in the form of a trigonometric series. This principle also led him and Euler to treat in a similar manner the problems of the vibration of a column of air and of an elastic rod. The problem of the vibration of a heavy string suspended from one end was also treated in the same manner by these mathematicians and deserves special mention here as in it Bessel's functions of the zeroth order appear for the first time.[3] In none of these cases, however, was any method given for determining the coefficients of the series.

This last remark also applies to the more complicated problems of the vibration of rectangular and circular membranes, which were discussed by Euler[4] in 1764, and in the last of which the general Bessel's functions of integral orders occur.

It is in problems connected with astronomy that the first completely successful application of the method here considered occurs. Legendre in a paper published in the Mémoires des Savants Étrangers for 1785, first introduced the zonal harmonics P_m and applied them to the determination of the attraction of solids of revolution. He was followed by Laplace, who in one of the most remarkable memoirs ever written[5] determined the potential of a solid differing but little from a sphere by means of the development according to the spherical harmonics Y_m.

Very closely related to this problem is Gauss's celebrated treatment of the theory of terrestrial magnetism,[6] which we will for that reason mention here, although it was not published until more than half a century later. This paper is particularly noteworthy as it contains a numerical application of the method on a larger scale than has ever been attempted before or since.

After the researches of Legendre and Laplace there was a pause of a quarter of a century until in 1812 Fourier's extensive memoir: *Théorie du mouvement*

---

[1] See preface.

[2] See two articles by Bernoulli and one by Euler in the Memoirs of the Academy of Berlin for this year.

[3] See the Transactions of the Academy of St. Petersburg for 1732-33, 1734 and 1781.

[4] Transactions of the Academy of St. Petersburg.

[5] "Théorie des attractions des sphéroïdes et de la figure des Planètes" Mémoires de l'académie des sciences 1782. This article, although bearing an earlier date than that of Legendre, was really inspired by it. It is here that "Laplace's equation" first appears, occurring, however, only in polar coördinates.

[6] Resultate aus den Beobachtungen des magnetischen Vereins im Jahre 1838. Leipzig, 1839. Reprinted in Gauss's collected works, Vol. V., p. 121.

de la chaleur dans les corps solides was crowned by the French Academy. Although not printed until the years 1824-26,[7] the manuscript of this work was in the meantime accessible to the other French mathematicians presently to be mentioned. The first part of this memoir, which was reproduced with but few alterations in the *Théorie analytique de la chaleur* (1822), contains a treatment of the following problems and of practically all of their special cases:

(*a*) The one dimensional flow of heat. (*b*) The two dimensional flow of heat in a rectangle. (*c*) The three dimensional flow of heat in a rectangular parallelopiped. (*d*) The flow of heat in a sphere when the temperature depends only on the distance from the centre. (*e*) The flow of heat in a right circular cylinder when the temperature depends only on the distance from the axis. In these problems not merely the simpler boundary conditions are considered but also the question of radiation into an atmosphere. In special cases of the first three problems just mentioned (when one or more dimensions become infinite) the series degenerate into "Fourier's integrals."

More important even than any of these special problems is the great advance which Fourier caused the theory of trigonometric series to make. In a posthumous paper Euler had given the formulae for determining the coefficients,[8] but Fourier was the first to assert and to attempt to prove that any function, even though for different values of the argument it is expressed by different analytical formulae, can be developed in such a series. The fact that the real importance of trigonometric series was thus for the first time shown justifies us in associating Fourier's name with them, although, as we have seen, they were known long before his day.

Fourier's results were extended by Laplace in 1820[9] to the general (unsymmetrical) case of the flow of heat in a sphere, and by Poisson[10] (1821) to the unsymmetrical flow of heat in a cylinder.

In 1835 Green published a paper[11] in which the method we are considering is employed to determine the potential of a heterogeneous ellipsoid. This paper, in which the analysis is performed at once for space of n dimensions, anticipates much that was subsequently done by others, but has failed to exert an influence proportional to its importance.

At about this time Lamé began a series of publications which have connected his name inseparably with the problem of the permanent state of temperature of an ellipsoid. In the first of these[12] the equation $\nabla^2 V = 0$ is transformed to ellipsoidal coördinates and is then broken up into three ordinary differential

[7] Mémoires de l'académie des sciences for 1819-20 and 1821-22.

[8] Lagrange had practically determined these coefficients long before but failed to notice what he had got.

[9] Connaissance des Temps pour l'an 1823.

[10] Journal de l'École Polytechnique, 19e Cahier. Although the final forms to which Poisson reduces his results are similar to Fourier's, his methods are very different.

[11] "On the determination of the exterior and interior attraction of ellipsoids of variable densities." Transactions of the Cambridge Philosophical Society.

[12] Mémoires des Savants Étrangers, Vol. V. Although the volume is dated 1838 this paper (which was reprinted in Liouville's Journal, 1837) must have appeared at least as early as 1835.

equations. The rest of the solution, however, is hardly touched upon. Lamé's most important work on this subject[13] was published in Liouville's Journal in 1839, and in it the complete solution of the problem is given. Lamé clearly shows in this paper how he arrived at his solution, by considering first the simpler case of a sphere where, instead of the polar coördinates θ and ϕ, the parameters of two families of confocal cones of the second degree are used as coördinates. This system of curvilinear coördinates, which, when applied to the complete sphere, merely gives the old results of Laplace in a new form, is barely mentioned in Lamé's later publications. In the same volume of Liouville's Journal Lamé published a second paper in which he applies his results to the special cases of ellipsoids of revolution.

These two papers form the starting-point for a series of articles on the same subject by Heine and Liouville. Heine in his doctor dissertation[14] (1842) determined the potential not merely for the interior of an ellipsoid of revolution when the value of the potential is given on the surface, but also for the exterior of such an ellipsoid and for the shell between two confocal ellipsoids of revolution. Even in the first of these problems, which is equivalent to that of Lamé, he simplified Lamé's solution materially by showing that the functions used may be reduced to spherical harmonics, while in the other two problems he introduced spherical harmonics of the second kind, which were then new. Shortly afterwards[15] Heine and Liouville published simultaneously two papers in which they arrived independently of each other at about the same results. In each of these papers attention is called to the fact that the product of two Lamé's functions is a spherical harmonic, and this fact is made use of to throw Lamé's solution of the problem of the permanent state of temperatures of an ellipsoid into a more elementary form. Besides this the second solution of Lamé's equation is introduced for the sake of solving the potential problem for the *exterior* of the ellipsoid.

In thus following up the theory of heat and the related potential problems, we have lost sight of the question of small vibrations, to which during the early part of the century a great deal of attention had been devoted by Poisson, who frequently made use of the method of development in series. In his memoirs[16] most of the problems left unfinished by Bernoulli and Euler are thoroughly treated, as well as various slight modifications of them. When, however, he attacked the problem of the vibration of an elastic plate he was unable to make much progress, owing in part to the erroneous form of his boundary conditions.

---

[13] "Sur l'équilibre des Températures dans un ellipsoïde à trois axes inégaux." An article by the same author on the two dimensional potential will be found in Vol. I. of this Journal.

[14] Reprinted in Crelle's Journal, Vol. 26 (1843).
In the same Journal for 1847 F. Neumann discussed the related problem of the magnetisation of a soft iron ellipsoid of revolution.

[15] *Heine:* Crelle's Journal, Vol. 29, 1845. *Liouville:* Liouville's Journal, Vol. X., 1845, and Vol. XI., 1846. For a treatment of the problem of the potential of an ellipsoidal shell by means of a development of $\frac{1}{r}$ in terms of Lamé's functions, see a paper by Heine in Crelle's Journal, Vol. 42, 1851.

[16] See especially the one in the Mémoires de l'académie des sciences, Vol. VIII., 1829.

He was, nevertheless, able to solve the problem of the *symmetrical* vibration of a free circular plate. The complete theory of the vibration of a free circular plate was first given by Kirchhoff.[17]

Passing now to a new subject, the theory of the equilibrium of an elastic spherical shell, we find a solution by Lamé in Liouville's Journal for 1854, and by Sir William Thomson (1862) in the Philosophical Transactions for 1863. Both of these papers consist of an application of the spherical harmonic analysis to this rather complicated problem. Thomson, however, considers besides Lamé's problem certain related questions and the form of his analysis is very different from Lamé's, being of the same nature as that used in the Appendix B of his Natural Philosophy of which we shall have to speak presently. These investigations form the starting point for a number of recent memoirs among which those of G. H. Darwin on cosmographical questions deserve special mention.

Closely related to this last mentioned problem is the theory of the small vibrations of an elastic sphere. While the simplest case of this problem was treated by Poisson in the memoir referred to above, the general solution has been only recently obtained by Jaerisch (1879)[18] and Lamb (1882).[19] The functions involved are the same as those which occur in the problem of the non-stationary flow of heat in a sphere as solved by Laplace.

The Appendix B of Thomson and Tait's Natural Philosophy,[20] to which we have already referred, deserves to be regarded as one of the most important contributions to the general theory. The way in which spherical harmonics are introduced (as homogeneous functions of the rectangular coördinates) was then new[21] and the solution of the potential problem for a variety of new solids was indicated; viz., for solids whose boundaries consist of concentric spheres, cones of revolution, and planes. We shall have more to say presently concerning the method employed for the solution of these problems.

Although connected only indirectly with the theory we are discussing, it will be well to mention at this point the method of electrical images which is also due to Sir William Thomson (1845). This method enables us to solve many potential problems for the inverse of any solid when once we have solved it for the solid itself. By means of this method most of the solutions of potential problems obtained by our method may be applied at once with very little modification to systems of curvilinear coördinates derived by inversion from those we have used. It will not be necessary to mention separately problems of this sort, as it is clearly immaterial whether they be solved directly or by means of the method of inversion.[22]

Returning now to the Continent, we find as the next important question

[17] Crelle's Journal, Vol. 40, 1850.

[18] Crelle's Journal, Vol. 88.

[19] Proc. Lond. Math. Soc.

[20] First edition, 1867. This appendix was evidently written as early as 1862, as Thomson refers to it in the memoir quoted above.

[21] The same method was used at about the same time by Clebsch.

[22] A case in point would be the potential problem for the shell between two non-intersecting eccentric spheres, since these spheres can be inverted into concentric spheres. This problem was treated directly by C. Neumann in a monograph published in Halle in 1862.

taken up the problem of the potential of an anchor ring. The first publication on this subject is a monograph by C. Neumann[23] (1864), but in Riemann's posthumous papers which were not published until 1876, ten years after his death, will be found a short fragment on this subject, which (*cf.* the last page of Hattendorf's edition of Riemann's lectures: "Partielle Differentialgleichungen") would appear to date back to the winter 1860-61. This fragment is of peculiar interest, as the opening paragraphs clearly show that Riemann had in mind an extended article on the fundamental principles of our subject.

We will next mention two papers by Mehler in which the functions known as "conal harmonics," which had already been introduced by Thomson in the Appendix B above mentioned, were applied to the solution of two problems in electrostatics. The first of these papers[24] (1868) deals with the solid bounded by two intersecting spheres, while in the second[25] (1870) the infinite cone of revolution is treated. Both of these problems are essentially different from those discussed in the "Appendix B," inasmuch as the infinite series which we usually have degenerate in these cases into definite integrals, just as they do in some simpler cases treated by Fourier. The later of the two papers just quoted also contains valuable information concerning the nature of the solution of similar problems for the hyperboloids and paraboloids of revolution. The solutions of these problems are not, however, given.

It remains, in order to close the history of this part of the subject, to mention a number of memoirs which although treating entirely new problems are of far less importance than most of those considered up to this point, partly because the solution is not brought to a point where it can be of much immediate use, and partly because most of the methods employed are such as could not fail to present themselves to any one attacking these problems.

Of these the first is a paper by Mathieu[26] on the vibration of an elliptic membrane (1868), in which the functions of the elliptic cylinder occur for the first time.

This was followed in the same year by a paper on closely allied subjects by H. Weber,[27] in which not merely the case of the complete ellipse is briefly considered, but also that in which the boundary consists of two arcs of confocal ellipses and two arcs of hyperbolas confocal with them. The special case in which the ellipses and hyperbolas become confocal parabolas is also considered, whereby the functions of the parabolic cylinder are for the first time introduced.

In Mathieu's "Cours de physique mathématique" (1873) the problem of the non-stationary flow of heat in an ellipsoid is touched upon, and an elaborate though not very satisfactory treatment of the special cases where we have ellipsoids of revolution is given. New functions appear in all of these problems.

Of late years C. Baer has supplied a number of missing links in the chain

[23] "Theorie der Elektricitäts- und Wärme-Vertheilung in einem Ringe." Halle.

[24] Crelle's Journal, Vol. 68, 1868.

[25] Jahresbericht des Gymnasiums zu Elbing.

[26] Liouville's Journal, Vol. XIII.

[27] "Ueber die Integration der partiellen Differentialgleichung $\frac{\delta^2 u}{\delta x^2} + \frac{\delta^2 u}{\delta y^2} + k^2 u = 0$." Math. Ann., Vol. I. No *physical* problem is mentioned in this paper.

of problems here considered by treating in succession the potential problem for the paraboloid of revolution,[28] the parabolic cylinder[29] and the general paraboloid.[30] In the first of these problems Bessel's functions occur, as had already been stated by Mehler, while in the last we find the functions of the elliptic cylinder. For each of the three systems of coördinates employed the same author also touches upon the more general problem of the non-stationary flow of heat, in which new functions occur.

Except in the case of the anchor ring we have found so far only such solids treated by our method as are bounded by surfaces of the first or second degree. Wangerin[31] (1875-76) considered in connection with the theory of the potential, more general systems of curvilinear coördinates than had previously been used in physical questions, namely, *cyclidic* coördinates.[32] He showed, however, merely how to break up Laplace's equation into three ordinary differential equations.[33]

An important branch of our theory which we have not yet touched upon dates back to the year 1836, when Sturm published a series of fundamentally important papers in the first two volumes of Liouville's Journal. The physical question which lies at the basis of these papers is the problem of the flow of heat in a heterogeneous bar.[34] The method here employed depends upon the fact that the functions which occur are characterized by the number of times they vanish in a certain interval. This same idea reappears in Thomson and Tait's Appendix B already referred to, but first finds its full expression in this more general field of the three dimensional potential in an article by Klein: "Ueber Körper welche von confocalen Flächen zweiten Grades begrenzt sind"[35] (1881). Still more recently (1889-90) Klein has in his lectures extended this theory to the treatment of solids bounded by six confocal cyclids, and has indicated how all the potential problems heretofore treated by our method are special cases of this one.[36]

Of late years, especially since the year 1880, the younger English mathematicians have done a vast amount of work in the theory we are here considering. Although much of this work is of great value, hardly any of it can be regarded as being a real *development* of the method; it is rather an application of it to

---

[28] "Ueber das Gleichgewicht und die Bewegung der Wärme in einem Rotationsparaboloid." Dissertation, Halle, 1881.

[29] "Die Funktion des parabolischen Cylinders," Gymnasialprogramm Cüstrin, 1883.

[30] "Parabolische Coordinaten," Frankfurt, 1888. See also a paper by Greenhill in the Proc. Lond. Math. Soc., Vol. XIX., 1889 (read Dec. 8, 1887). Also a posthumous paper by Lamé in Liouville's Journal for 1874, Vol. XIX.

[31] Preisschriften der Jablanowski'schen Gesellschaft, No. XVIII., and Crelle's Journal, Vol. 82. See also, concerning a still further extension, the Berliner Monatsberichten for 1878.

[32] Cyclids are a kind of surface of the fourth order (see Salmon's Geom. of three Dimensions, p. 527). In his first memoir Wangerin considers only cyclids of revolution.

[33] See also a paper by this author in Grünert's Archiv for 1873, where the problem of the equilibrium of elastic solids of revolution is treated.

[34] The similar problem of the vibration of a heterogeneous string under the action of an external force was treated by Maggi (Giornale di Matematiche, 1880). Several special cases are also considered here in detail.

[35] Math. Ann., 18.

[36] For an exposition of this theory see the treatise: Ueber die Reihenentwickelungen der Potentialtheorie, Leipsic, Teubner, 1894, by the writer of the present chapter.

a great variety of problems. We must therefore content ourselves with giving a mere list of a few of the more important of these papers.

Niven: On the Conduction of Heat in Ellipsoids of Revolution. Phil. Trans., 1880.

Niven: On the Induction of Electric Currents in Infinite Plates and Spherical Shells. Phil. Trans., 1881.

Hicks: On Toroidal Functions. Phil. Trans., 1881.

Hicks: On the Steady Motion and Small Vibrations of a Hollow Vortex. Phil. Trans., 1884, 1885.

Lamb: On Ellipsoidal Current Sheets. Phil. Trans., 1887.

Chree: The Equations of an Isotropic Elastic Solid in Polar and Cylindrical Coördinates, their Solution and Application. Camb. Phil. Soc. Trans., XIV., 1889.

Hobson: On a Class of Spherical Harmonics of Complex Degree with Applications to Physical Problems. Camb. Phil. Soc. Trans., XIV., 1889.

Chree: On some Compound Vibrating Systems. Camb. Phil. Soc. Trans., XV., 1891.

Niven: On Ellipsoidal Harmonics. Phil. Trans., 1892.

The historical sketch we have just given would naturally require as a supplement some account of the work that has been done on the question of the convergence of the various series which occur. This, however, would carry us too far, and we will content ourselves with mentioning the two fundamental memoirs by Dirichlet in Crelle's Journal, one in 1829 on Fourier's series, and one, which has been criticised to some extent by subsequent mathematicians, in 1837 on Laplace's spherical harmonic development.

Another subject which naturally presents itself here is the theory of the various new functions we have met. Those properties of these functions, however, which the physicist needs have usually been investigated by the physicists themselves in the papers mentioned above; while any thorough account of the development of the theory of these functions would lead us into the vast region of the modern theory of linear differential equations.

We will therefore close by merely giving a list of books which will be found useful by those wishing to continue their study of the subject further.

We begin with the books relating directly to physical questions:

Fourier: Théorie Analytique de la Chaleur, 1822.

Lamé: Leçons sur les Functions inverses des Transcendantes et les Surfaces isothermes, 1857.

Lamé: Leçons sur les Coordonnées Curvilignes et leurs diverses Applications, 1859.

Mathieu: Cours de Physique Mathématique, 1873.

Riemann: Partielle Differentialgleichungen, und deren Anwendung auf physikalische Fragen (edited by Hattendorf), third edition. 1882.

F. Neumann: Theorie des Potentials und der Kugelfunktionen (edited by C. Neumann), 1887.

Thomson and Tait: Natural Philosophy, second edition, 1879.

Rayleigh: Theory of Sound, 1877.

Basset: Hydrodynamics, 1888.

Love: Theory of Elasticity, 1892.

Heine: Handbuch der Kugelfunktionen (second edition), 1878-81.

Ferrers: Spherical Harmonics, 1881.

Haentzschel: Reduction der Potentialgleichung auf gewöhnliche Differentialgleichungen, 1893.

These last three books would also belong in the following list of books relating to the theory of the various functions we use:

Todhunter: The Functions of Laplace, Lamé and Bessel, 1875.

Lommel: Studien über die Bessel'schen Funktionen, 1868.

F. Neumann: Beiträge zur Theorie der Kugelfunktionen, 1878.

And finally concerning the question of convergence:

C. Neumann: Über die nach Kreis-, Kugel- und Cylinder-Functionen fortschreitenden Entwickelungen, 1881.

APPENDIX.

TABLES.

Table I., a table of Surface Zonal Harmonics (Legendrians), gives the values of the first seven Harmonics $P_1(\cos\theta)$, $P_2(\cos\theta)$, $\cdots P_7(\cos\theta)$ for the argument θ in degrees. It is taken from the Philosophical Magazine for December, 1891, and was computed by Messrs. C. E. Holland, P. R. James, and C. G. Lamb, under the direction of Professor John Perry.

Table II., a table of Surface Zonal Harmonics (Legendrians), gives the values of the first seven Harmonics $P_1(x)$, $P_2(x)$, $\cdots P_7(x)$ for the argument x. It is reduced from the Tables of Legendrian Functions computed under the direction of Dr. J. W. L. Glaisher, and published in the Report of the British Association for the Advancement of Science for the year 1879.

Table III., the table of Hyperbolic Functions, gives the values of e^x, e^{-x}, $\sinh x$, $\cosh x$, and $\operatorname{gd} x$ (Gudermannian of x) for values of x from 0.00 to 1.00; and the values of $\log\sinh x$ and $\log\cosh x$ for values of x from 1.00 to 10.0. The values of $\operatorname{gd} x$, $\log\sinh x$, and $\log\cosh x$ are taken from the Mathematical Tables prepared by Professor J. M. Peirce (Boston: Ginn & Co.).

The $\log\sinh x$ and $\log\cosh x$ for values of x between 0.00 and 1.00 can be obtained from the values given for the Gudermannian of x in the table by the aid of the relations

$$\log\sinh x = \log\tan(\operatorname{gd} x) \qquad \log\cosh x = \log\sec(\operatorname{gd} x).$$

Table IV. gives the first twelve roots of $J_0(x) = 0$ and $J_1(x) = 0$ each divided by π. The table is taken from Lord Rayleigh's Sound, Vol. I., page 274, and is due to Professor Stokes, Camb. Phil. Trans., Vol. IX., page 186.

Table V. gives the first nine roots of $J_0(x) = 0$, $J_1(x) = 0$, $\cdots J_5(x) = 0$. The table is taken from Rayleigh's Sound, Vol. I., page 274, and is due to Professor J. Bourget, Ann. de l'Ecole Normale, T. III., 1866, page 82.

Table VI., the table of Bessel's Functions, gives the values of the Bessel's Functions $J_0(x)$ and $J_1(x)$ for the argument x from $x = 0$ to $x = 15$. It is taken from Rayleigh's Sound, Vol. I., page 265, and from Lommel's Bessel'sche Functionen.

TABLE I. — SURFACE ZONAL HARMONICS

θ	$P_1(\cos\theta)$	$P_2(\cos\theta)$	$P_3(\cos\theta)$	$P_4(\cos\theta)$	$P_5(\cos\theta)$	$P_6(\cos\theta)$	$P_7(\cos\theta)$
0°	1.0000	1.0000	1.0000	1.0000	1.0000	1.0000	1.0000
1	.9998	.9995	.9991	.9985	.9977	.9967	.9955
2	.9994	.9982	.9963	.9939	.9909	.9872	.9829
3	.9986	.9959	.9918	.9863	.9795	.9713	.9617
4	.9976	.9927	.9854	.9758	.9638	.9495	.9329
5	.9962	.9886	.9773	.9623	.9437	.9216	.8961
6	.9945	.9836	.9674	.9459	.9194	.8881	.8522
7	.9925	.9777	.9557	.9267	.8911	.8476	.7986
8	.9903	.9709	.9423	.9048	.8589	.8053	.7448
9	.9877	.9633	.9273	.8803	.8232	.7571	.6831
10	.9848	.9548	.9106	.8532	.7840	.7045	.6164
11	.9816	.9454	.8923	.8238	.7417	.6483	.5461
12	.9781	.9352	.8724	.7920	.6966	.5892	.4732
13	.9744	.9241	.8511	.7582	.6489	.5273	.3940
14	.9703	.9122	.8283	.7224	.5990	.4635	.3219
15	.9659	.8995	.8042	.6847	.5471	.3982	.2454
16	.9613	.8860	.7787	.6454	.4937	.3322	.1699
17	.9563	.8718	.7519	.6046	.4391	.2660	.0961
18	.9511	.8568	.7240	.5624	.3836	.2002	.0289
19	.9455	.8410	.6950	.5192	.3276	.1347	−.0443
20	.9397	.8245	.6649	.4750	.2715	.0719	−.1072
21	.9336	.8074	.6338	.4300	.2156	.0107	−.1662
22	.9272	.7895	.6019	.3845	.1602	−.0481	−.2201
23	.9205	.7710	.5692	.3386	.1057	−.1038	−.2681
24	.9135	.7518	.5357	.2926	.0525	−.1559	−.3095
25	.9063	.7321	.5016	.2465	.0009	−.2053	−.3463
26	.8988	.7117	.4670	.2007	−.0489	−.2478	−.3717
27	.8910	.6908	.4319	.1553	−.0964	−.2869	−.3921
28	.8829	.6694	.3964	.1105	−.1415	−.3211	−.4052
29	.8746	.6474	.3607	.0665	−.1839	−.3503	−.4114
30	.8660	.6250	.3248	.0234	−.2233	−.3740	−.4101
31	.8572	.6021	.2887	−.0185	−.2595	−.3924	−.4022
32	.8480	.5788	.2527	−.0591	−.2923	−.4052	−.3876
33	.8387	.5551	.2167	−.0982	−.3216	−.4126	−.3670
34	.8290	.5310	.1809	−.1357	−.3473	−.4148	−.3409
35	.8192	.5065	.1454	−.1714	−.3691	−.4115	−.3096
36	.8090	.4818	.1102	−.2052	−.3871	−.4031	−.2738
37	.7986	.4567	.0755	−.2370	−.4011	−.3898	−.2343
38	.7880	.4314	.0413	−.2666	−.4112	−.3719	−.1918
39	.7771	.4059	.0077	−.2940	−.4174	−.3497	−.1469
40	.7660	.3802	−.0252	−.3190	−.4197	−.3234	−.1003
41	.7547	.3544	−.0574	−.3416	−.4181	−.2938	−.0534
42	.7431	.3284	−.0887	−.3616	−.4128	−.2611	−.0065
43	.7314	.3023	−.1191	−.3791	−.4038	−.2255	.0398
44	.7193	.2762	−.1485	−.3940	−.3914	−.1878	.0846
45°	.7071	.2500	−.1768	−.4062	−.3757	−.1485	.1270

TABLE I. — SURFACE ZONAL HARMONICS

θ	$P_1(\cos\theta)$	$P_2(\cos\theta)$	$P_3(\cos\theta)$	$P_4(\cos\theta)$	$P_5(\cos\theta)$	$P_6(\cos\theta)$	$P_7(\cos\theta)$
45°	.7071	.2500	−.1768	−.4062	−.3757	−.1485	.1270
46	.6947	.2238	−.2040	−.4158	−.3568	−.1079	.1666
47	.6820	.1977	−.2300	−.4252	−.3350	−.0645	.2054
48	.6691	.1716	−.2547	−.4270	−.3105	−.0251	.2349
49	.6561	.1456	−.2781	−.4286	−.2836	.0161	.2627
50	.6428	.1198	−.3002	−.4275	−.2545	.0563	.2854
51	.6293	.0941	−.3209	−.4239	−.2235	.0954	.3031
52	.6157	.0686	−.3401	−.4178	−.1910	.1326	.3153
53	.6018	.0433	−.3578	−.4093	−.1571	.1677	.3221
54	.5878	.0182	−.3740	−.3984	−.1223	.2002	.3234
55	.5736	−.0065	−.3886	−.3852	−.0868	.2297	.3191
56	.5592	−.0310	−.4016	−.3698	−.0510	.2559	.3095
57	.5446	−.0551	−.4131	−.3524	−.0150	.2787	.2949
58	.5299	−.0788	−.4229	−.3331	.0206	.2976	.2752
59	.5150	−.1021	−.4310	−.3119	.0557	.3125	.2511
60	.5000	−.1250	−.4375	−.2891	.0898	.3232	.2231
61	.4848	−.1474	−.4423	−.2647	.1229	.3298	.1916
62	.4695	−.1694	−.4455	−.2390	.1545	.3321	.1571
63	.4540	−.1908	−.4471	−.2121	.1844	.3302	.1203
64	.4384	−.2117	−.4470	−.1841	.2123	.3240	.0818
65	.4226	−.2321	−.4452	−.1552	.2381	.3138	.0422
66	.4067	−.2518	−.4419	−.1256	.2615	.2996	.0021
67	.3907	−.2710	−.4370	−.0955	.2824	.2819	−.0375
68	.3746	−.2896	−.4305	−.0650	.3005	.2605	−.0763
69	.3584	−.3074	−.4225	−.0344	.3158	.2361	−.1135
70	.3420	−.3245	−.4130	−.0038	.3281	.2089	−.1485
71	.3256	−.3410	−.4021	.0267	.3373	.1786	−.1811
72	.3090	−.3568	−.3898	.0568	.3434	.1472	−.2099
73	.2924	−.3718	−.3761	.0864	.3463	.1144	−.2347
74	.2756	−.3860	−.3611	.1153	.3461	.0795	−.2559
75	.2588	−.3995	−.3449	.1434	.3427	.0431	−.2730
76	.2419	−.4112	−.3275	.1705	.3362	.0076	−.2848
77	.2250	−.4241	−.3090	.1964	.3267	−.0284	−.2919
78	.2079	−.4352	−.2894	.2211	.3143	−.0644	−.2943
79	.1908	−.4454	−.2688	.2443	.2990	−.0989	−.2913
80	.1736	−.4548	−.2474	.2659	.2810	−.1321	−.2835
81	.1564	−.4633	−.2251	.2859	.2606	−.1635	−.2709
82	.1392	−.4709	−.2020	.3040	.2378	−.1926	−.2536
83	.1219	−.4777	−.1783	.3203	.2129	−.2193	−.2321
84	.1045	−.4836	−.1539	.3345	.1861	−.2431	−.2067
85	.0872	−.4886	−.1291	.3468	.1577	−.2638	−.1779
86	.0698	−.4927	−.1038	.3569	.1278	−.2811	−.1460
87	.0523	−.4959	−.0781	.3648	.0969	−.2947	−.1117
88	.0349	−.4982	−.0522	.3704	.0651	−.3045	−.0735
89	.0175	−.4995	−.0262	.3739	.0327	−.3105	−.0381
90°	.0000	−.5000	.0000	.3750	.0000	−.3125	.0000

TABLE II.—SURFACE ZONAL HARMONICS.

x	$P_1(x)$	$P_2(x)$	$P_3(x)$	$P_4(x)$	$P_5(x)$	$P_6(x)$	$P_7(x)$
0.00	0.0000	−.5000	0.0000	0.3750	0.0000	−.3125	0.0000
.01	.0100	−.4998	−.0150	.3746	.0187	−.3118	−.0219
.02	.0200	−.4994	−.0300	.3735	.0374	−.3099	−.0436
.03	.0300	−.4986	−.0449	.3716	.0560	−.3066	−.0651
.04	.0400	−.4976	−.0598	.3690	.0744	−.3021	−.0862
.05	.0500	−.4962	−.0747	.3657	.0927	−.2962	−.1069
.06	.0600	−.4946	−.0895	.3616	.1106	−.2891	−.1270
.07	.0700	−.4926	−.1041	.3567	.1283	−.2808	−.1464
.08	.0800	−.4904	−.1187	.3512	.1455	−.2713	−.1651
.09	.0900	−.4878	−.1332	.3449	.1624	−.2606	−.1828
.10	.1000	−.4850	−.1475	.3379	.1788	−.2488	−.1995
.11	.1100	−.4818	−.1617	.3303	.1947	−.2360	−.2151
.12	.1200	−.4784	−.1757	.3219	.2101	−.2220	−.2295
.13	.1300	−.4746	−.1895	.3129	.2248	−.2071	−.2427
.14	.1400	−.4706	−.2031	.3032	.2389	−.1913	−.2545
.15	.1500	−.4662	−.2166	.2928	.2523	−.1746	−.2649
.16	.1600	−.4616	−.2298	.2819	.2650	−.1572	−.2738
.17	.1700	−.4566	−.2427	.2703	.2769	−.1389	−.2812
.18	.1800	−.4514	−.2554	.2581	.2880	−.1201	−.2870
.19	.1900	−.4458	−.2679	.2453	.2982	−.1006	−.2911
.20	.2000	−.4400	−.2800	.2320	.3075	−.0806	−.2935
.21	.2100	−.4338	−.2918	.2181	.3159	−.0601	−.2943
.22	.2200	−.4274	−.3034	.2037	.3234	−.0394	−.2933
.23	.2300	−.4206	−.3146	.1889	.3299	−.0183	−.2906
.24	.2400	−.4136	−.3254	.1735	.3353	.0029	−.2861
.25	.2500	−.4062	−.3359	.1577	.3397	.0243	−.2799
.26	.2600	−.3986	−.3461	.1415	.3431	.0456	−.2720
.27	.2700	−.3906	−.3558	.1249	.3453	.0669	−.2625
.28	.2800	−.3824	−.3651	.1079	.3465	.0879	−.2512
.29	.2900	−.3738	−.3740	.0906	.3465	.1087	−.2384
.30	.3000	−.3650	−.3825	.0729	.3454	.1292	−.2241
.31	.3100	−.3558	−.3905	.0550	.3431	.1492	−.2082
.32	.3200	−.3464	−.3981	.0369	.3397	.1686	−.1910
.33	.3300	−.3366	−.4052	.0185	.3351	.1873	−.1724
.34	.3400	−.3266	−.4117	−.0000	.3294	.2053	−.1527
.35	.3500	−.3162	−.4178	−.0187	.3225	.2225	−.1318
.36	.3600	−.3056	−.4234	−.0375	.3144	.2388	−.1098
.37	.3700	−.2946	−.4284	−.0564	.3051	.2540	−.0870
.38	.3800	−.2834	−.4328	−.0753	.2948	.2681	−.0635
.39	.3900	−.2718	−.4367	−.0942	.2833	.2810	−.0393
.40	.4000	−.2600	−.4400	−.1130	.2706	.2926	−.0146
.41	.4100	−.2478	−.4427	−.1317	.2569	.3029	.0104
.42	.4200	−.2354	−.4448	−.1504	.2421	.3118	.0356
.43	.4300	−.2226	−.4462	−.1688	.2263	.3191	.0608
.44	.4400	−.2096	−.4470	−.1870	.2095	.3249	.0859
.45	.4500	−.1962	−.4472	−.2050	.1917	.3290	.1106
.46	.4600	−.1826	−.4467	−.2226	.1730	.3314	.1348
.47	.4700	−.1686	−.4454	−.2399	.1534	.3321	.1584
.48	.4800	−.1544	−.4435	−.2568	.1330	.3310	.1811
.49	.4900	−.1398	−.4409	−.2732	.1118	.3280	.2027
.50	.5000	−.1250	−.4375	−.2891	.0898	.3232	.2231

TABLE II.—SURFACE ZONAL HARMONICS.

x	$P_1(x)$	$P_2(x)$	$P_3(x)$	$P_4(x)$	$P_5(x)$	$P_6(x)$	$P_7(x)$
.50	.5000	−.1250	−.4375	−.2891	.0898	.3232	.2231
.51	.5100	−.1098	−.4334	−.3044	.0673	.3166	.2422
.52	.5200	−.0944	−.4285	−.3191	.0441	.3080	.2596
.53	.5300	−.0786	−.4228	−.3332	.0204	.2975	.2753
.54	.5400	−.0626	−.4163	−.3465	−.0037	.2851	.2891
.55	.5500	−.0462	−.4091	−.3590	−.0282	.2708	.3007
.56	.5600	−.0296	−.4010	−.3707	−.0529	.2546	.3102
.57	.5700	−.0126	−.3920	−.3815	−.0779	.2366	.3172
.58	.5800	.0046	−.3822	−.3914	−.1028	.2168	.3217
.59	.5900	.0222	−.3716	−.4002	−.1278	.1953	.3235
.60	.6000	.0400	−.3600	−.4080	−.1526	.1721	.3226
.61	.6100	.0582	−.3475	−.4146	−.1772	.1473	.3188
.62	.6200	.0766	−.3342	−.4200	−.2014	.1211	.3121
.63	.6300	.0954	−.3199	−.4242	−.2251	.0935	.3023
.64	.6400	.1144	−.3046	−.4270	−.2482	.0646	.2895
.65	.6500	.1338	−.2884	−.4284	−.2705	.0347	.2737
.66	.6600	.1534	−.2713	−.4284	−.2919	.0038	.2548
.67	.6700	.1734	−.2531	−.4268	−.3122	−.0278	.2329
.68	.6800	.1936	−.2339	−.4236	−.3313	−.0601	.2081
.69	.6900	.2142	−.2137	−.4187	−.3490	−.0926	.1805
.70	.7000	.2350	−.1925	−.4121	−.3652	−.1253	.1502
.71	.7100	.2562	−.1702	−.4036	−.3796	−.1578	.1173
.72	.7200	.2776	−.1469	−.3933	−.3922	−.1899	.0822
.73	.7300	.2994	−.1225	−.3810	−.4026	−.2214	.0450
.74	.7400	.3214	−.0969	−.3666	−.4107	−.2518	.0061
.75	.7500	.3438	−.0703	−.3501	−.4164	−.2808	−.0342
.76	.7600	.3664	−.0426	−.3314	−.4193	−.3081	−.0754
.77	.7700	.3894	−.0137	−.3104	−.4193	−.3333	−.1171
.78	.7800	.4126	.0164	−.2871	−.4162	−.3559	−.1588
.79	.7900	.4362	.0476	−.2613	−.4097	−.3756	−.1999
.80	.8000	.4600	.0800	−.2330	−.3995	−.3918	−.2397
.81	.8100	.4842	.1136	−.2021	−.3855	−.4041	−.2774
.82	.8200	.5086	.1484	−.1685	−.3674	−.4119	−.3124
.83	.8300	.5334	.1845	−.1321	−.3449	−.4147	−.3437
.84	.8400	.5584	.2218	−.0928	−.3177	−.4120	−.3703
.85	.8500	.5838	.2603	−.0506	−.2857	−.4030	−.3913
.86	.8600	.6094	.3001	−.0053	−.2484	−.3872	−.4055
.87	.8700	.6354	.3413	.0431	−.2056	−.3638	−.4116
.88	.8800	.6616	.3837	.0947	−.1570	−.3322	−.4083
.89	.8900	.6882	.4274	.1496	−.1023	−.2916	−.3942
.90	.9000	.7150	.4725	.2079	−.0411	−.2412	−.3678
.91	.9100	.7422	.5189	.2698	.0268	−.1802	−.3274
.92	.9200	.7696	.5667	.3352	.1017	−.1077	−.2713
.93	.9300	.7974	.6159	.4044	.1842	−.0229	−.1975
.94	.9400	.8254	.6665	.4773	.2744	.0751	−.1040
.95	.9500	.8538	.7184	.5541	.3727	.1875	.0112
.96	.9600	.8824	.7718	.6349	.4796	.3151	.1506
.97	.9700	.9114	.8267	.7198	.5954	.4590	.3165
.98	.9800	.9406	.8830	.8089	.7204	.6204	.5115
.99	.9900	.9702	.9407	.9022	.8552	.8003	.7384
1.00	1.0000	1.0000	1.0000	1.0000	1.0000	1.0000	1.0000

TABLE III.—HYPERBOLIC FUNCTIONS.

x	e^x	e^{-x}	$\sinh x$	$\cosh x$	$\operatorname{gd} x$
0.00	1.0000	1.0000	0.0000	1.0000	0°.0000
.01	1.0100	0.9900	.0100	1.0000	0.5729
.02	1.0202	.9802	.0200	1.0002	1.1458
.03	1.0305	.9704	.0300	1.0004	1.7186
.04	1.0408	.9608	.0400	1.0008	2.2912
.05	1.0513	.9512	.0500	1.0012	2.8636
.06	1.0618	.9418	.0600	1.0018	3.4357
.07	1.0725	.9324	.0701	1.0025	4.0074
.08	1.0833	.9231	.0801	1.0032	4.5788
.09	1.0942	.9139	.0901	1.0040	5.1497
.10	1.1052	.9048	.1002	1.0050	5.720
.11	1.1163	.8958	.1102	1.0061	6.290
.12	1.1275	.8869	.1203	1.0072	6.859
.13	1.1388	.8781	.1304	1.0085	7.428
.14	1.1503	.8694	.1405	1.0098	7.995
.15	1.1618	.8607	.1506	1.0113	8.562
.16	1.1735	.8521	.1607	1.0128	9.128
.17	1.1853	.8437	.1708	1.0145	9.694
.18	1.1972	.8353	.1810	1.0162	10.258
.19	1.2092	.8270	.1911	1.0181	10.821
.20	1.2214	.8187	.2013	1.0201	11.384
.21	1.2337	.8106	.2115	1.0221	11.945
.22	1.2461	.8025	.2218	1.0243	12.505
.23	1.2586	.7945	.2320	1.0266	13.063
.24	1.2712	.7866	.2423	1.0289	13.621
.25	1.2840	.7788	.2526	1.0314	14.177
.26	1.2969	.7711	.2629	1.0340	14.732
.27	1.3100	.7634	.2733	1.0367	15.285
.28	1.3231	.7558	.2837	1.0395	15.837
.29	1.3364	.7483	.2941	1.0423	16.388
.30	1.3499	.7408	.3045	1.0453	16.937
.31	1.3634	.7334	.3150	1.0484	17.484
.32	1.3771	.7261	.3255	1.0516	18.030
.33	1.3910	.7189	.3360	1.0549	18.573
.34	1.4049	.7118	.3466	1.0584	19.116
.35	1.4191	.7047	.3572	1.0619	19.656
.36	1.4333	.6977	.3678	1.0655	20.195
.37	1.4477	.6907	.3785	1.0692	20.732
.38	1.4623	.6839	.3892	1.0731	21.267
.39	1.4770	.6771	.4000	1.0770	21.800
.40	1.4918	.6703	.4108	1.0811	22.331
.41	1.5068	.6636	.4216	1.0852	22.859
.42	1.5220	.6570	.4325	1.0895	23.386
.43	1.5373	.6505	.4434	1.0939	23.911
.44	1.5527	.6440	.4543	1.0984	24.434
.45	1.5683	.6376	.4653	1.1030	24.955
.46	1.5841	.6313	.4764	1.1077	25.473
.47	1.6000	.6250	.4875	1.1125	25.989
.48	1.6161	.6188	.4986	1.1174	26.503
.49	1.6323	.6126	.5098	1.1225	27.015
0.50	1.6487	0.6065	0.5211	1.1276	27°.524

TABLE III.—HYPERBOLIC FUNCTIONS.

x	e^x	e^{-x}	$\sinh x$	$\cosh x$	$\text{gd}\, x$
0.50	1.6487	0.6065	0.5211	1.1276	27°.524
.51	1.6653	.6005	.5324	1.1329	28.031
.52	1.6820	.5945	.5438	1.1383	28.535
.53	1.6989	.5886	.5552	1.1438	29.037
.54	1.7160	.5827	.5666	1.1494	29.537
.55	1.7333	.5770	.5782	1.1551	30.034
.56	1.7507	.5712	.5897	1.1609	30.529
.57	1.7683	.5655	.6014	1.1669	31.021
.58	1.7860	.5599	.6131	1.1730	31.511
.59	1.8040	.5543	.6248	1.1792	31.998
.60	1.8221	.5488	.6367	1.1855	32.483
.61	1.8404	.5433	.6485	1.1919	32.965
.62	1.8589	.5379	.6605	1.1984	33.444
.63	1.8776	.5326	.6725	1.2051	33.921
.64	1.8965	.5273	.6846	1.2119	34.395
.65	1.9155	.5220	.6967	1.2188	34.867
.66	1.9348	.5169	.7090	1.2258	35.336
.67	1.9542	.5117	.7213	1.2330	35.802
.68	1.9739	.5066	.7336	1.2402	36.265
.69	1.9937	.5016	.7461	1.2476	36.726
.70	2.0138	.4966	.7586	1.2552	37.183
.71	2.0340	.4916	.7712	1.2628	37.638
.72	2.0544	.4867	.7838	1.2706	38.091
.73	2.0751	.4819	.7966	1.2785	38.540
.74	2.0959	.4771	.8094	1.2865	38.987
.75	2.1170	.4724	.8223	1.2947	39.431
.76	2.1383	.4677	.8353	1.3030	39.872
.77	2.1598	.4630	.8484	1.3114	40.310
.78	2.1815	.4584	.8615	1.3199	40.746
.79	2.2034	.4538	.8748	1.3286	41.179
.80	2.2255	.4493	.8881	1.3374	41.608
.81	2.2479	.4449	.9015	1.3464	42.035
.82	2.2705	.4404	.9150	1.3555	42.460
.83	2.2933	.4360	.9286	1.3647	42.881
.84	2.3164	.4317	.9423	1.3740	43.299
.85	2.3396	.4274	.9561	1.3835	43.715
.86	2.3632	.4232	.9700	1.3932	44.128
.87	2.3869	.4190	.9840	1.4029	44.537
.88	2.4109	.4148	.9981	1.4128	44.944
.89	2.4351	.4107	1.0122	1.4229	45.348
.90	2.4596	.4066	1.0265	1.4331	45.750
.91	2.4843	.4025	1.0409	1.4434	46.148
.92	2.5093	.3985	1.0554	1.4539	46.544
.93	2.5345	.3946	1.0700	1.4645	46.936
.94	2.5600	.3906	1.0847	1.4753	47.326
.95	2.5857	.3867	1.0995	1.4862	47.713
.96	2.6117	.3829	1.1144	1.4973	48.097
.97	2.6379	.3791	1.1294	1.5085	48.478
.98	2.6645	.3753	1.1446	1.5199	48.857
.99	2.6912	.3716	1.1598	1.5314	49.232
1.00	2.7183	0.3679	1.1752	1.5431	49°.605

TABLE III.—HYPERBOLIC FUNCTIONS.

x	$l \sinh x$	$l \cosh x$	x	$l \sinh x$	$l \cosh x$	x	$l \sinh x$	$l \cosh x$
1.00	0.0701	0.1884	1.50	0.3282	0.3715	2.00	0.5595	0.5754
1.01	.0758	.1917	1.51	.3330	.3754	2.01	.5640	.5796
1.02	.0815	.1950	1.52	.3378	.3794	2.02	.5685	.5838
1.03	.0871	.1984	1.53	.3426	.3833	2.03	.5730	.5880
1.04	.0927	.2018	1.54	.3474	.3873	2.04	.5775	.5922
1.05	.0982	.2051	1.55	.3521	.3913	2.05	.5820	.5964
1.06	.1038	.2086	1.56	.3569	.3952	2.06	.5865	.6006
1.07	.1093	.2120	1.57	.3616	.3992	2.07	.5910	.6048
1.08	.1148	.2154	1.58	.3663	.4032	2.08	.5955	.6090
1.09	.1203	.2189	1.59	.3711	.4072	2.09	.6000	.6132
1.10	.1257	.2223	1.60	.3758	.4112	2.10	.6044	.6175
1.11	.1311	.2258	1.61	.3805	.4152	2.11	.6089	.6217
1.12	.1365	.2293	1.62	.3852	.4192	2.12	.6134	.6259
1.13	.1419	.2328	1.63	.3899	.4232	2.13	.6178	.6301
1.14	.1472	.2364	1.64	.3946	.4273	2.14	.6223	.6343
1.15	.1525	.2399	1.65	.3992	.4313	2.15	.6268	.6386
1.16	.1578	.2435	1.66	.4039	.4353	2.16	.6312	.6428
1.17	.1631	.2470	1.67	.4086	.4394	2.17	.6357	.6470
1.18	.1684	.2506	1.68	.4132	.4434	2.18	.6401	.6512
1.19	.1736	.2542	1.69	.4179	.4475	2.19	.6446	.6555
1.20	.1788	.2578	1.70	.4225	.4515	2.20	.6491	.6597
1.21	.1840	.2615	1.71	.4272	.4556	2.21	.6535	.6640
1.22	.1892	.2651	1.72	.4318	.4597	2.22	.6580	.6682
1.23	.1944	.2688	1.73	.4364	.4637	2.23	.6624	.6724
1.24	.1995	.2724	1.74	.4411	.4678	2.24	.6668	.6767
1.25	.2046	.2761	1.75	.4457	.4719	2.25	.6713	.6809
1.26	.2098	.2798	1.76	.4503	.4760	2.26	.6757	.6852
1.27	.2148	.2835	1.77	.4549	.4801	2.27	.6802	.6894
1.28	.2199	.2872	1.78	.4595	.4842	2.28	.6846	.6937
1.29	.2250	.2909	1.79	.4641	.4883	2.29	.6890	.6979
1.30	.2300	.2947	1.80	.4687	.4924	2.30	.6935	.7022
1.31	.2351	.2984	1.81	.4733	.4965	2.31	.6979	.7064
1.32	.2401	.3022	1.82	.4778	.5006	2.32	.7023	.7107
1.33	.2451	.3059	1.83	.4824	.5048	2.33	.7067	.7150
1.34	.2501	.3097	1.84	.4870	.5089	2.34	.7112	.7192
1.35	.2551	.3135	1.85	.4915	.5130	2.35	.7156	.7235
1.36	.2600	.3173	1.86	.4961	.5172	2.36	.7200	.7278
1.37	.2650	.3211	1.87	.5007	.5213	2.37	.7244	.7320
1.38	.2699	.3249	1.88	.5052	.5254	2.38	.7289	.7363
1.39	.2748	.3288	1.89	.5098	.5296	2.39	.7333	.7406
1.40	.2797	.3326	1.90	.5143	.5337	2.40	.7377	.7448
1.41	.2846	.3365	1.91	.5188	.5379	2.41	.7421	.7491
1.42	.2895	.3403	1.92	.5234	.5421	2.42	.7465	.7534
1.43	.2944	.3442	1.93	.5279	.5462	2.43	.7509	.7577
1.44	.2993	.3481	1.94	.5324	.5504	2.44	.7553	.7619
1.45	.3041	.3520	1.95	.5370	.5545	2.45	.7597	.7662
1.46	.3090	.3559	1.96	.5415	.5587	2.46	.7642	.7705
1.47	.3138	.3598	1.97	.5460	.5629	2.47	.7686	.7748
1.48	.3186	.3637	1.98	.5505	.5671	2.48	.7730	.7791
1.49	.3234	.3676	1.99	.5550	.5713	2.49	.7774	.7833
1.50	0.3282	0.3715	2.00	0.5595	0.5754	2.50	0.7818	0.7876

TABLE III.—HYPERBOLIC FUNCTIONS.

x	$l \sinh x$	$l \cosh x$	x	$l \sinh x$	$l \cosh x$	x	$l \sinh x$	$l \cosh x$
2.50	0.7818	0.7876	2.75	0.8915	0.8951	3.0	1.0008	1.0029
2.51	.7862	.7919	2.76	.8959	.8994	3.1	1.0444	1.0462
2.52	.7906	.7962	2.77	.9003	.9037	3.2	1.0880	1.0894
2.53	.7950	.8005	2.78	.9046	.9080	3.3	1.1316	1.1327
2.54	.7994	.8048	2.79	.9090	.9123	3.4	1.1751	1.1761
2.55	.8038	.8091	2.80	.9134	.9166	3.5	1.2186	1.2194
2.56	.8082	.8134	2.81	.9178	.9209	3.6	1.2621	1.2628
2.57	.8126	.8176	2.82	.9221	.9252	3.7	1.3056	1.3061
2.58	.8169	.8219	2.83	.9265	.9295	3.8	1.3491	1.3495
2.59	.8213	.8262	2.84	.9309	.9338	3.9	1.3925	1.3929
2.60	.8257	.8305	2.85	.9353	.9382	4.0	1.4360	1.4363
2.61	.8301	.8348	2.86	.9396	.9425	4.1	1.4795	1.4797
2.62	.8345	.8391	2.87	.9440	.9468	4.2	1.5229	1.5231
2.63	.8389	.8434	2.88	.9484	.9511	4.3	1.5664	1.5665
2.64	.8433	.8477	2.89	.9527	.9554	4.4	1.6098	1.6099
2.65	.8477	.8520	2.90	.9571	.9597	4.5	1.6532	1.6533
2.66	.8521	.8563	2.91	.9615	.9641	4.6	1.6967	1.6968
2.67	.8564	.8606	2.92	.9658	.9684	4.7	1.7401	1.7402
2.68	.8608	.8649	2.93	.9702	.9727	4.8	1.7836	1.7836
2.69	.8652	.8692	2.94	.9746	.9770	4.9	1.8270	1.8270
2.70	.8696	.8735	2.95	.9789	.9813	5.0	1.8704	1.8705
2.71	.8740	.8778	2.96	.9833	.9856	6.0	2.3047	2.3047
2.72	.8784	.8821	2.97	.9877	.9900	7.0	2.7390	2.7390
2.73	.8827	.8864	2.98	.9920	.9943	8.0	3.1733	3.1733
2.74	.8871	.8907	2.99	.9964	.9986	9.0	3.6076	3.6076
2.75	0.8915	0.8951	3.00	1.0008	1.0029	10.0	4.0419	4.0419

TABLE IV.—ROOTS OF BESSEL'S FUNCTIONS.

	$\frac{x}{\pi}$ for $J_0(x)=0$	$\frac{x}{\pi}$ for $J_1(x)=0$		$\frac{x}{\pi}$ for $J_0(x)=0$	$\frac{x}{\pi}$ for $J_1(x)=0$
1	0.7655	1.2197	7	6.7519	7.2448
2	1.7571	2.2330	8	7.7516	8.2454
3	2.7546	3.2383	9	8.7514	9.2459
4	3.7534	4.2411	10	9.7513	10.2463
5	4.7527	5.2428	11	10.7512	11.2466
6	5.7522	6.2439	12	11.7511	12.2469

TABLE V.—ROOTS OF $J_n(x)=0$.

	$n=0$	$n=1$	$n=2$	$n=3$	$n=4$	$n=5$
1	2.405	3.832	5.135	6.379	7.586	8.780
2	5.520	7.016	8.417	9.760	11.064	12.339
3	8.654	10.173	11.620	13.017	14.373	15.700
4	11.792	13.323	14.796	16.224	17.616	18.982
5	14.931	16.470	17.960	19.410	20.827	22.220
6	18.071	19.616	21.117	22.583	24.018	25.431
7	21.212	22.760	24.270	25.749	27.200	28.628
8	24.353	25.903	27.421	28.909	30.371	31.813
9	27.494	29.047	30.571	32.050	33.512	34.983

TABLE VI.— BESSEL'S FUNCTIONS.

x	$J_0(x)$	$J_1(x)$	x	$J_0(x)$	$J_1(x)$	x	$J_0(x)$	$J_1(x)$
0.0	1.0000	0.0000	5.0	−.1776	−.3276	10.0	−.2459	.0435
0.1	.9975	.0499	5.1	−.1443	−.3371	10.1	−.2490	.0184
0.2	.9900	.0995	5.2	−.1103	−.3432	10.2	−.2496	−.0066
0.3	.9776	.1483	5.3	−.0758	−.3460	10.3	−.2477	−.0313
0.4	.9604	.1960	5.4	−.0412	−.3453	10.4	−.2434	−.0555
0.5	.9385	.2423	5.5	−.0068	−.3414	10.5	−.2366	−.0789
0.6	.9120	.2867	5.6	.0270	−.3343	10.6	−.2276	−.1012
0.7	.8812	.3290	5.7	.0599	−.3241	10.7	−.2164	−.1224
0.8	.8463	.3688	5.8	.0917	−.3110	10.8	−.2032	−.1422
0.9	.8075	.4060	5.9	.1220	−.2951	10.9	−.1881	−.1604
1.0	.7652	.4401	6.0	.1506	−.2767	11.0	−.1712	−.1768
1.1	.7196	.4709	6.1	.1773	−.2559	11.1	−.1528	−.1913
1.2	.6711	.4983	6.2	.2017	−.2329	11.2	−.1330	−.2039
1.3	.6201	.5220	6.3	.2238	−.2081	11.3	−.1121	−.2143
1.4	.5669	.5419	6.4	.2433	−.1816	11.4	−.0902	−.2225
1.5	.5118	.5579	6.5	.2601	−.1538	11.5	−.0677	−.2284
1.6	.4554	.5699	6.6	.2740	−.1250	11.6	−.0446	−.2320
1.7	.3980	.5778	6.7	.2851	−.0953	11.7	−.0213	−.2333
1.8	.3400	.5815	6.8	.2931	−.0652	11.8	.0020	−.2323
1.9	.2818	.5812	6.9	.2981	−.0349	11.9	.0250	−.2290
2.0	.2239	.5767	7.0	.3001	−.0047	12.0	.0477	−.2234
2.1	.1666	.5683	7.1	.2991	.0252	12.1	.0697	−.2157
2.2	.1104	.5560	7.2	.2951	.0543	12.2	.0908	−.2060
2.3	.0555	.5399	7.3	.2882	.0826	12.3	.1108	−.1943
2.4	.0025	.5202	7.4	.2786	.1096	12.4	.1296	−.1807
2.5	−.0484	.4971	7.5	.2663	.1352	12.5	.1469	−.1655
2.6	−.0968	.4708	7.6	.2516	.1592	12.6	.1626	−.1487
2.7	−.1424	.4416	7.7	.2346	.1813	12.7	.1766	−.1307
2.8	−.1850	.4097	7.8	.2154	.2014	12.8	.1887	−.1114
2.9	−.2243	.3754	7.9	.1944	.2192	12.9	.1988	−.0912
3.0	−.2601	.3391	8.0	.1717	.2346	13.0	.2069	−.0703
3.1	−.2921	.3009	8.1	.1475	.2476	13.1	.2129	−.0489
3.2	−.3202	.2613	8.2	.1222	.2580	13.2	.2167	−.0271
3.3	−.3443	.2207	8.3	.0960	.2657	13.3	.2183	−.0052
3.4	−.3643	.1792	8.4	.0692	.2708	13.4	.2177	.0166
3.5	−.3801	.1374	8.5	.0419	.2731	13.5	.2150	.0380
3.6	−.3918	.0955	8.6	.0146	.2728	13.6	.2101	.0590
3.7	−.3992	.0538	8.7	−.0125	.2697	13.7	.2032	.0791
3.8	−.4026	.0128	8.8	−.0392	.2641	13.8	.1943	.0984
3.9	−.4018	−.0272	8.9	−.0653	.2559	13.9	.1836	.1166
4.0	−.3972	−.0660	9.0	−.0903	.2453	14.0	.1711	.1334
4.1	−.3887	−.1033	9.1	−.1142	.2324	14.1	.1570	.1488
4.2	−.3766	−.1386	9.2	−.1367	.2174	14.2	.1414	.1626
4.3	−.3610	−.1719	9.3	−.1577	.2004	14.3	.1245	.1747
4.4	−.3423	−.2028	9.4	−.1768	.1816	14.4	.1065	.1850
4.5	−.3205	−.2311	9.5	−.1939	.1613	14.5	.0875	.1934
4.6	−.2961	−.2566	9.6	−.2090	.1395	14.6	.0679	.1999
4.7	−.2693	−.2791	9.7	−.2218	.1166	14.7	.0476	.2043
4.8	−.2404	−.2985	9.8	−.2323	.0928	14.8	.0271	.2066
4.9	−.2097	−.3147	9.9	−.2403	.0684	14.9	.0064	.2069
5.0	−.1776	−.3276	10.0	−.2459	.0435	15.0	−.0142	.2051